RESEARCH IN
PROTOZOOLOGY

RESEARCH IN PROTOZOOLOGY

In Four Volumes

EDITED BY

TZE-TUAN CHEN

Professor of Zoology
University of Southern California
Los Angeles, California

VOLUME 4

PERGAMON PRESS

OXFORD · NEW YORK · TORONTO
SYDNEY · BRAUNSCHWEIG

Pergamon Press Ltd., Headington Hill Hall, Oxford
Pergamon Press Inc., Maxwell House, Fairview Park, Elmsford,
New York 10523
Pergamon of Canada Ltd., 207 Queen's Quay West, Toronto 1
Pergamon Press (Aust.) Pty. Ltd., 19a Boundary Street,
Rushcutters Bay, N.S.W. 2011, Australia
Vieweg & Sohn GmbH, Burgplatz 1, Braunschweig

First edition 1972
Library of Congress Catalog Card No. 66–22364

Printed in Germany
08 016437 4

CONTENTS

5

PREFACE

VOLUME 4 is the final volume in this series and we hope to have achieved our original aim of covering the progress being made in the field of Protozoology.

We are grateful to many protozoologists who report that the first three volumes have served them as excellent sources of reference and in some cases as texts.

I am indebted to the Board of Consultants including Drs. William Balamuth, A. C. Giese, R. F. Kimball, Norman D. Levine and William Trager for their continuous assistance in making the difficult decisions.

I want to express my personal gratitude to Dr. Ruth S. Lynch for her help in editing, to my secretarial assistant, Miss Ryle Sonduck, and to the staff of Pergamon Press in New York and Oxford for their usual helpful co-operation throughout.

Los Angeles, California   T. T. CHEN

SYNCHRONIZED CELL DIVISION IN PROTOZOA

Erik Zeuthen and Leif Rasmussen

*Biological Institute of the Carlsberg Foundation,
16 Tagensvej, 2200 Copenhagen N, Denmark*

CONTENTS

I. INTRODUCTION

Much knowledge about the course of cellular changes during a cell generation cycle has been obtained from studies on single cells of known ages, with the aid of refined cytological techniques. Such methods will continue to be of use. However, detailed quantitative investigations at the biochemical level require material in amounts far exceeding that which one cell can supply. This has led to a demand for cultures in which the cells engage simultaneously in division. In the present paper it has been our purpose to discuss the mechanisms underlying experimentally induced synchronous division in protozoa; when these are understood it may be possible to extend synchronization techniques to cover a wider range of biological systems.

Following Pardee, let us here make a distinction between *synchronous* and *synchronized* cells and in this introduction deal with the two in that order.

When a cell divides the two daughter cells begin their independent lives at exactly the same moment. In the following generation cycle these two may also divide simultaneously or within a very short interval of each other, and in this way phasing of division activity among the progeny can be retained for some time. Ultimately, however, such phasing disappears. In spite of this, *small* clones do approach the ideal of a *naturally synchronous* population, and for decades they have been exploited successfully in studies requiring only few cells. It is not well understood why the natural synchrony of such clones decays. Part of the explanation may perhaps be traced to non-equivalence of the products of a cell division; for example, in ciliates it is known that the proter and opisthe daughter cells are not identical in respect of cytoplasmic structures, the former containing the original oral apparatus and the latter a newly formed one. Protozoa which possess a molecular mosaic in their surface reflected at the microscope level, are useful objects for study because the structures can serve as markers of existing cytoplasmic differences. Such differences may be present, mostly undetected, in all cells.

There is one instance of a naturally synchronous system where considerable amounts of experimental material are readily obtainable. We refer to the case of the myxomycete *Physarum polycephalum* in which numerous nuclei share a large cytoplasm thus forming a plasmodium of gigantic size. In this system the nuclei undergo mitosis in perfect synchrony, because they are controlled by factors uniformly distributed in a cytoplasm which is well stirred by cyclosis.

In cultures in the exponential phase of population increase the cells are generally completely out of phase with one another with respect to any specific activity which takes place during only a fraction of the generation time. Any phasing which can be observed is the result of some disturbance inflicted on the culture, as, for example, through a change of temperature, medium, illumination, etc. In the early days of tissue culture the chance observation of a partial phasing of cell behavior led to planned efforts to increase the number of mitoses in progress at a particular time. Also long ago the botanists, and in particular the algologists, discovered that cell division increased greatly during the hours of darkness. However, neither of these observations invited further exploration then because present deep interest in the study of cellular activities through the generation cycle was not yet developed.

Examples of *synchronized* cultures are batches of eggs which are fertilized simultaneously in the laboratory, and cell suspensions in which the organisms are *forced* to align mutually their activities in the cell cycle, in response to appropriate external influences. The acute need for phased cultures for biochemical studies was alleviated around 1953–4 when successful synchronization techniques were published independently for bacteria,[111] protozoa[225] and algae,[223] and opened new fields of research in each case. Early hopes that cyclic temperature changes, effective in protozoa, might be used to synchronize mammalian cells were not fulfilled. Instead tissue culturists have settled for techniques involving either (i) separation of cells undergoing mitosis, (ii) collection of cells at the metaphase stage through addition of colcemid, or (iii) collection of cells just prior to the period of DNA synthesis through starvation for one or more deoxynucleotide precursors of DNA. For still other cell types, e.g. yeast, different synchronizing methods have been worked out. In brief, there is no unifying principle among the methods used to obtain synchronized cell populations, because no single technique can be applied with equal success to all cell types. Some bacteriologists have issued warnings against the use of synchronization methods which in any way distort cell life, and today many people prefer separation techniques which are claimed to produce more "normal" cells. In our group we hold to the view that synchronized cell systems are worthy of further study as long as they can be reproduced at will and provided that any alterations which they may bring about in the cells' biological characteristics appear to be fully reversible. Our cycling temperature synchronizing method fulfils these requirements. However, temperature shifts affect the rates of all biochemical reactions and also have profound physical effects on membranes and cytoplasm. Hence in using such a multi-target instrument, we assign ourselves a very difficult task when we try to establish by which mechanisms the synchrony is induced. Assisted by others we have studied this system for many years, and we believe that we have now gained enough insight into the mechanisms involved to be able to improve the synchronization obtainable with temperature shifts in *Tetrahymena* and perhaps in other cells. We persevere with this technique because temperature shifts can be more easily effected and controlled than any other interference with cell life. Moreover, we have now reached the point where emphasis can be placed on biochemical studies of the type carried out on the naturally synchronous *Physarum* system.

In this review we shall first deal extensively with the process of synchronization as such, in populations of the ciliate *Tetrahymena pyriformis* and of the heterotroph flagellate *Astasia longa*, and then proceed to a discussion of studies on the mitotic cycle in *Physarum*, especially as these relate to the results obtained with *Tetrahymena*. We have decided not to include material on synchronized autotrophs; these also have been reviewed frequently by others. Already two books[33, 267] containing collected papers on synchronous and synchronized cells have been published. To a large extent the view points developed here incorporate and build upon ideas presented in two previous reviews from this laboratory[263, 268] which are still acceptable today. While this is so, we have been at pains to avoid unnecessary duplication of earlier texts. Reviews with different emphasis from ours are also available.[215, 217, 222]

This chapter was concluded in June 1969.

II. TETRAHYMENA PYRIFORMIS (CILIATA, HOLOTRICHA)

Today several methods are available by which degrees of phasing of the cell cycle can be obtained in large cultures of *Tetrahymena*. All are based on temporary interference with normal cell proliferation by means of agents having either broad or narrow spectra of action on cell metabolism. With any technique the purpose is first to collect all cells in a common situation and then to release them simultaneously for joint progression through the cell cycle. Broad-spectrum interference with cell activity has been made with intermittent temperature shifts either to the cold or to the warm side of the growth temperature. Narrow-spectrum interference includes attempts to phase cells by specific starvation for thymine. Between these two extremes are the relatively non-specific effects of interference with RNA synthesis, energy metabolism, etc. Behind the use of broad-spectrum interference lies the thought that among a multitude of effects those most directly related to cell division will stand out and give a useful degree of division synchrony. The information available to date on the heat synchronized *Tetrahymena* system, which has been integrated in this review, indicates that the crucial effect of the heat shocks is to produce synchronous cell division, and that this in turn brings synthesis of DNA into synchrony. In contrast, the specific interference resulting from temporary starvation of cells for thymine is directed primarily against DNA replication and produces a useful degree of synchronization thereof, with synchronization of cell division as a secondary phenomenon.

Tetrahymena pyriformis was the first animal cell used in deliberate synchronization studies. In our laboratory it has served as a model for such studies with the emphasis on the effects of temperature shocks. In heat synchronized *Tetrahymena* as in other cells, DNA replication and cell division are mutually exclusive processes. However, in some of these cells the cessation of DNA synthesis prior to the first synchronous division occurs dangerously late if division is to distribute a fully replicated DNA complement between the daughter cells. This dilemma may be tolerated by *Tetrahymena* by virtue of its polygenomic nature, and it could account for imperfections in the division synchrony achieved so far. Other eukaryotic cells might not be so indifferent in the above situation. At any rate, the synchronized *Tetrahymena* system invites further study of the relation between DNA replication and cell division. Some introductory work along these lines will be described; one approach has been to inhibit and release DNA replication and cell division independently of each other in the same population, and another involves varying parameters in the cycling temperature synchronization procedure. Experiments with the latter technique are in progress, but not yet far enough advanced to be included in this review.

The subsection entitled "Mechanisms in the synchronization of cell division by set-back reactions" is mainly based on research conducted in this laboratory; it contains a unifying hypothesis concerning these mechanisms which is presented for the first time here, and which we believe can resolve some of the differences existing between workers in this field. In general, the hypothesis states that the differential regression, called set-back, in cells' progress towards division, which is brought about by heat shocks amongst other things, is the direct consequence of a collapse of dynamic and highly labile structures somehow essential for division. A main point in the hypothesis is that this collapse is the result of a limitation of the supply of specific proteins. Such limitation might be due, at least in part, to a general block for synthesis of messenger RNAs, some of which are required for the synthesis of division-relevant proteins.

1. Synchrony of Cell Division Induced by Multiple Heat Shocks†

A. DESCRIPTION OF THE SYSTEM: FROM LOGARITHMIC GROWTH PHASE, THROUGH SYNCHRONOUS DIVISION AND BACK TO NORMAL GROWTH

Studies with *Tetrahymena pyriformis*, amicronucleate strain GL, produced a cell system in which, at constant temperature, a first highly synchronous cell division is followed by other, less synchronous divisions through which

† When the population is given a treatment (e.g. a heat or a cold shock) which elicits a partially synchronous division, and a subsequent similar treatment after this division burst has taken place, and so on, we speak of a "one shock per generation" procedure. In contrast, when the population is given several treatments prior to the first synchronous division burst we speak of a "multi-shock" procedure.

Abbreviations, etc., used in this review: IH denotes the time of initiation of heat treatment; EH: the end of heat treatment (if this designation has a suffix it indicates the number of heat shocks applied); EC: the end of cold treatment (sometimes also with a suffix); TCA: trichloroacetic acid. All temperatures are given in degrees Celsius.

the system reverts to exponential and asynchronous growth and division. The first summarizing report on these studies appeared as an editorial in *Science*[1] here quoted:

An important problem in microbiology is the study of the variation in properties that must occur during the stages in the division cycle of cells. Such studies in the populations of growing cells are difficult because of the random occurrence of cell division which results in a smooth exponential increase in the total amounts of cell components and metabolic activities. Recognition of this problem has resulted in a number of recent papers devoted to techniques for the synchronization of cell division and to the properties of synchronized cultures.

Starting with single *Tetrahymena* cells Zeuthen[262] was able to study synchronized growth through four division cycles. He made the rather surprising observation that the respiratory activity per cell increased linearly rather than exponentially during the interval between cell divisions in which the cells were increasing in size. This suggested that growth, when divorced from cell division, is not autocatalytic. Scherbaum and Zeuthen[225] then developed a method for the synchronization of cell division in mass cultures of *Tetrahymena*. A growing culture was raised from the optimum growth temperature of 29° to the sublethal temperature of 34° for ½ hour. This interrupted cell growth and the initiation of cell division, but it did not prevent the completion of divisions already in progress. On the return of the culture to 29°, synthesis of protoplasm resumed, but there was no cell division. During 8 hours of these cyclic temperature changes, the cells increased three- to four-fold in size without cell division. If the culture was then held at 24°, there was a lag period of about 90 minutes, after which about 85 per cent of the cells divided nearly simultaneously. Two additional division cycles followed at less than 2-hour intervals, the proportion of synchronized cell divisions gradually decreasing.

Synchrony induction was in the early experiments monitored by observation of the single outwardly conspicuous event in the cell cycle, namely the fission of the cytoplasm, cytokinesis. Other morphological phenomena also became synchronized, and thus displayed themselves in a more or less normal time relation to cell division. They are seen in the following order in *Tetrahymena pyriformis*, strain GL: oral morphogenesis from 40 to 80 minutes after the last heat shock; cytokinesis, beginning with fission line formation, from 60 to 100 minutes; macronuclear stretching and fission from 65 to 90 minutes; a "division maximum" (a maximum of about 90 per cent of the cells showing furrowing) at 85 minutes. Also synthetic activities become synchronized, e.g. macronuclear DNA synthesis begins at 90 and ends at 140 minutes. In micronucleate strains the micronucleus is slightly ahead of the macronucleus with respect to synchronous division. The described series of synchronized events repeats itself several times with degenerating synchrony. This returns the system to the fully asynchronous state of exponential growth and division. Heat synchronization has no effect on cell viability.

a. *Population dynamics*

Figure 1 shows some of the features of the population dynamics before, during and after the synchronizing treatment. *Before* the initiation of the heat shocks, at a constant temperature of 29°, a straight-line relationship

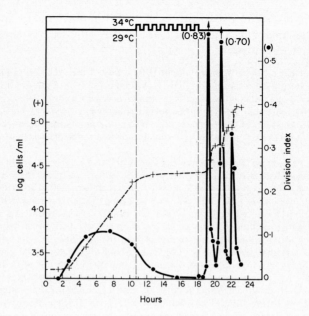

FIG. 1. Population density and division index (crosses and filled circles, respectively) before, during and after eight temperature shocks which induce synchronous divisions in a culture of *Tetrahymena pyriformis*. Effects of inoculation of the culture may be observed during the first 4 to 5 hours. From Zeuthen and Scherbaum[272] by courtesy of Butterworths Scientific Publications.

between the logarithm of the cell density (ordinate) and time (abscissa) is seen. Furthermore, the observed division index (fraction of cells dividing) is rather constant at 0.10 (10 per cent), and this indicates a balance between numbers of cells entering and completing division. Thus division activity in the culture occurs randomly, or, expressed in other words, the cells divide asynchronously. *During* the induction period the temperature is shifted every half hour between 29 and 34°, with the result that cell multiplication ceases. The division index decreases from about 0.10 to near zero during the first two to three heat shocks, and in reflection of this the cell density curve levels off, usually at a value 30 per cent above that observed at the initiation of the heat shocks. For several hours cell multiplication ceases altogether, but growth continues, see Figs. 11 and 12. *After* the end of the synchronizing treatment cell divisions are still suppressed for more than 1 hour. Then begins the period of synchronous cell division. At the time of the first synchronous cell division (maximal at 85 ± 5 minutes after EH) division indices as high as 0.95 may be observed and cell counts show that almost all cells divide. Almost 2 hours later, at EH + 180 minutes, the second synchronous division takes place. The highest division index observed is 0.70, and also this division results in a near doubling of the cell number. Between the two synchronous cell divisions few cells divide, and indices close to zero are observed. The third division maximum, which in this case had an index of 0.33, occurs later. As here described the synchrony deteriorates in time, and this happens at almost the same rate as in small cultures established with selected single dividing cells.[187] In due time a synchronized culture reverts to normal exponential and asynchronous growth, provided nutritional factors do not become limiting.

Three points need further comment. Firstly, as indicated in Fig. 1 the population was synchronized by shifts between 29 and 34°. In later work the temperature shifts were often made between 28 and 34°. In reality, in the correct temperature range (31–33°) a change of one degree, or perhaps even less, is sufficient to precipitate the critical events leading to synchrony of division (local experience). Secondly, it was mentioned that the cells increase in dry mass during the synchronizing treatment. It has been shown that the cells do not quite double their mass between either divisions 1 and 2, or divisions 2 and 3. This means that the once oversized cells reduce their mass in the course of their synchronous divisions and in due time attain the size characteristic of the exponentially multiplying cells. Thirdly, other authors have often used the term *balanced growth* to characterize cell populations in which each chemical parameter exactly doubles within a cell generation time. It has been shown that *Tetrahymena* cultures inoculated with stationary phase cells require 15 hours of growth at 28° after inoculation before the balanced state is obtained.[99] If it is considered advantageous that the synchronizing treatment is applied to populations in balanced growth, appropriate measures must be taken to achieve this.[183]

In summary, cells in division are not seen after the first two to three heat shocks, but increase in cell mass continues as more shocks are applied. At suitable constant temperature the cell system reverts through a series of synchronous cell divisions to normal growth.

b. *Cellular structures*

Later we shall deal with the mechanisms by which heat shocks induce synchronous division in *Tetrahymena*. It will be suggested that effects on cell structures and particularly on their earliest stages of development are deeply and directly involved. Therefore it is considered essential that this account of the synchronized *Tetrahymena* cell system is introduced by a description of some structures which we think may be implicated. The reader is referred also to a review by Frankel and Williams.[81]

Between two divisions cells from exponentially growing populations double their structural components. Examples of these are somatic and oral kinetosomes, other mouth structures, contractile vacuole pores, nuclei, etc. In the case of some structures this doubling activity extends through the entire cell cycle and may be shown to take place in all cells at all times. Doubling of other structures, however, may be limited to a restricted part of the cell cycle and cells exhibiting such developing structures appear with characteristic frequencies in the culture. Understandably, synchrony induction and synchrony alter normal frequency patterns.

A cilium is attached to a basal body, a kinetosome, located just below the cilium's point of emergence from the cell cortex. The somatic kinetosomes are arranged in n (mostly 17 or 18) longitudinal rows, called kineties, of which most extend like meridians from pole to pole of the cell. It is not yet known how new kinetosomes arise. The most widely held view is that a new kinetosome is formed in close connection with an "old" one,[12] and it arises as a probasal body.[4] In their study of morphological events in the normal cell cycle, Williams and Scherbaum[255] divided the entire interval between two successive cell divisions into four developmental stages. They found in all stages non-ciliated kinetosomes and short cilia. This suggests that kinetosome reproduction and ciliary synthesis can occur at any time during the cell generation. Direct counts of kinetosomes at the time of cell division are difficult to obtain, but the authors have evidence which suggests that the rate of kinetosomal multiplication possible is highest during cell division. This is based on the following observations: Prior to the synchronizing treatment there was a mean of forty kinetosomes along a certain kinety, with a range from twenty nine to sixty; during the induction period the mean number increased and reached sixty (range 48 to 69) prior to synchronous division, and it dropped to thirty-four immediately after division. The last two mean figures are suggestive of an increase from about sixty to about sixty-eight somatic kinetosomes immediately prior to separation of the daughter cells—see also a report by Allen.[4] Randall and Disbrey's[190] results for *T. pyriformis*, strain S, show that the number of kinetosomes along a certain kinety increases rather evenly from thirty to sixty during the heat treatment (in this case seven shocks), and is again reduced to thirty after division. No range of counts is given. The counts were made on fairly uniform cells from a population which had previously been synchronized, allowed to go through one division, and then treated with a new series of seven heat shocks. In conclusion, it appears that formation of kinetosomes and growth of cilia take place throughout the cell cycle, as well as throughout the entire synchronizing treatment.

The kineties have been investigated in heat-treated and normal cells. No change in number of kineties per cell was found, and in essence, also their morphology was unchanged.[255] The distance between the kineties, however, was clearly greater in synchronized cells, which is to be expected on the basis of the cell's greater width.

The oral area is located near the anterior end of the cell. It is seen in Fig. 2, left part, and it consists of a buccal cavity which on the animal's right

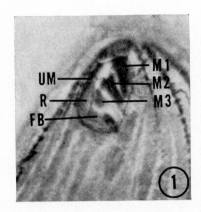 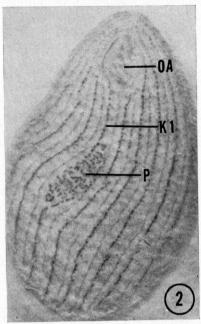

FIG. 2. Part 1. The oral area of *Tetrahymena pyriformis*. UM: undulating membrane; R: oral ribs; FB: part of the deep fiber bundle; M 1, M 2 and M 3: The three membranelles. Magnification approximately 1650×. Part 2. The oral primordium (P) is the fairly irregular field of kinetosomes located adjacent to the mid-region of kinety 1 (K 1). The anterior oral area (OA) is visible out of focus. Magnification 1130×. From Frankel[76] by courtesy of the Danish Science Press.

side has an undulating membrane and on its left three membranelles. The membranelles and the undulating membrane are compound ciliary structures with densely packed rows of kinetosomes at their bases (three rows in each of the membranelles and two rows, only one ciliated, in the undulating membrane). Two of the kineties terminate at the posterior margin of the oral area, and not near the apex of the cell like the rest. On the right side of the buccal cavity, apparently attached to the kinetosomes of the undulating membrane, is a set of oral ribs. Continuous with this is a fiber bundle which extends posteriorly deep into the cell.

Prior to division the cell develops a second mouth destined for the posterior daughter cell. The new oral area appears near the equator of the cell, and in the space between the neighboring kineties number 1 and no. *n*, of which the former is the most right of the two post-oral kineties. In this region kinetosomes have been seen to reproduce.[4] First, an apparently disorganized array of about 180 non-ciliated kinetosomes appears adjacent

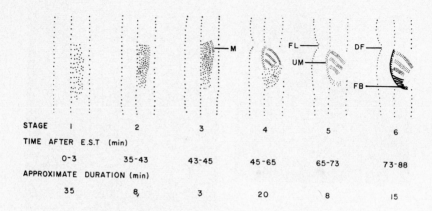

FIG. 3. Summary of the stages of oral development in *Tetrahymena pyriformis*. In these diagrams only the developing oral primordium and the region immediately surrounding it are shown. The approximate time at which each of these stages is observed in synchronized cells (maintained at 28°) is indicated beneath the diagrams (E.S.T. is in this case identical to EH). These stages may be characterized as follows:

Stage 1. Loose "anarchic field" with indefinite margins.

Stage 2. Compact field, with clearly defined outer margin.

Stage 3. Beginning of membranelle formation; anterior parts of one or two membranelles (*M*) visible, as double files of kinetosomes.

Stage 4. All three membranelles visible; beginning of undulating membrane (*UM*) formation (*UM* less than half complete).

Stage 5. Membranelles complete and now consist of triple files of kinetosomes; undulating membrane more than half complete.

Stage 6. Membranelles sink into shallow buccal cavity; oral ribs appear; deep fiber bundle, (*FB*) forms.

The fission line (*FL*) appears as an equatorial zone of discontinuities in the kineties in stage 5; the division furrow (*DF*) constricts along the fission line during stage 6. From Frankel[79] by courtesy of the Rockefeller University Press.

to kinety 1, see Fig. 2, right part. The area in which they are collected, the "anarchic field", soon becomes more compact; then kinetosomes become arranged into the rows described above, and essentially in the order in which the membranelles are numbered in Fig. 2, left part, thus from anterior to posterior and from the cell's left to the cell's right. After this process is well under way the second row of kinetosomes is added to the undulating membrane, and third rows to each of the membranellar bases 1–3. The new oral area is completed when ciliation has taken place. Finally, the deep fiber bundle develops and at the same time the oral structure sinks into the definitive buccal cavity.

The morphological development is continuous, but may be divided into six discrete stages as shown in Fig. 3.[79] This figure also gives information about the duration in minutes of each single stage. There is no major difference in the formation of the oral structures in synchronized and normal cells. The times required for each stage in Fig. 3 apply to synchronized cells. In exponentially growing cells under similar growth conditions development is slightly slower.

The most extensive study on the distribution of developmental stages in *Tetrahymena* cultures has been reported by Frankel.[77] He has found that in normally growing cultures, i.e. before the synchronizing treatment, 61 per cent of the cells had no visible oral primordia. Of the remaining 39 per cent, 14 per cent were in the anarchic field stage (Fig. 3, stage 1), whereas stages 2–6 had 4–7 per cent each. In their study on the effects of the heat treatment on morphogenesis in synchronized *T. pyriformis* Williams and Scherbaum[255] reported a slightly higher value, 18 per cent, for the frequency of cells having an anarchic field. No change in this value occurred in the interval from the *initiation* of the synchronizing treatment until the second heat shock. From this time on there was a steady increase until the third shock, when *all* cells had anarchic fields. Significantly, direct correlation was found between the fraction of cells in the anarchic field stage and the fraction which can participate in the first synchronous division.[255] The cells remained in the anarchic field stage through shocks 6, 7 and 8 until EH + 45 minutes. This is in accordance with earlier observations[110] on synchronous division in *T. pyriformis* WH − 6 that subsequent to the synchronizing treatment all cells had large anarchic fields. Parallel with the decrease of division stages during the first couple of heat shocks, developmental stages further advanced than the anarchic field stage disappeared from the culture.

Although somewhat different temperature regimens will produce synchronous divisions of equal quality, the populations may go through slightly different morphological developments during the synchronizing treatments. Frankel[77] studied morphogenesis in *T. pyriformis* GL under conditions which differed from the method used by Williams and Scherbaum[255] in this one respect: the heat shocks lasted 20 minutes and the interval between

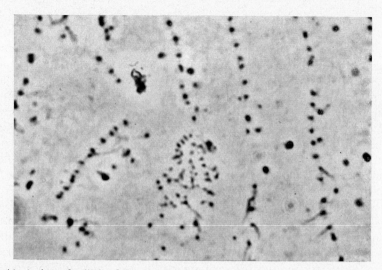

FIG. 4A. A piece of pellicle of *Tetrahymena pyriformis* bearing an oral primordium in the anarchic field stage, stage 1 of Fig. 3. Few cilia are seen because most cilia have been stripped off the kinetosomes in the somatic kineties and the anarchic field kinetosomes are non-ciliated at this stage; phase contrast micrograph; magnification 1500×. From Williams and Zeuthen[256] by courtesy of the Danish Science Press.

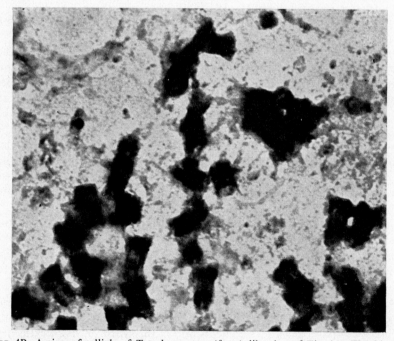

FIG. 4B. A piece of pellicle of *Tetrahymena pyriformis* like that of Fig. 4A. The kinetosomes appear as elongated cylinders; note absence of fibrous connections between the kinetosomes; electron micrograph; magnification 15,000×. From Williams and Zeuthen[256] by courtesy of the Danish Science Press.

them 40 minutes. Frankel states that towards the end of the interval between the shocks, membranelle development starts in a considerable fraction of the cells; as a result of the shocks these membranelles are resorbed. This development does not occur in cultures synchronized by means of heat shocks lasting 30 minutes, i.e. as by Williams and Scherbaum, because it requires a longer duration of shock-interval to manifest itself. On the other hand, if the intervals are appreciably longer than 40 minutes, cell division occurs during the shock treatment, and then effective synchronization is not possible. The conclusion thus emerges that both *formation* and *maintenance* of the oral primordium are incompatible with the elevated temperature of 34°.[77]

Oral kinetosomes are collected in the primordium during heat treatment; no definitive order is established among them as long as the heat shocks continue. Isolated anarchic oral primordia show non-ciliated kinetosomes without connecting fibers as viewed in the electron microscope (Fig. 4B). In contrast, in stages 3 and later, kinetosomes are criss-cross connected with filaments and fibers ranging in thickness from 4 to more than 20 mμ (Fig. 5), and there is evidence of progressive bundling of fibers, many of which have microtubular dimensions. The suggestion has been made that a heat shock breaks or depolymerizes filaments and fibers at an early stage of the development towards a heat-stable configuration in which dense bundling is a characteristic feature.[256, 273]

The primordium and the finished oral apparatus can be isolated from lysed cells. The finished oral apparatus as isolated in 1 M hexylene glycol and viewed in the electron microscope may be seen in Fig. 6, and a schematic figure based on studies of such isolates is shown in Fig. 7.

The isolated oral apparatus is a fibrillar organelle system essentially devoid of cilia. Structurally, it consists primarily of cellular membranes, many kinetosomes and associated microtubular fibers[253] and microfilaments (Dr. Jason Wolfe, personal communication). Anatomical details can be seen from Figs. 6 and 7. From the undulating membrane on one side and from membranelle number 3 on the other arises a fiber bundle which extends downwards to the vicinity of the macronucleus. The upper ramifications of this bundle which are connected to the undulating membrane support the rib wall of the buccal cavity and its deeper parts may line a channel running from the cytopharynx deep into the cytoplasm[169] The deeper parts of the fiber bundle regress from 60 to 80 minutes after EH. At mid-cytokinesis the old and the new OA are structurally identical, and shortly after both show extended deep fibers. Regression of the deep fiber is correlated with cessation of feeding.[41]

Non-ciliated kinetosomes and mitotic centrioles share common features. They act as centers upon which cytoplasmic fibers converge, and these fibers may be instrumental in establishing a specific spatial arrangement of both kinetosomes and centrioles, the former in ciliate oral morphogenesis,

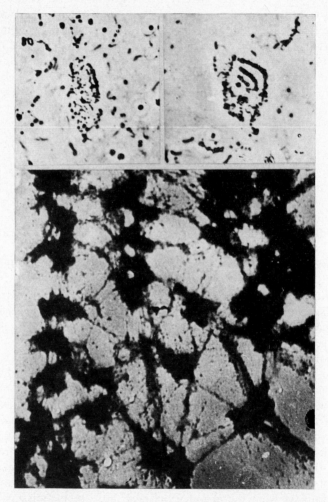

FIG. 5. Two upper frames: phase contrast micrographs of oral primordia of *Tetrahymena pyriformis* in morphogenetic stages 3 (left) and 4 (right) isolated with tertiary butyl alcohol. Note the progressive development of the membranelles from stage 3 to stage 4; magnification 1200×. Lower part: electron micrograph of a late stage 3 oral primordium. Well-defined fiber bundles connect kinetosomes and take part in the organization of kinetosomes into membranelles; magnification 15,000×. From Williams and Zeuthen[256] by courtesy of the Danish Science Press.

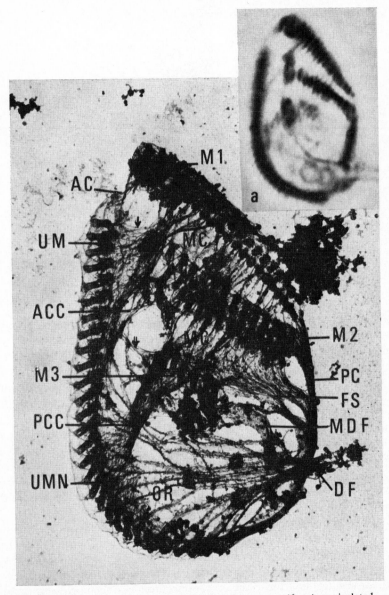

FIG. 6. An intact oral apparatus of *Tetrahymena pyriformis* as isolated with hexylene glycol. Fibrillar material connects various parts of the undulating membrane and the three membranelles. (a) Phase contrast micrograph of an oral apparatus from the same isolate. From Forer et al.[74] by courtesy of the Danish Science Press.

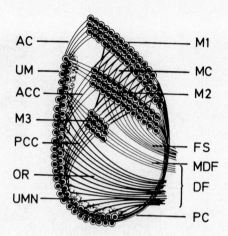

FIG. 7. Schematic drawing of the oral apparatus of *Tetrahymena pyriformis*. Note the extensive pattern of fibrils connecting the subunits of the oral structures. From Forer *et al.*[74] by courtesy of the Danish Science Press.

the latter in the polarization of the mitotic figure. To stress this parallel the term centrioles has been applied to the non-ciliated kinetosomes of the oral primordium in *Tetrahymena*.[273] According to this description this cell is highly polycentriolar.

It is not difficult to accept that subpellicular fibers are essential for oral morphogenesis. It is not as easy to see a connection between oral morphogenesis and cell division, and what happens in the oral field may be only an indicator of essentially similar structural events elsewhere in the cell which are in more direct control of nuclear and cellular divisions. The observation of Holz *et al.*[110] that micronuclear mitosis is blocked by heat shocks in *T. pyriformis*, strain WH − 6, may fit in here, because blocked anaphase movement of the chromosomes and primordium development resume at the same time.

The fission line first appears as equatorial discontinuities in the kineties immediately above the anterior end of the oral primordium.[76] The zone of discontinuities spreads in both directions around the cell, and forms a complete ring by the beginning of stage 5. The division furrow begins as a small local notch anterior to the oral primordium and then proceeds around the cell, following the preformed fission line. The notch is often present in stage 5 cells, but most of the actual cleavage of the cell takes place during stage 6, see Fig. 3.

Ciliates have nuclear dimorphism: typically, each cell has both a micronucleus and a macronucleus. Micronuclei divide by normal meiosis and

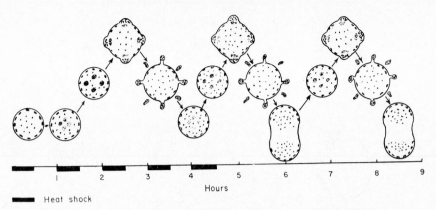

Fig. 8. A schematic representation of the macronuclear events in *Tetrahymena pyriformis* during the heat-shock treatment and the subsequent division cycles. See text for details. From Elliott *et al.*[69] by courtesy of the Rockefeller University Press.

mitosis; a micronucleus may give rise to a macronucleus by endomitotic polyploidization. The existence of macronuclear chromosomes has never been demonstrated, and the process by which a macronucleus stretches and pinches into two is denoted *amitosis*, but is by and large not understood. Some strains of *Tetrahymena*, including *T. pyriformis* GL are amicronucleate, i.e. they have lost the micronucleus, and consequently multiply asexually.

Macronuclear components increase during the synchronizing treatment as will be described later. This is reflected in macronuclear volume. Williams and Scherbaum[255] found that the mean macronuclear volume increased 4 times from logarithmic growth phase to immediately before the synchronous division. At the same time the cell volume increased 3.7 times. In a minor fraction of the cells the macronucleus swelled excessively before the division; however, during division, when the average nucleus had a volume less than half that before division, the range of volume variations became reduced. Shortly after division the macronuclear volume again increased.

Macronuclear events have been followed in the electron microscope in synchronously dividing cells. Elliott *et al.*[68, 69] synchronized *T. pyriformis*, strain WH − 6 (variety 1, mating type I) with multiple heat shocks after Holz *et al.*[110] by applying five heat shocks at 42.8° alternating with half-hour intervals at the growth temperature of 35°. In the procedure stationary phase cells were transferred to fresh growth medium and temperature cycling was immediately begun. The following is a description of Fig. 8, which depicts their results. During the first heat shock (initiated at time 0) the interphase macronucleus contains peripheral crescent-shaped nucleoli and evenly distributed chromatin bodies. Large centrally located RNA bodies, composed of fibers, appear 1–2 hours later. After 2–2½ hours these

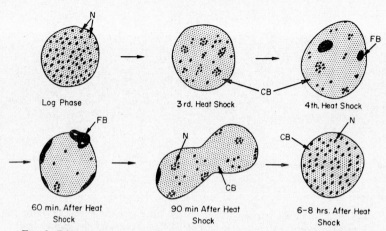

Fig. 9. Diagrammatic representation of macronuclear events during heat shock synchronization of exponentially growing *Tetrahymena pyriformis*. The last three drawings show events which occur after the end of the last heat shock. FB: fusion bodies; N: nucleoli; CB: chromatin bodies. From Cameron *et al.*[34] by courtesy of the Society of Protozoologists.

bodies move to the periphery and disintegrate; the nucleoli aggregate and form blebs which protrude into the cytoplasm into which they may liberate their contents. These nuclear events are repeated during the 4th and 5th hours. The shocks are terminated at $4\frac{1}{2}$ hours, and cytokinesis occurs at about 6 hours, with participation of 80 per cent of the cells. The fine structure of the dividing macronucleus fits the description of the macronucleus seen in the original stationary phase cells, at the start of the first heat shock. Between synchronous divisions 1 and 2, the described events repeat themselves.

Cameron and Guile[31] have suggested that the RNA bodies reported by Elliott *et al.* were late fusion products of nucleoli. Such fusion bodies are present in stationary phase cells transferred to fresh medium, but are absent in logarithmically multiplying cells. Whatever the origin of the bodies described by Elliott *et al.*, it is quite possible that their cyclic dispersion and reaggregation was due to the experiments having been made on cultures newly inoculated with stationary phase cells, rather than to the heat shocks applied. In a further study Cameron *et al.*[34] established synchronous division, by exposing *T. pyriformis*, strain HSM, to five heat shocks at 39°, separated by 25-minute intervals at the growth temperature of 29°. A division maximum of 60 per cent was observed at EH + 75 minutes. When the synchronizing treatment was applied to logarithmically multiplying cells they found nucleolar fusion in response to the heat shocks, but no rhythmic disintegration of the fusion bodies occurred during the shock treatment, cf. Fig. 9. The following pattern of macronuclear events was observed after the heat shocks. The fusion bodies moved to the periphery, flattened against the nuclear envelope, disaggregated and began to spread out. Nuclear bulges were noted, but no blebs were seen to separate from the nucleus.

The state of the chromatin was followed by Holz[108] using Feulgen stained preparations. His conclusions are valid for synchronized *T. pyriformis*, strain WH − 6, and apply also to strain GL, except for statements about the micronucleus. Holz observed that immediately after cell division the macronucleus assumed a moderately dense and homogeneous granular appearance. At the time when the micronucleus migrated to the periphery of the organism, the macronucleus increased in size and granularity. As the micronucleus divided, the macronucleus elongated, and at the time when the cleavage furrow separated the daughter cells, it divided into approximately equal halves, which again assumed a moderately dense and granular appearance. Small portions of material occasionally separated from the main mass of the macronucleus during its division[82, 108, 223] and were later resorbed. We consider that increased granulation in early stages of synchronous cytokinesis may account for great increase in self-absorption in about half of the cells at this time, as reflected in autoradiographic results with tritiated thymidine (cf. p. 35).

The chromatin condensation which takes place during division in logarithmically multiplying cells seems to be more pronounced in amicronucleate than in micronucleate strains;[73, 202] this difference was first pointed out by Roth and Minick[202] who compared *T. pyriformis*, strains W and HAM 3. It is generally accepted that chromatin condensation slows transcription and one would expect the state of the chromatin to be reflected at the levels of RNA synthesis measured by uridine incorporation, and of enzyme activities, e.g. as expressed in the cells' respiration rate. For *Tetrahymena*, as for other cells, the question whether[9] or not[188] there is a provisional stop for RNA synthesis during cell division has been much debated, as has the issue whether[262, 272] or not[144] there is a provisional stop for increase in respiration rate during division. The available evidence indicates that in the synchronized population there is predivision chromatin condensation,[114] slow-down of RNA synthesis[9] and stop for increase in cellular respiration, but opinions differ on these points when we turn to normal cells. Probably the specific growth conditions are important. Clones of *T. pyriformis* GL started in fresh medium with cells from an early stationary phase culture showed plateaus in respiration rate prior to each division when the cell number increased from 1 to 8 or from 1 to 16,[262] but this was infrequently observed in clones started with cells from the exponential growth phase.[144]

Recently Wolfe[258] studied the macronuclear structures in *Tetrahymena*. Isolated macronuclei were spread on an air–water interface after chelation of the divalent ions. The contents of the burst nuclei were examined with the electron microscope. Chromatin spheres remained attached to one another by 10-mμ fibrils, each containing two 2-mμ-thick strands, presumably two DNA double helices. Wolfe suggests that the 10-mμ fibrils are the sites of DNA replication as well as of transcription, and that they represent nucleoprotein in a state of flux from one chromatin sphere to another. During nuclear division some chromatin spheres move to one sister nucleus, some to the other. Several chromatin spheres are attached to each other, which indicates that distribution of chromatin units is not random.

Amitosis resembles mitosis in major respects. In both the chromatin condenses, an elongation process separates it to opposite poles and oriented microtubules seem to be involved in this. Spindle-like intra-macronuclear fibrils tend to form near the constriction of the dividing macronucleus of *Tetrahymena*. This was shown for the micronucleate strain HAM 3 by Roth and Minick,[202] and bundles of microtubuli which fit this description have also been observed in the amicronucleate strain GL by Nilsson (personal communication). In addition, Nilsson found parallel cytoplasmic microtubuli in the macronuclear constriction zone. As far as is known the nuclear envelope is continuous during the division of the macronucleus,[202] but it is more richly supplied with pores than the nuclear membrane of any other cell studied to date.[261]

c. *Macromolecules*

A shift from 28 to 34° in a population of *T. pyriformis* GL affects growth parameters to different degrees. Figure 10 shows that such a shift in an

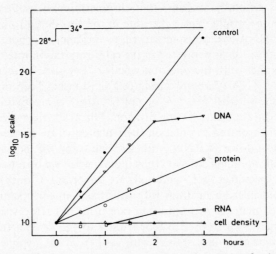

Fig. 10. The effect on growth parameters per unit volume of culture of a shift of temperature from 28 to 34°. DNA: after Burton; [21] RNA: O.D. at 260 mμ; protein: after Lowry *et al.*[142] The control curve (filled circles) is representative of the increase which takes place in each of the parameters shown when the temperature is maintained at 28° throughout. Rasmussen, unpublished experiments.

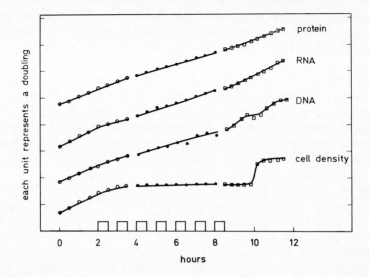

FIG. 11. Population parameters expressed per unit volume of culture before, during and after the synchronizing treatment. Protein content was measured according to Lowry,[142] RNA by the orcinol method[226] and DNA by the method of Burton.[21] Curves from three separate experiments are shown fitted end to end. In fact, the actual numerical values obtained in these analyses vary from culture to culture notwithstanding efforts to maintain constant conditions between cultures; for example, in thirty-three experiments carried out on exponentially multiplying *T. pyriformis* the parameters in question varied as indicated by the following mean values (in $\mu\mu g$ per cell) with their standard deviations (in the same units and as a percentage of the mean) and ranges; protein 2650 $\pm$ 330 (12 per cent), range 2010 to 3450; RNA: 270 $\pm$ 27 (10 per cent), range 212 to 340; DNA: 12.7 $\pm$ 3.3 (26 per cent), range 6.5 to 21.0. It is obvious that the DNA content of the cells is more variable than the content of either RNA or protein. In these experiments cell density varied from 24,000 to 64,000 cells per ml, but there was no correlation between the values of DNA per cell and cell density. The following table shows some of the numerical values (in $\mu\mu g$ per cell) from which the fitted curves of Fig. 11 were constructed.

	Experiment 1		Experiment 2		Experiment 3	
Time, hours	0	3.5	4	8	8.5	11.3
Cells per ml	24,000	45,500	65,000	65,000	44,500	87,000
Protein	3350	3790	3260	6050	6550	5550
RNA	275	314	300	580	625	607
DNA	18	18.4	8.15	15.8	13.0	11.5

Villadsen and Zeuthen, unpublished experiments.

exponentially growing population immediately and fully inhibits RNA synthesis, and that this is paralleled by inhibition of cell separation. Syntheses of protein and of DNA are also inhibited, the latter fully and abruptly after 2 hours at 34°.

During the synchrony-induction period macromolecular syntheses continue, but at reduced rates as seen in Fig. 11. The components DNA, RNA and protein roughly double in amount during a synchronizing treatment consisting of seven heat shocks, but would increase to a lesser or greater degree according as fewer or more heat shocks were applied. The doubling which occurs in the case of seven shocks is incidental and should not be over-interpreted.

Other data indicate that some parameters undergo periodic variations during the heat shocks. It was seen in Fig. 10 that RNA synthesis is more severely affected by the temperature shifts than is protein synthesis. This could be anticipated from previous results which showed cessations first of phosphorus accumulation, then of increase in dry matter in response to each heat shock.[99] Furthermore, glycogen seems to be accumulated faster than other macromolecules during the synchronizing treatment,[221] and this is probably reflected in the change in the ratio between total phosphorus and dry matter. This ratio becomes low during the heat treatments, but returns to normal values in the course of three synchronous divisions.[99]

In many instances constant values for ratios between biochemical parameters, measured at distinct phases of the synchronizing program and of the synchronous division cycles, have been reported. As an example we mention that the relative distribution of phosphorus in nucleic acid (of which 95 per cent is RNA), acid soluble phosphorus and phospholipids was constant at all time points studied,[99] and in the approximate proportions 60 : 30 : 10. Similarly, Scherbaum[214] found constant ratios between DNA, RNA and dry weight from logarithmic growth, through division-blocked to synchronous division stages. We are left with an impression of a system in which the cells' content of various components can vary widely, while the ratios between them remain fairly constant. Indeed, the only dramatic feature is the blockage of divisions during the period with heat shocks. On this background it is not surprising that the cells tolerate that the growth medium is replaced with an inorganic medium or even with distilled water, e.g. after the sixth heat shock, without any consequence at least for the first synchronous division.

After the period with heat shocks for some hours the cells divide more rapidly than they grow.[225, 272] During this time synthesis of DNA is periodic like cell division, while synthesis of RNA and protein are not clearly so (see Fig. 11). The finer analysis obtained from measurements of the rate at which heat synchronized cells incorporate ^{3}H-uridine into acid precipitable substances, supplies details not accessible from the data of Fig. 11. Thus the curve of this rate versus time was shown to describe two maxima

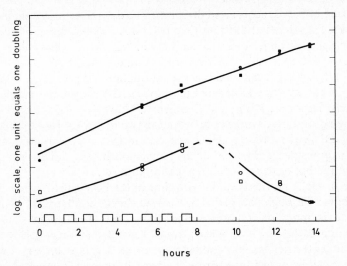

Fig. 12. Cell volume and amount of DNA, both expressed per unit of culture (filled symbols), and volume and DNA, both expressed per cell (open symbols) during and after the synchronizing treatment in *Tetrahymena pyriformis*. Data from Zeuthen and Scherbaum.[272]

prior to synchronous division 1,[162, 274] of which the first preceded a maximum in the rate of incorporation of ^{14}C-phenylalanine under comparable conditions.[274]

DNA. That DNA per cell can vary widely in *Tetrahymena* under various conditions is well known, and is exemplified in particular by cultures subjected to a synchronizing heat treatment. Thus DNA per cell increased by a factor 4 during the period with heat shocks as shown in Fig. 12.[272] Later it decreased in consequence of the synchronous divisions, although total DNA per unit volume of the culture kept rising throughout the experiment. Changes in amounts of DNA are closely paralleled by changes in cell volume and this agrees with Kamiya's[123] finding of constant ratios between DNA and total nitrogen from IH through division 1. Variation in the average size of the newly synchronized cells is often seen from one experiment to another. In the original work, as depicted in Fig. 12, growth during the period with heat shocks was more intensive than that recorded in later publications. Thus Scherbaum *et al.*[224] reported far less growth and a much smaller increase in DNA per average cell (1.8-fold) from IH to $EH_{(7)}$. Scherbaum[214] also found that the base ratios of the macronuclear DNA were unchanged in heat synchronized cells as compared with exponentially multiplying cells.

The above information prompts the question: how do the high amounts of DNA per cell become reduced to the *pre*-heat treatment values? Hardin *et al.*[101] recently measured amounts of DNA in *Tetrahymena* cultures during the period of heat synchronized divisions. The determinations were made on macronuclei isolated at different time points. At the beginning of division 2 the average macronucleus contained only 85 per cent of the value expected if DNA had doubled after division 1. The authors suggested that this reduction in DNA per cell reflected that only a fraction (70 per cent) of the chromatin in the average macronucleus duplicated between the synchronous divisions. However, Hjelm and Zeuthen[103] have pointed out that not all cells engage in DNA synthesis between the two divisions. Hence, the reduction in DNA per average cell which occurs in the synchronous division steps, may result from partial or no replication between divisions in some cells on the background of full replication in most.

Rate of DNA synthesis varies with time in the heat synchronized culture, as follows: Through the period with heat shocks until EH + 40 minutes the rate is more or less constant; between EH + 40 minutes and EH + 80 minutes the rate is very low; from EH + 80 minutes to EH + 130 to 140 minutes the rate is high, and before division 2, it is again very low. This sequence can be pieced together on the basis of reports from various workers.[101, 103, 143, 224] Scherbaum et al.[224] analysed the period up to and including division 1. They found evidence of accelerated synthesis immediately before and during the synchronous division, and they suggested that this synthesis triggers this division; we shall later present evidence for the view that this synthesis rather represents the beginning of the S period which relates functionally to the second division. As will be pointed out below, present methods for DNA determinations lack accuracy. Consequently neither our description above, nor the reported values for amounts of DNA should be overrated.

In summary, DNA per cell increases during the period with heat shocks; this increase is a function of the growth conditions in the particular experiment and of the number of heat shocks applied to induce the synchrony. During the synchronous divisions, the total DNA in the culture increases, while DNA per average cell approaches the lower *pre*-heat treatment values of exponentially growing cells.

With some analytical methods the ease with which DNA can be extracted from cells or from precipitated pellets of macromolecules affects the chemical determination of DNA. This may be of minor significance when the cultures can be considered uniform from sample to sample, as in *Tetrahymena* grown exponentially. However, serious difficulties arise when the configuration of the deoxyribonucleo-proteins can be assumed to change from sample to sample as in a synchronized population. That this difficulty exists in studies with *Tetrahymena* is clearly evident from the results of Holm.[105-7] In his hands the measured amount of DNA changes periodically with the heat shocks, but this may wholly or in part reflect differences in the time necessary to hydrolyze the deoxyribonucleo-protein which condenses visibly during the heat shocks (Dr. J. R. Nilsson, personal communication). Whether or not this influences the determination of DNA in samples of synchronously dividing cells is not known at the moment, but warnings have been issued on the subject.[143] Qualitative differences in DNA from normal and synchronized cells have also been reported.[143, 156] Similarly, the state of the chromatin at the time when cells are fixed for autoradiography of DNA synthesized in the presence of ³H-thymidine greatly influences the grain count over a macronucleus, and to the extent that the presence in the macronucleus of previously synthesized DNA can be masked by self-absorption, probably at the level of the chromatin granules. Lack of appreciation of this fact led Hjelm and Zeuthen[103] to propose that much newly made DNA is lost from some macronuclei just prior to synchronous division. The work has now been repeated with ¹⁴C-thymidine, and this tracer provides no evidence of such loss.[114] That a small loss does in fact take place can be seen only cytologically.[223]

So far we have talked about DNA measured on a per average cell basis; a more detailed description of the DNA synthesis requires information on the behavior of each single cell. Tritium labeled thymidine and autoradiographic techniques have been of some help for this purpose[38, 103] and one question put forward was this: how long is the period which must pass from any particular moment until all macronuclei in a culture have engaged in synthesis of DNA? First let us follow an exponentially growing culture to which labeled thymidine was added at time zero in so high amount that it was not used up to any appreciable degree during the experiment. Subsequently, samples were removed and assessed for labeled macronuclei. As is shown in Fig. 13 the 100 per cent labeling level was reached about 2 hours after ³H-thymidine was added. It was concluded from this experiment that macronuclei entered DNA synthesis asynchronously, and that 2 hours were needed before all had done so. Let us next consider a culture exposed to the

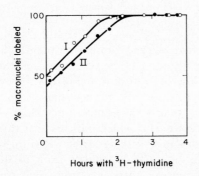

FIG. 13. Rate at which *Tetrahymena pyriformis* cells of an exponentially multiplying, asynchronous population enter DNA synthesis. Tritium-labeled thymidine was added at time zero. From Hjelm and Zeuthen[103] by courtesy of the Danish Science Press.

synchronizing treatment. Figure 14 shows the rate with which macronuclei in two aliquots of a population of *T. pyriformis* became labeled with ³H-thymidine added at IH and IH + 3 hours, i.e. at the beginning of the first and the fourth shocks, respectively. In both cases all macronuclei were labeled only after several hours. In the first aliquot (I, open circles) the level of 100 per cent labeling was obtained after 4–5 hours, in the second (II, filled circles) the level of 85 per cent labeling was reached after 4 hours. It was concluded from the data presented in Fig. 14 that the macronuclei engaged asynchronously in DNA synthesis during the period with heat shocks; that they did so more slowly than during exponential multiplication;

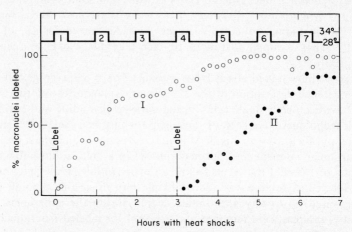

FIG. 14. Rate at which the macronuclei in two aliquots of a population of *Tetrahymena pyriformis* become labeled with tritium-labeled thymidine added at the times indicated by arrows, and present for the remainder of each experiment. Both aliquots are exposed to heat shocks as indicated at the top of the figure. From Hjelm and Zeuthen[103] by courtesy of the Danish Science Press.

and that cells which were blocked for division and which had already synthesized DNA since the beginning of the synchronizing treatment could resume DNA synthesis at a later time prior to the first synchronous division. It was found by Cerroni and Zeuthen[38] that macronuclei engaged asynchronously in DNA synthesis for some time after EH, and their observation has been confirmed in later experiments which indicate that this interval lasts for a period of between 20 and 55 minutes after EH (Fig. 3 of Hjelm and Zeuthen[103]). From this time (60–30 minutes prior to the first synchronous division maximum) onwards there was decreasing participation in DNA synthesis, and around EH + 75–80 minutes extremely few macronuclei were engaged in DNA synthesis. In contrast, between 80 and 110 minutes after EH, that is, during and immediately after the first division, 55–65 per cent of all macronuclei again entered DNA synthesis. In most of these cells the two macronuclei became labeled before the daughter cells separated. Thus they essentially by-passed the normal G_1 period. Finally, let us follow a culture through the synchronous divisions. Figure 15 shows an experiment in which cell divisions were synchronized as shown by the upper curve. Three aliquots from this population received ^{3}H-thymidine at the times indicated by the arrows, and curves I, II and III show the degree of synchrony with which the macronuclei entered DNA replication during and after each of the three synchronous divisions. As indicated by the slope of the curves this synchrony degenerated as did the synchrony of the divisions themselves. Furthermore, not all nuclei took part in this synchronous DNA synthesis. The participation was around 65 per cent after division 1, and slightly higher, 75 per cent, after division 2. Failure of some cells to synthesize DNA between subsequent divisions may be an important factor in the regulation of the amount of DNA per cell which takes place during the synchronous division steps. The data of Fig. 16 do not permit us to decide whether those cells which failed to replicate DNA between two divisions also failed to

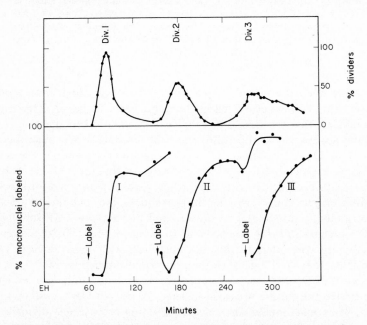

FIG. 15. Rate at which *Tetrahymena pyriformis* cells from a heat synchronized population enter DNA synthesis. Tritium-labeled thymidine was added at the times indicated by the arrows to three separate aliquots of a culture. The upper curve shows the fraction of cells in division, estimated visually, versus time; the lower curves indicate the rate at which the macronuclei became labeled. From Hjelm and Zeuthen[103] by courtesy of the Danish Science Press.

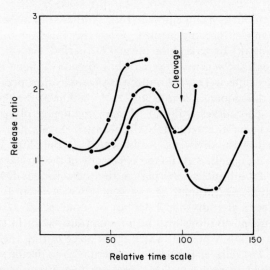

FIG. 16. Ratio between radioactivities (in counts per minute) of protein which can be released with ATP, and protein remaining on the ribosomes from synchronized *Tetrahymena pyriformis* as a function of time after EH (zero point of abscissa). See text for details. One hundred relative time units equal 80 minutes. From Plesner[179] by courtesy of the Long Island Biological Association.

replicate between the next divisions, or whether this block for a normal function shifted from cell to cell in the population. However, later experiments indicate that the second possibility is the more likely one.

During and for some time after the heat treatment each macronucleus shows cyclic DNA synthesis. Each nucleus continues essentially the course taken in the exponentially growing population, so that the nuclei remain out of phase with each other. Therefore, when visible preparation for synchronous division gets under way after the heat shocks, the macronuclei are more or less randomly distributed between S, and non-S phases. Morphological post-EH preparations for synchronous division, in phase in 90 per cent of the cells, are correlated in the following way with macronuclear DNA synthesizing activity: First, when the oral primordium gets reactivated new engagement in DNA replication apparently ceases; secondly, as the macronuclei stretch the culture becomes depleted for cells in DNA synthesis; and thirdly, when macronuclei and cells are dividing or have just divided two-thirds of the cells engage rapidly in synthesis of DNA. Hjelm and Zeuthen[103] have suggested that cells in synchronous division emit signals for initiation of DNA replication, to which two-thirds of all macronuclei in a division synchronized population respond because they have passed a critical point in the extended time between subsequent S periods. One-third or so of the cells do not respond, because they have not yet passed this point of maturation. Hjelm and Zeuthen summarized their description with the words: "The heat shocks have synchronized macronuclear and cellular division, and these events have in turn synchronized DNA synthesis in the larger part of the population." It is of interest here that in *Stentor* cytoplasmic factors are involved in the initiation of DNA replication.[235]

A recent investigation undertaken in our laboratory by Andersen *et al.*,[5] and yet unpublished, confirmed and extended the conclusions which were based on the autoradiographic experiments. Bromodeoxyuridine and radioactive phosphate were added to cultures at various times and became incorporated into DNA synthesized subsequently. After isolation the DNA was separated by centrifugation in cesium chloride gradients and assayed. The following conclusions could be drawn: (i) DNA synthesis is nearly linear from IH up to EH + 30 minutes; (ii) full replication of all the DNA (± 5 per cent) present at IH requires 4–5 hours; (iii) DNA synthesis is almost blocked from EH + 30 to EH + 75 minutes; (iv) high rates of DNA synthesis are measured during and immediately after division 1; (v) prior to division 2 there is a second pause in DNA synthesizing activity; (vi) synchrony of DNA synthesis is observed also during and after division 2; (vii) within errors of measurement, it can be stated that no second round of DNA replication begins until the foregoing round has been completed. This is true for exponentially multiplying as well as for synchronized cells. Thus in the single cell, as in the whole population, the DNA must be fully replicated before a new round of replication can be initiated.

We shall see later that cell divisions are synchronized through reversal of morphogenetic processes. There is no evidence for the occurrence of a similar reversal of DNA synthesis.

RNA. Knowledge is limited about roles played by RNA during the induction and course of synchronous division, but we have comments on the information supplied in Figs. 10 and 11. Scherbaum[214] found constant base ratios in total RNA from cells in logarithmic, division-blocked and synchronous division stages, whereas Christensson[47] reported differences in the relative amounts of two major RNA-subfractions, one present in the aqueous, the other in the phenol phase after extraction according to Kirby.[125] Neither of these fractions has been well characterized. Christensson speculated that the phenol phase RNA fraction may be mostly macronuclear which he thinks would agree with cytochemical observations by Agrell[2] of an RNA fraction which is tightly bound to protein and to DNA. In *Tetrahymena* most macronuclear RNA has a sedimentation constant of 25 S,[137] and this is a precursor of the 50 S ribosomal subparticle.[138] It has also been reported that there is increased ribosome formation after the synchronized cell division.[138]

Plesner[179–80] has studied the ribosome composition of synchronized *Tetrahymena*. He found that in the presence of 0.025 M magnesium stable ribosomes of classes 70 S, 80 S and 100 S could be separated from the cells

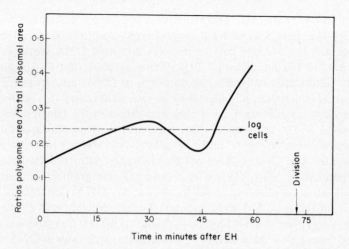

FIG. 17. The ratios of polysome to total ribosome contents (expressed as areas under sucrose gradient profiles) of synchronized *Tetrahymena pyriformis* are plotted as a function of time after EH. From Zimmerman[274] by courtesy of Academic Press.

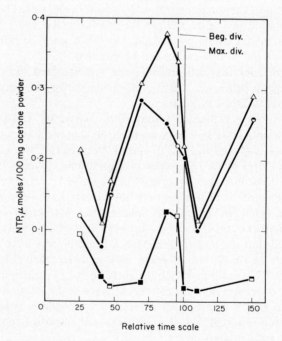

FIG. 18. The content of ATP and GTP in acetone powders of synchronized *Tetrahymena*, from EH through division 1. Circles: ATP; squares: GTP; triangles: sum of ATP and GTP. Abscissa as in Fig. 16. From Plesner[181] by courtesy of the Danish Science Press.

at all stages between EH and division, but that at low magnesium concentration (less than 10^{-4} M) only the 70 S ribosomes from cells which had reached a time point between 15 and 30 minutes prior to the synchronous division maximum remained stable. In the absence of magnesium ions these stable 70 S ribosomes liberated newly synthesized protein upon incubation with 5×10^{-4} M ATP. In this connection it is interesting that cellular ATP + GTP reaches a high maximum roughly 10–15 minutes prior to the division maximum and a minimum 10 minutes later (see Fig. 18). Figure 16 shows the change in the ratio of radioactivity in newly synthesized protein (released as described above) to radioactivity in protein remaining in the ribosomes. Further work has been done on *Tetrahymena* ribosomes by Leick and Plesner,[135-6, 139] but this mostly concerns exponentially multiplying cells.

In Zimmerman's laboratory measurements have been made of the extent to which the ribosomes exist in aggregate form.[274] The results were expressed as the ratio of the amount of polysome material (120 S) to the total amount of ribosome + polysome material (84 S + 120 S)—see Fig. 17. This ratio increases with time for 30 minutes after EH, then decreases and again increases around division. With respect to the ups and downs in the ratio curve there is fair harmony with the curves for rate of protein synthesis,[274] increase in cell mass[99] and in respiratory activity, a measure of enzyme activities.[272]

Protein. The protein content of *T. pyriformis* varies with the culture conditions. In exponentially multiplying populations values ranging from 0.74 to 3.2 mg protein per million cells have been published.[46, 134-5, 221, 227] *Throughout the synchronizing heat treatment* a general increase in protein content (expressed per unit volume of culture) takes place, as shown in Fig. 11; this is in agreement with previous reports.[46, 134, 219, 223, 227] So far, detailed information is scarce with respect to the effect of a single heat shock in the synchronizing program on the synthesis of various fractions of cell protein. However, there are two reports dealing with the rate at which total protein is synthesized during the heat treatment.[155, 245] Although the experimental protocols are different the same conclusion can be drawn in each case, namely, that the rate of protein synthesis during the eighth heat shock is only 20 per cent or less, of the rate observed before the first heat shock. This is based on comparisons of radioactivity incorporated into protein after standard pulses with radioactive amino acids. The first attempt to measure differential increases in the cell proteins during the period with heat shocks was made by Christensson.[46] At various time points in the synchronizing program he separated the cell proteins into 0.1 M KCl–0.011 M acetic acid soluble and insoluble fractions. Synthesis of the soluble proteins was then seen to lag behind synthesis of the insoluble. *After the heat treatment* protein content of the culture increases, but more slowly than cell number. This is seen in Fig. 11 and confirms previous findings.[99]

Changes in the protein composition before and during cell division have been demonstrated by Watanabe and Ikeda,[246-7] who worked with heat synchronized *T. pyriformis*, strain W, harvested at three different time points after the heat treatment: at EH, EH + 45 minutes, and EH + 65 minutes. They separated two protein fractions, soluble in water and 0.6 M KCl respectively, and fractionated each further on DEAE-cellulose columns, obtaining ten subfractions from the water soluble and three from the KCl soluble proteins. In the water-soluble material the amount of a particular subfraction, peak 7, reached a maximum at EH + 45 minutes. Of the subfractions from the KCl soluble protein peak 1 increased in size with time, whereas peak 2 decreased. Peak 3 was relatively large at all three time points, but greatest at EH + 45 minutes.

Watanabe and Ikeda[246-7] have also demonstrated differential protein synthesis in the heat-treated cells by other means. They have shown that ^{3}H-leucine and ^{3}H-phenylalanine are incorporated at decreasing rates during the interval from EH to division 1, whereas the rate of incorporation of ^{3}H-methionine increases within this period. Furthermore, they have investigated the variation in the number of SH groups per unit weight of protein in all of the ten subfractions of the water soluble and three subfractions of the KCl-soluble proteins at EH, EH + 45 minutes and EH + 67 minutes respectively.[115] Among the water-soluble protein subfractions at EH + 45 minutes, peaks 6, 7 and 8 are much richer in SH groups than are the other peaks. Of the KCl-soluble subfractions, peaks 2 and 3 contain many more SH groups than does peak 1 which has few particularly in the later part of the period studied. Finally, Watanabe and Ikeda[246] have observed that a heat shock which blocks cell division inhibits incorporation of ^{3}H-phenylalanine into all subfractions of the water soluble proteins, and most strongly in the case of peak 7. Sodium fluoride added at EH in a concentration which blocks cell division, inhibits the described increase at EH + 45 minutes in the amounts of the water-soluble peak 7 and KCl-soluble peak 3 fractions. Lowe and Zimmerman[141, 274] have been unable to confirm some of these findings, but it should be pointed out that their attempts were made on a different strain of *T. pyriformis*, namely GL, and using a different DEAE cellulose.

Several workers have examined the composition of histones and basic proteins isolated from whole *Tetrahymena* cells and from cell organelles.[48, 101, 117, 220] All conclude that macronuclear histones composition stays essentially constant as the cells multiply exponentially, and throughout all stages of synchronized population growth, the histone to DNA ratio being approximately 1. According to Iwai *et al.*[117] this ratio rises somewhat from the value of 1 at the time of the first synchronous division and becomes maximal at the beginning of the S period, after which it falls again to 1 around the second division. Christensson[48] has measured the amount of macronuclear basic protein per unit of DNA during and after the synchronization and reports a constant value of 1.88 at all stages studied. Also the amino acid composition of the histones[117] and basic proteins[48] is unchanged or shows only minor variations upon synchronization. Electrophoretic separation produces composite patterns consisting of fifteen[117] or nineteen[220] bands in the case of histone material, and of eighteen[48] bands in the case of total basic proteins, and these patterns too are essentially unaltered by the experimental heat synchronizing treatment. Incorporation of ^{14}C-lysine suggests that the synthesis of all component bands extends throughout the S period.[220] It can be concluded that the macronuclei of *Tetrahymena* are equipped with the same complement of basic proteins at all stages of the cell cycle and that the rise in the amount of these substances within the macronucleus parallels or slightly precedes the synthesis of DNA.

Carbohydrate polymers, aggregates and lipids. With respect to these compounds we have found little information having direct relevance to the topics covered in this paper. Accumulation of glycogen in response to the heat shocks was demonstrated by Scherbaum and Levy[221] and of more complex carbohydrates, some rich in amino sugars, by Lindh and Christensson.[47, 140] The latter authors deal extensively with aggregates of carbohydrates and RNA, while Rosenbaum et al. present thought-provoking ideas about the significance of membrane lipids in the cells' response to elevated temperatures.[199]

d. Low molecular weight compounds

Nucleotides and derivatives. Studies on free nucleotides and related low molecular weight substances in heat synchronized *Tetrahymena* were initiated by Plesner. He developed the large-scale techniques for growing, synchronizing and harvesting the cells for biochemical analysis[178, 181, 183] which have been useful also to other workers.[219, 252] At the time of sampling the cells were quickly collected by high-speed centrifugation, the supernatant was decanted and the dense pellet was stirred with a large volume of acetone at $-20°$. After several washings in cold acetone and ether the residues were dried and stored in the deep-freeze for later extraction, chromatographic separation, and chemical or enzymic analysis and characterization. Scherbaum adopted different procedures for the isolation of the pool of TCA-soluble compounds from such large-scale cultures. With Chou[42-45] he washed the cells centrifugally at 4°, first several times in cold 0.4 per cent NaCl, then once in 0.001 N HCl (and one may speculate as to what happens to the pool during these steps). They then extracted the washed cells with 15 per cent TCA and chromatographed the lyophilized extracts after removal of glycogen and TCA.

The collection of chromatograms published shows a perplexing number of nucleotides and other phosphorus-containing compounds.[42-45, 181] For many of these the absolute and relative concentrations appeared to

2a*

depend on the cultural growth stage. Chou and Scherbaum found that the block for growth and division inflicted on exponentially multiplying cells through a shift in temperature from 28° to 34° greatly increased the cells' content of two[43] or three[44] phosphorus-containing deoxy-sugars. The increase leveled off after 2–3 hours at the higher temperature. Intermittent heat shocks, as used in the synchronization program, had a similar though smaller effect, and new concentration levels tended to be established after the first two heat shocks. Subsequent to the heat treatment there was gradual return to lower concentrations. Labeling experiments with ^{32}P suggested that the compounds in question were liberated from TCA-precipitable material and not synthesized *de novo*.[44] A later report showed that they were in fact complexes containing ethanolamine, glycerol and phosphoric acid probably originating from lipid-containing structural elements in the normal cell. The presence of deoxy-sugars in the complexes was not confirmed.[45] Also the pyridine nucleotides, DPNH and TPNH, were found to change significantly in amount during the heat shocks and first synchronous division.[170, 217] When comparison is made with the cells' RNA, the pool of free nucleotides is seen to be rich in cytidylic and uridylic acids.[47] In the amount of nucleoside triphosphates there was increase and decrease in response to each heat shock and shock interval,[218] reflecting the oscillations in growth observed earlier.[99] Uracil derivatives labeled with ^{14}C-uracil added externally, also increased when a heat shock was inflicted on growing cells.[162] When starved, cells from both exponentially multiplying[50, 51] and synchronized[37, 181] cultures degrade RNA and excrete purines and pyrimidines or their derivatives. In the synchronized cell this excretion was stimulated by a heat shock, while chilling caused reuptake.[37]

Plesner firmly established that there was a strong cyclic behaviour of the nucleoside triphosphates in synchronously dividing cells held at constant temperature, see Fig. 18. Only ATP and GTP were involved. They oscillated in parallel, and their sum, designated NTP, showed a minimum at 40 relative time units, a maximum at 80–90 units (division at 100 units), a new minimum at 110 units, and a second maximum prior to division 2. These oscillations were demonstrated in both growing and starving cells, and were thus highly correlated with synchronous division. Watanabe[244] measured the activity of the enzyme ATPase in synchronized *T. pyriformis* W cells, and found that this showed fluctuations similar to those just described for the substrate. Plesner clearly recognized the danger of extrapolating from acetone powders to the living state. Consequently, he checked and confirmed his results by another method. Small samples of cells were squirted into a solution containing luciferin-luciferase together with digitonin which caused lysis and immediate liberation of NTP. Light emitted in response was measured and interpreted in terms of NTP released (ATP + triphosphates which can donate phosphate to ADP). Plesner's results were prematurely criticized,[29, 216] but have been confirmed by the critics.[218]

At the moment it is not easy to fit these scattered biochemical findings into a coherent picture. An attempt to do this has been made by Plesner[179-82] who sees the accumulation of NTP prior to a division as a consequence of reduction in growth rate, and as essential for stripping the ribosomes of newly synthesized, division-related proteins, whereby the system can proceed in time.

Sugars. Thus far few investigators have attempted to study how division-synchronizing treatments affect cellular content of free sugars. Early experiments of Kamiya[123] established that acid soluble sugars accumulated in *Tetrahymena* exposed to heat shocks. His results are shown in Fig. 19; they include values for total sugar measured by two independent methods,

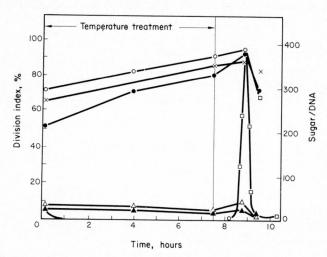

FIG. 19. Total sugar, hexose, pentose and ketosugar, all expressed per unit of DNA (right ordinate) in *Tetrahymena geleii* W, during and after the synchronizing treatment. Open circles: total sugar measured with the reagents α-naphthol and sulphuric acid; filled circles: total sugar with indole–sulphuric acid; crosses: hexose with cysteine–sulphuric acid; filled triangles: pentose with orcinol; open triangles: ketohexose with cysteine carbazole–sulphuric acid. The course of the synchronous division in terms of division indices (left ordinate) is indicated by the small filled circles.
From Kamiya[123] by courtesy of the Japanese Biochemical Society.

and for pentose and ketosugar. From the figure it can be seen that total sugar, of which more than 80 per cent was aldohexose, increased both during and after the heat treatment, until the time of the first synchronous division when it showed a sudden decrease. Pentose and ketohexose were present in much smaller amounts (about 10 per cent each of total sugar); they remained essentially constant in amount until after the heat treatment, then increased slightly, and finally decreased at division as did hexose. Kamiya chose to refer his sugar concentrations to the amount of DNA present in the samples, and in doing so he partially concealed the extent to which the cells actually accumulated sugars, because DNA increases in amount throughout the experimental period. Similarly the abrupt rise in DNA content, which is known to follow synchronous division 1, alone could account for the coincident drop in sugar concentration, expressed per unit of DNA.

For detailed information the reader is referred to a more recent paper by Lindh and Christensson[140] describing the results of a very extensive study of the content of both free and combined carbohydrates in synchronized *Tetrahymena*.

Free amino acids. Wu and Hogg[260] studied the amino acid composition of *T. pyriformis*, strain E, cultured in both peptone-containing and chemically-defined media, and showed that no correlation existed between cell composition and medium composition. In addition they found[259] that the ratio of the total amount of an amino acid in the whole cell (i.e. free, conjugated and incorporated within protein) to the amount in protein alone was fairly constant, ranging from 1.4 to 2.1 for the different amino acids excluding glycine, for which it had the value 3.0. Their work indicates that not only the protein, but also the non-protein amino acid composition is rather carefully regulated in these cells.

Evidence of genetic control of amino acid composition has been obtained in another line of research. Studying *T. pyriformis*, variety 6, mating types I and II, Wells[248] established differences between the two types in the ratio of free arginine to alanine, and in free isoleucine to leucine. A progeny clone, derived from conjugation of the two types, resembled the one parent with respect to the one ratio and the other in regard to the second. She concluded that the composition of the pool of free amino acids is under genetic control in *Tetrahymena* as in other organisms.

Pertinent to the case of synchronized *Tetrahymena* cells is the question: are the protein, peptide and amino acid compositions altered during the treatment with heat shocks? If the answer is affirmative, then gene actions or physiological mechanisms may have been disrupted, or set differently by the temperature changes, so that the composition patterns of synchronously dividing cells reflect a regulation back to normal values, superimposed upon any variations which may normally occur in the course of the cell cycle. We have already discussed the claim that groups of proteins appear in characteristic sequences prior to the synchronized division. In the following pages we shall review observations on the extent to which the amino acid pool composition varies during and after the synchronization treatment.

Both Christensson[46] and Scherbaum et al.[219] have reported values for amounts of individual amino acids in the synchronized cells. The work of Christensson, which is the more detailed, will be discussed first. His results include observations on populations growing exponentially before the heat shocks, on cells before and after heat shocks numbers 4, 6 and 8, and on cells before and after synchronous division. We have chosen to present

Christensson's data graphically—see Fig. 20 in which the data from his Table 1 have been plotted arbitrarily on logarithmic scales. Values for cells after synchronous division have been corrected to the values which would have been obtained per cell had the cells not divided. The correction factor, 47/26, is the ratio deduced from Christensson's curves, between the amounts of protein per cell before and after division. We have arbitrarily plotted Christensson's results for cells "before division" and "after division" at 1 and 2 hours respectively after $EH_{(8)}$.

The free amino acids can be divided into three groups according to how they accumulate with time during and after the synchronizing treatment. The first group comprises amino acids which increase in amount roughly in parallel with the increase in protein (see Fig. 20, top). Within this group the overall regulation of the ratio of free amino acids (FAA) to protein is unaffected by the heat shocks and by the division, but as pointed out by Christensson the amino acid concentrations show increases and decreases with the heat shocks and shock intervals, reflecting inhibitions and resumptions of protein synthesis. The second group (shown at the bottom of Fig. 20 above the curve for cell protein) consists of amino acids whose amounts increase much during the early heat shocks (lysine, histidine) and much less or not at all during the later shocks (lysine). Amino acids of the third group (shown at the bottom of Fig. 20 below the curve for protein) follow the reverse course; among these proline represents the extreme, staying constant per cell during the whole period with heat shocks. Perhaps the size of the free amino acid pool is regulated differently for each of these groups of amino acids, but nothing is known so far about the mechanisms involved. From the data used to plot the curves of Fig. 20 we can calculate that the ratio of free amino acid plus protein amino acid to protein amino acid, (FAA + PAA)/PAA, is around 1.15 at all stages; this is lower than the figure arrived at by Wu and Hogg. However, peptides, which are probably present in considerable amounts[259] were excluded in Christensson's analysis, and this is one of several reasons why direct comparison can not be made with Wu and Hogg's work on normal cells.

Scherbaum et al.[219] extracted the free amino acid pool from synchronized cells at four different stages of development, and measured the amounts of its components individually or in groups (see Table 1). Careful comparison of Table 1 and Fig. 20 will disclose both similarities and discrepancies between their findings and Christensson's. Concurrently they determined the amino acid composition of protein hydrolysates from parallel samples. From the bottom lines of Table 1 it can be seen that the ratio (FAA + PAA)/ PAA calculated from their data ranges from 1.13–1.14, which is in agreement with values based on Christensson's results. In contrast, this ratio is decreased in synchronized cells transferred to an inorganic starvation medium; for example, a value of 1.08 can be calculated from results of Crockett et al.,[52] who found that 4.4×10^5 cells, with a protein content of 1.96 μmg

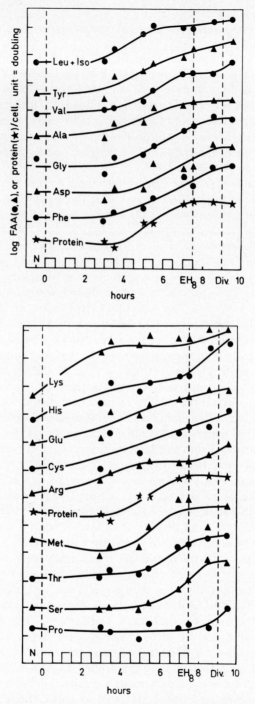

FIG. 20. Amounts of free amino acid per cell in *Tetrahymena* cultures before, during and after eight synchronizing heat shocks. Amounts of protein per cell are shown for comparison. Ordinate: logarithmic scale, arbitrary units; see text for details. Data from Christensson.[46]

TABLE 1. AMINO ACID COMPOSITION OF *Tetrahymena pyriformis*

Determinations were carried out at four growth stages, namely: (1) during exponential multiplication; (2) at the end of heat treatment; (3) prior to synchronous division; (4) during stationary phase. FAA: free amino acids, determined in the ethanolsoluble fraction of the cells; PAA: protein amino acids. Values are expressed in micromoles per gram dry weight of cells. From Scherbaum et al.[219] by courtesy of the Wistar Institute.

	1		2		3		4	
	FAA	PAA	FAA	PAA	FAA	PAA	FAA	PAA
Alanine	45.3	205.9	41.3	196.7	34.3	179.7	36.3	187.0
Arginine	31.5	70.5	27.0	89.2	38.2	85.9	30.1	81.7
Asparagine	7.9		7.3		6.2		6.9	
Aspartic acid	17.3	339.6	21.6	381.1	27.1	356.4	18.8	346.5
Cystine	8.3	32.5	11.9	40.0	10.0	17.5	8.6	45.0
Glutamic acid	34.5	402.2	45.0	415.0	37.6	440.4	39.2	419.0
Glutamine	12.8		12.1		18.0		13.8	
Glycine	33.4	213.2	29.6	216.1	26.8	234.5	29.9	215.6
Leucine / Isoleucine	34.0	415.3	48.4	471.2	60.3	447.8	46.0	466.5
Phenylalanine	35.8	182.5	24.4	214.1	29.8	206.2	28.4	194.5
Lysine Histidine	12.1	85.5	11.3	77.0	9.9	80.0	10.6	81.0
Proline	18.7	129.2	18.3	137.6	21.8	143.0	19.1	136.1
Serine	17.5	175.1	18.7	202.3	20.8	176.5	18.7	183.7
Threonine	14.1	93.7	20.2	86.4	16.2	82.2	17.2	83.8
Tyrosine	16.6	229.1	22.4	222.0	26.4	225.8	21.1	236.1
Valine Methionine								
Sum of PAA		2574.3		2748.7		2675.9		2676.8
Sum of FAA	339.8		359.5		383.4		344.7	
(FAA + PAA)/PAA		1.13		1.13		1.14		1.13

per cell, contained 11.2 μg ninhydrin-reactive amino nitrogen, extractable with cold perchloric acid.

The pool of free amino acids reflects fine balances between uptake, incorporation into protein, catabolism and excretion of all the individual components. The work reviewed above strongly suggests that heat shocks interfere with normal regulation of the size and composition of this pool in *Tetrahymena*. This interference can be at any level between the genes and their physiological expression.

As mentioned earlier transfer of heat-treated cells to a starvation medium does not prevent development of synchronous division, as mentioned earlier. In these circumstances, however, protein turnover occurs and this is accompanied by a leakage of ninhydrin-positive material into the medium.[52] Cann[37] showed that 90 per cent of this material is free L-alanine, and that the leakage is stimulated by an increase of temperature from 28° to 34°, and blocked at or below 7.5°. Excretion of free amino acids into the extracellular fluid has also been demonstrated for cells cultured in nutrient medium, after maximum growth has been attained.[260]

In sum the results reported above show interesting, but not well-understood changes in the composition of the pool of free amino acids in *Tetrahymena* exposed to the adverse temperature conditions which induce synchronous division. Most striking is the observation that some free amino acids increase in concentration while others decrease during the early temperature shocks. These changes tend to reverse during the later shocks and after the shocks. The two amino acids which initially increase most in concentration are histidine and especially lysine, the lysine to protein ratio rising by a factor 3 before heat shock 4. Here it may be relevant that L-lysine in high concentration is known to interfere with division, both in cells multiplying exponentially in growth medium (unpublished observations of Hoffmann, Rasmussen and Zeuthen) and in synchronized populations in starvation medium.[195]

B. MECHANISMS IN THE SYNCHRONIZATION OF CELL DIVISION BY SET-BACK REACTIONS

It is obvious that synchronization of cell division with heat shocks implies that the cells are treated rather violently and in ways which put regulation mechanisms under considerable stress. However, the end result, synchronous cell division, suggests that the stressed mechanisms have operated rather successfully after all. The situation would have been simple if the heat shocks had blocked the cells at a particular point in a normal cycle, and progression in the cycle had been resumed from this point after the heat treatment. This is not entirely so as will appear. A somewhat more complicated situation had existed, if Scherbaum and Zeuthen[225, 272] had been right in the hypothesis on which they initiated work on *Tetrahymena*. This hypothesis stated that division-synchrony could be obtained, if the asynchronously dividing cells could be differentially affected merely in their *progression* through the cell cycle. This may be what happens in some of the other synchronized systems, and this situation has attracted builders of mathematical models of cell synchronization and desynchronization.[35, 70] However, it was commented in the original paper by Scherbaum and Zeuthen[225] and clearly demonstrated later that things must be more complicated at least in the case of *Tetrahymena* in which both heat and cold shocks also cause differential *regression* ("set-back") in the cell cycle, except in those cells which stand just before division and which continue to *progress*.[237] These regressions and progressions phase the cells for morphogenesis and for cell division, two closely coupled phenomena in this organism. While this case is mathematically manageable,[36] it is still too simple to account for the additional experimental finding that cortical structures regress while at the same time macronuclear chemical events progress within the same cell. As already described, the division-blocked cells perform periodic DNA replication; the synthesis is asynchronous, and this may perhaps be true for any other specific compound which is synthesized as part of a sequential program in the normal cell cycle. Macromolecules to be used in division and in oral morphogenesis may also belong in this category, as we have evidence that oral precursor proteins are synthesized during the period with heat shocks. The puzzling observation is that after the first synchronous division a program which involves DNA replication falls in phase and follows synchronous division fairly normally, even though until shortly before the division it ran its course asynchronously. Therefore regulation mechanisms are stressed not only while the cells are being phased for synchronous divisions but also during the period when these divisions occur.

a. *Physiological mechanisms in the induction of synchronous division with heat shocks*

Mechanisms in the induction of synchronous division have been studied at physiological and biochemical levels. The first approximation towards an understanding of the mechanisms involved in synchronization with heat shocks was based on measurements of the time which passed between consecutive divisions in single cells from a normal population hit by a heat shock between the two divisions. In these studies the age of the cells since the foregoing division was varied in order to mimic the situation which exists when a normal asynchronously multiplying population receives the first heat shock in a series. Cell response was found to increase with cell age, and the results supply a basis for interpretation of the degrees of phasing which can be obtained with one or several heat shocks applied to an exponentially growing culture. Measurements of division delays produced in response to environmental influences have indicated fundamental similarity in behavior between normal and heat synchronized cells. Both of these can therefore be used, in the presence or absence of perturbing factors, to supply information about physiological and biochemical processes involved in preparation of a cell division. This information supplies a basis on which a model can be proposed for the mechanism of synchronization with heat shocks. The model derives support from kinetic data of various kinds, and especially from evidence which suggests that synchronization is brought about by damage to macromolecular systems. In brief, we suggest that heat shocks cause structural damage, that this brings cells into phase, and that repair of damage done is mediated through processes which occur sequentially in the normal cell cycle. Thus the physiologically relevant targets of the synchronizing heat shocks may be proteins or aggregates of proteins whose presence is necessary for cell division to occur. We shall gradually approach the question whether damage is direct or results from primary disturbances of translational or transcriptional events. What needs to be stated here is that repair must involve the metabolic level at which damage is inflicted and perhaps antecedent levels also.

In the detailed discussion to follow the material has, for convenience, been divided into sections numbered (i)–(ix), the last of which is both summarizing and integrating in nature.

(i) The response of a cell population to a heat shock can be analyzed by applying the shock to single cells of known ages, age being defined as time elapsed since the moment of cell separation in the foregoing division. This has been done by Thormar[237] who exposed single *T. pyriformis* GL cells maintained at 28.5° to the elevated temperature of 34.1° for 15 minutes and noted the resultant generation times. From these he could calculate the intervals by which division was delayed in treated cells as compared with controls. Thormar's results are depicted in Fig. 21, curve A_1, in which delay of division (right ordinate) is plotted against the age of the cell upon treatment (abscissa). From this figure it can be seen that in very young cells the treatment produced a division delay of about 50 minutes; then with increasing age the delay increased up to a maximum of about $2\frac{1}{2}$ hours which was produced in cells about 2 hours old. Cells older than this, i.e. cells just about to divide or in the process of dividing, were only slightly delayed in completing their cell cycles. On the basis of these findings some cells would be expected to be insensitive and divide during the first heat shock of the standard synchronization program, namely those which are close to division at the beginning of heat treatment, and this is borne out in practice. Moreover, the division-delaying effect of a temporary treatment (a shock) on the other sensitive cells which comprise the larger part of a population, is greater the older the cells. It is in fact this age-dependent response, which underlies the good synchrony produced by a series of shocks. Because most cells are delayed *in excess* of the duration of a treatment, the term *excess delay* has been introduced; it denotes the delay of cell division minus the duration of the treatment. At the end of a heat shock the treated cell finds itself further separated in time from the forthcoming division than it was when the shock

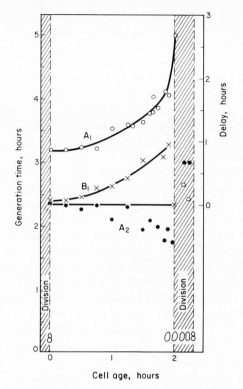

Fig. 21. Effect of various treatments on the generation time of single *Tetrahymena pyriformis* cells. Curves A_1 and B_1: durations of the treated cell generation cycle (left ordinate) as a function of the age of the cell upon exposure to 34.1° for 15 minutes or to 31.1° for 20 minutes, respectively —the time in hours (including the duration of the heat shock) by which the treated cell generation cycle is delayed compared to that of control cells kept at 28.5° is shown on the right ordinate. Filled circles (observations denoted A_2): duration of the cell cycle in the subsequent untreated progeny generation as a function of the age of the parent cell upon commencement of the treatment used to produce the response shown in curve A_1. From Thormar[237] by courtesy of the Danish Science Press.

was initiated, as indicated by observation of division times in two sister cells of which the one has been heat treated after the previous division. The treated cell can thus be said to have been set back in its cycle. The terms *excess delay* and *set-back*[263] are equivalent; they measure identical intervals.

Cell multiplication in a culture of *T. pyriformis* can be phased by application of a single heat shock; additional shocks improve the degree of phasing obtained. Thus an exponentially growing culture exposed to a heat shock (i.e. transferred to 34.1° for 30 minutes and then back to 28.5°) shows a majority of 25 per cent of the cells in division about 2 hours after the shock.[265] This partially phased culture responds to a second heat shock in the same manner as it did to the first, whereby the degree of phasing is increased. Gradually three synchronous division maxima build up as a function of the number of heat shocks applied—see Fig. 6 in review by Zeuthen.[268] However, some cells eventually begin to divide again in spite of the heat treatment, if this consists of fifteen shocks or more. This shows that adaptation to the temperature changes is possible.

Thormar[237] has also reported that individual cells subjected to a temperature shock during one generation cycle take a shorter time to go through the following cycle. It is unknown whether only a specific part of the cell cycle is shortened, or whether all parts are affected. However, we know that after the eight standard heat shocks the interval between synchronized divisions 1 and 2 is about 110 minutes and this is considerably shorter than the normal generation time which is about 150 minutes under comparable conditions. As already stated a majority of the division-synchronized cells either by-pass the G_1 phase and go directly into the S phase or show a very short G_1 phase. In the normal cell cycle, there is a G_1 period of about 15–20 minutes under the growth conditions employed here.[32] In conclusion, when a randomly dividing culture of T. pyriformis is exposed to a heat shock, its cells are affected to different degrees according to age. This results in a phasing of cell division in the culture. The phasing which arises from one heat shock is not spectacular. However, if the first shock is followed by others according to a certain program the phasing is greatly improved, and we may speak about synchronized cell divisions.

(ii) Normal and heat synchronized cells respond to temperature shocks in the same general way. Like normal cells heat synchronized cells go through a cycle in which they have first increasing then suddenly disappearing capacity to respond with delayed division to either a heat or a cold shock. Transition from maximal to no delay is sharper in the single cell than in the synchronized population. In heat synchronized cells this transition occurs at EH + 50–55 minutes,[76] i.e. about 35 minutes before the separation of daughter cells takes place at a maximal rate. In normal cells the transition point is reached 25 minutes before cell separation.[237] The similarity between normal and synchronized cells is not limited to their response to temperature shocks but extends to their reaction towards chemical agents, which is discussed below.

(iii) Division of heat synchronized cells is readily blocked with inhibitors of protein synthesis. Rasmussen and Zeuthen[194–6, 264] and Holz et al.[109] used metabolic inhibitors and structural analogues in an attempt to characterize the anabolism which the cells must perform after the heat shocks in order to divide later. To put it differently, they tried to find out which division-relevant macromolecular syntheses were either suppressed during the period with heat shocks, or somehow made useless or insufficient for later cell divisions. In these studies the cells were transferred to an inorganic medium between the last two heat shocks, under which conditions two synchronous divisions develop as shown by Hamburger and Zeuthen.[98] Inhibitors or analogues were added at EH, and delays of synchronous division were measured. This system showed itself to be highly sensitive to amino acid analogues, notably p-fluorophenylalanine and canavanine,[39, 194-6] and to certain agents interfering with sterol or lipid metabolism, but essentially insensitive to purine and pyrimidine analogues[109] and to three established inhibitors of thymidine incorporation in this system, 5-fluoro-2-deoxyuridine, deoxyguanosine and deoxyadenosine,[38] cf. Table 2. A

TABLE 2. PERCENTAGE INHIBITION OF MACROMOLECULAR SYNTHESES

	6-MP 1.6×10^{-3} M	5-FUDR 5×10^{-3} M	dG 2×10^{-4} M	dA 2×10^{-4} M	Puromycin 1×10^{-3} M	DL-p-FPhe 1.8×10^{-4} M
1. Cell division	no division	normal	normal	normal	no division	no division
2. DNA	—	88	88	90	—	—
3. RNA	80	86	45	45	75	—2
4. Protein	45	—	5	1	93	40

6-MP: 6-methyl purine; 5-FUDR: 5-fluoro-2′-deoxyuridine; DL-p-FPhe: DL-para-fluorophenylalanine; dG: deoxyguanosine; dA: deoxyadenosine. From Cerroni and Zeuthen[39] by courtesy of Academic Press.

transition point was found to occur at EH + 40 minutes with 0.8 mM p-fluorophenylalanine; this was believed to reflect the same physiological event in the cell cycle as the transition points observed with heat and cold shocks. Responses to pulse exposures to p-fluorophenylalanine increased linearly from EH to the time of transition. There is, however, one significant difference between the effects of a temperature shock and a pulse treatment with p-fluorophenylalanine in comparable experiments. A heat shock sets the cells back only to the time EH. The amino acid analogue sets the cells further back, to 30 minutes prior to EH. Thus contamination of newly made proteins with p-fluorophenylalanine causes more extensive damage to the cells than does heat.[136, 271] The evidence from these studies, complemented with results of experiments on the effects of inhibitors of protein synthesis, e.g. puromycin and cycloheximide[79] and of energy metabolism,[97, 98, 191, 194, 244] supports the notion that proteins synthesized during the heat shocks are insufficient to sustain synchronized division. Accordingly, division-relevant proteins must be made after the shocks.

Pulse treatments with heat,[76] with hydrostatic pressure[146-8, 274] and with the quickly penetrating reducing agent β-mercaptoethanol[150] produced results resembling those obtained with p-fluorophenylalanine pulses, and in addition the experiments with β-mercaptoethanol revealed that the position of the transition point is a function of the strength of the treatment having an asymptote at EH + 50–55 minutes.

Among the ciliates, such age-dependent set-back reactions are not confined to *T. pyriformis*. When single *Paramecium aurelia* cells were exposed to short-term treatments with various noxious agents at different points in the cell cycle set-backs reminiscent of those known for *T. pyriformis* were found.[192–3] The treatments included exposure to *p*-fluorophenyl-alanine, puromycin, chloramphenicol, 5-fluoro-2-deoxyuridine, 2,4-di-nitrophenol and short-term starvation.

(iv) Synthesis of division-relevant proteins in excess of normal requirements might account for adaptation phenomena. Normal cells adapt to continuously elevated temperature (Fig. 22) and, as briefly commented

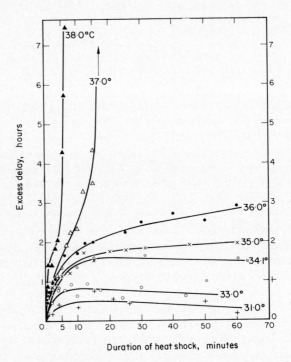

Duration of heat shock, minutes

FIG. 22. Prolongations of the generation cycle in single *Tetrahymena pyriformis* cells as functions of the durations of applied heat shocks of various temperatures. The response is given as excess delay (ordinate) which is the total delay measured minus the period of exposure to the shock. The cells, which represented an exponentially growing population, were cultivated at 28.5° and shocked when they had reached a standard age of 100 to 105 minutes after completion of cell division. A maximum on the delay curve indicates that the cells adapt to the treatment in question. From Thormar[237] by courtesy of the Danish Science Press.

above, heat-shocked cells can adapt to continued cycling of the temperature so that eventually after becoming gigantic they divide during the treatment. Heat synchronized cells also adapt to the continued presence of both *p*-fluorophenylalanine in inorganic medium (Fig. 23) and *β*-mercaptoethanol in growth medium (Fig. 24). In all cases of adaptation, the cells are initially set back by the agent, but later they make new progress towards division in its presence. In the case of *β*-mercaptoethanol Mazia and Zeuthen[150] proposed that the set-backs obtained resulted from reduction of S—S bonds, and that new progress in the presence of the agent occurred when protein required for division became available at the lowered redox potential in greater than normal amount. With proper modifications this interpretation might also be applied to adaptation to *p*-fluorophenylalanine. In this case it can be envisaged that an existing pool of protein, which is essential for, but not yet sufficient to support division, becomes contaminated with

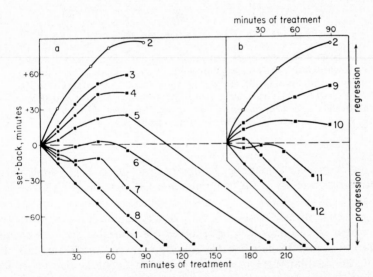

FIG. 23. Prolongations of the cell generation cycle (given as set-backs which are equivalent to excess delays) in synchronized *Tetrahymena* as functions of the duration of exposures (started at EH) to various concentrations of *p*-fluorophenylalanine. Adaptation to the treatment occurs when the set-back passes through a maximum value.

Part *a*: curves 1–8 were obtained with the following millimolar concentrations of fluorophenylalanine: **1**, 0.0; **2**, 0.8; **3**, 0.16; **4**, 0.08; **5**, 0.04; **6**, 0.02; **7**, 0.01; **8**, 0.003.

Part *b*: curves 1, 2 and 9–12 were obtained with the following millimolar concentrations of fluorophenylalanine (FPhe) and L-methionine (Met) in combination: **1**, 0.0 FPhe + 0.0 Met; **2**, 0.8 FPhe + 0.0 Met; **9**, 0.8 FPhe + 1.0 Met; **10**, 0.8 FPhe + 5 Met; **11**, 0.8 FPhe + 25 Met; **12**, 0.8 FPhe + 50 Met.

From Rasmussen and Zeuthen[196] by courtesy of Academic Press.

58 ERIK ZEUTHEN AND LEIF RASMUSSEN

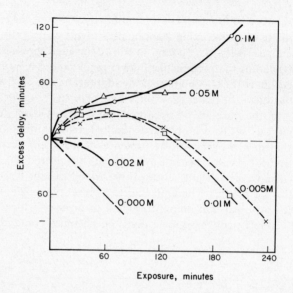

Fig. 24. Prolongations of the cell generation cycle (given as excess delays) in synchronized *Tetrahymena* as functions of the duration of exposures to various concentrations of mercaptoethanol. In all cases exposure to the drug was begun at EH + 30 minutes. Adaptation to the treatment occurs when the excess delay passes through a maximum value. From Mazia and Zeuthen[150] by courtesy of the Danish Science Press.

newly made protein containing the analogue in place of phenylalanine. If this protein as a result of the contamination has less efficacy for division it may have to be made in larger amounts before the cells can eventually divide in the presence of the analogue. In cells kept at supraoptimal temperatures, division-relevant proteins may be rendered less utilizable due to a changed configuration, and so here too have to be made in larger amounts than normal, for adaptation (expressed as ability to divide at the higher temperature) to take place. Adaptation to heat occurs only for temperatures up to *ca* 33°, i.e. slightly lower than the shock temperature used in the heat synchronizing program. In the case of *p*-fluorophenylalanine the adaptation has been observed to take place in starving cells in which new synthesis depends on turnover of cell components.[52] This indicates that the division-relevant proteins synthesized during adaptation must be small in amount. We shall later put forward a hypothesis according to which developments essential for division involve structures in which proteins are replaced at considerable rates with precursor proteins from the cytoplasm around them. The suggestion that adaptation to heat requires elevated precursor concentration derives support from the observation that heat shocked cells carry increased amounts of such proteins (cf. p. 65).

(v) Various models have been proposed to account for synchronization by set-back reactions. Figure 25 shows two schemes put forward by Rasmussen and Zeuthen.[194] The schemes are based on observed set-backs in normal and synchronized cells in response to heat, cold, and direct or indirect inhibition of the synthesis of division-relevant proteins. As previously outlined, the division machinery requires protein synthesis *before*, but not *after* a transition point. According to scheme I, the division-relevant protein P_1 increases in amount prior to the transition and this occurs in a balance between anabolic reactions (*a*) and catabolic side reactions (*c*) which compete with the division machinery for use of P_1. The balance between *a* and *c*

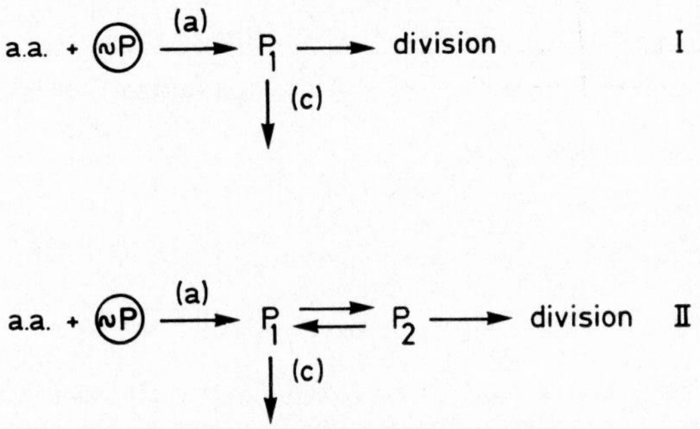

FIG. 25. Two schemes which can account, in a purely formal way, for the set-backs observed in *Tetrahymena* cells, subjected to some division-inhibiting influence. The schemes depict the build-up from amino acids (a.a.) and a source of energy ($\sim$ P) of a complex protein-containing product P_1, necessary for cell division. Anabolic reactions (c) and catabolic side reactions (c) both determine the rate at which P_1 accumulates. See text for further details. From Rasmussen and Zeuthen[194] by courtesy of the Danish Science Press.

favors the production of P_1 at 28°, but not at 34° and not at 28° when reactions *a* are inhibited in one of the ways described in paragraph (iii). In other words, heat and cold shocks and inhibition of essential syntheses result in removal of products of the type P_1 which must accumulate before the process of division can be precipitated. This precipitation takes place at the physiological transition point before a division. What happens at this time is obscure. Sudden stabilization of P_1 against reactions *c* must take place. According to scheme I this stabilization encompasses the whole amount of P_1 present just before the transition. Scheme II leaves the possibility open that only a part, P_2, of a more complex system, $P_1 + P_2$, becomes stabilized; in other words, in scheme II from the time of structural stabilization the reaction P_1 to P_2 is no longer reversible. P_2 may be part of a larger complex.

Cann[36] made these views the basis of a kinetic model which "predicts" (this word is nowadays used in retrospect) relevant experimental data obtained in this laboratory (see Fig. 26). According to Cann's model, division is precipitated when the product, C, of two sequential reactions is present in a critical amount. Division depletes the cell of C and thus starts it out in a new cycle. This model dispenses with the notion that c undergoes "stabilization", but proposes that the first reaction product, B, is subject to removal

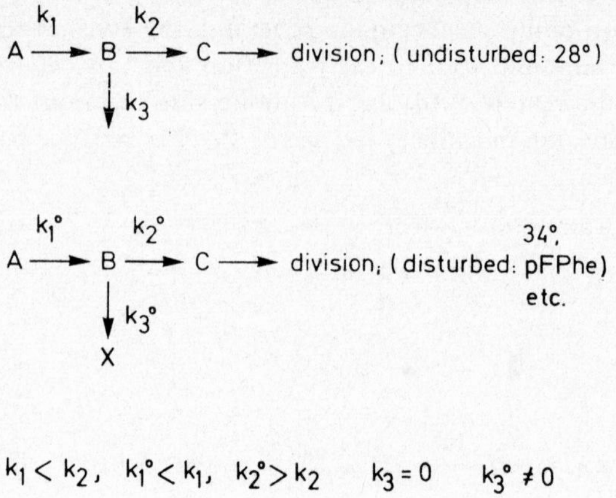

$$A \xrightarrow{k_1} B \xrightarrow{k_2} C \longrightarrow \text{division; (undisturbed: 28°)}$$
$$\downarrow k_3$$

$$A \xrightarrow{k_1°} B \xrightarrow{k_2°} C \longrightarrow \text{division; (disturbed: pFPhe)} \quad \substack{34° \\ \text{etc.}}$$
$$\downarrow k_3°$$
$$X$$

$$k_1 < k_2, \quad k_1° < k_1, \quad k_2° > k_2 \quad k_3 = 0 \quad k_3° \neq 0$$

FIG. 26. Cann's kinetic model. Cell division takes place when C is present in a critical amount. The normal reaction constants (k_1, k_2 and k_3) attain new values ($k_1°$, $k_2°$ and $k_3°$) under adverse conditions. From Cann[36] by courtesy of the Danish Science Press.

in a side reaction, B to X, which takes place to a significant extent only in disturbed cells. As a result of Cann's computations our data with heat[76] and p-fluorophenylalanine[194] can now be interpreted on the basis of identical assumptions with respect to change from normal of three rate constants k_1, k_2 and k_3. He assumes that the side reaction B to X (c in the schemes of Fig. 25) is normally blocked or running only slowly, but opens up in response to heat and p-fluorophenylalanine.

In combination, the work of Rasmussen and Zeuthen and of Cann provides a solid foundation for the idea that short-term heating, chilling and qualitative or quantitative interference with protein synthesis, make proteins already set aside for use in later division useless for this purpose. When a previously asynchronous culture is subjected to one of the above treatments, the cells are brought into a common situation from which new preparations for division occur in synchrony once the disturbing influence has been removed. The views of Rasmussen and Zeuthen have received support from pressure studies by MacDonald who stresses similarities between *Tetrahymena* and the sea urchin egg.[147-8]

(vi) In the following we shall present evidence for the standpoint that the damage produced by heat shocks requires high activation energy, and that set-back reflects this damage. Thormar[237] treated single normal cells with

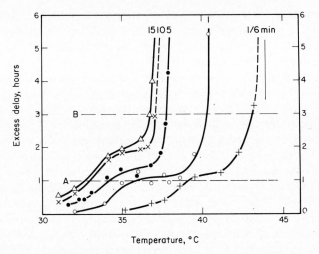

FIG. 27. Prolongation of the generation cycle (given as excess delay) in single *Tetrahymena pyriformis* cells representing an exponentially growing population after heat shocks administered in various combinations of shock temperature and period of exposure (indicated in minutes at the top of each curve). The cells were treated 100 to 105 minutes after completion of division. From Thormar[237] by courtesy of the Danish Science Press.

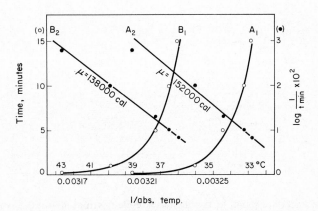

FIG. 28. A replot of the observations corresponding to the fixed excess delays of 1 and 3 hours, which are indicated by the dotted lines A and B in Fig. 27, and which are each assumed to reflect standard damage. Curves A_1 and B_1 of Fig. 28 show the relationship between shock temperatures used and the period of exposure (left ordinate) necessary to produce these fixed excess delays. Curves A_2 and B_2 referred to the right ordinate are Arrhenius plots of the same data. From Thormar[237] by courtesy of the Danish Science Press.

heat shocks of different heights and durations. The cells were first kept at the optimal growth temperature of 28.5°, and 100–105 minutes after they had divided they were exposed to the higher temperature. At this age the cells' response to a heat shock is maximal. As seen in Fig. 27, when the duration of the treatment is varied from 15 minutes to 10 seconds, standard set-backs of 1 and 3 hours require temperatures in the ranges 32.5–39° and 36.5–43°, respectively. The short delay of 1 hour (A) is fully reversible, in as far as the first division is followed by later divisions. The longer delays of 3 hours (B) are obtained only with treatments which affect the viability of the cells. The lines A_2 and B_2 of Fig. 28 are the Arrhenius plots which correspond to the temperature conditions that give the 1- and 3-hour delays. It appears that both temperature effects have high activation energies, 152,000 and 138,000 cal., respectively, indicating heat damage to complex systems: macromolecules, macromolecular aggregates or cellular structures. From the following it will appear that the high activation energies are unlikely to reflect simple heat denaturation of one or more proteins.

Division-delaying effects of temperature shocks can be traced on both sides of the optimal temperature, and even the smallest temperature changes have some effect. This is seen in Fig. 29 in which are brought together three

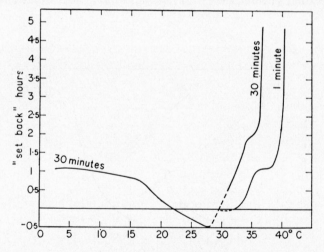

FIG. 29. Prolongation of the generation cycle (given as set-back) in single *Tetrahymena* cells exposed to temperature shocks for either 30- or 1-minute periods, as a function of the shock temperature in the range 3 to 37°. Before and after the treatment, which was commenced 100 to 110 minutes after completion of cell division, the cells were kept at 28.5°. With shock temperatures between 22° and 30° the set-backs obtained are negative, which indicates that the cells advance towards division during the treatments concerned. From Zeuthen[263] based on data by Thormar by courtesy of Academic Press.

curves obtained by Thormar. This figure shows division delays (ordinate) resulting from 30-minute exposures of normal cells, previously grown at 28–29°, to any of the various temperatures between 3° and 37° (abscissa). It appears that the cells continue to advance towards division during exposure to temperatures between 22° and 30°. In the interval between these two temperatures advance made depends on the shock temperature, but forward reactions always outbalance reactions which tend to direct the cell away from division. At 22° and 30° precisely, the two reactions cancel out, so that progression towards division is blocked. In other words, it is only when shock temperatures above 30° or below 22° are employed that the balance favours the side reactions, so that the cells are set back in time, as required for the described temperature synchronization of *Tetrahymena*. In comparable experiments on mass cultures[268] exponentially multiplying cells were chilled or heated for 1 or 2 hours and then returned to and observed at the original growth temperature. Only when the cells were heated to 32° (or more) or chilled to 16° (or less) did the fraction of dividers first decrease, subsequently increase and finally return to normal as compared with controls. It may be significant that both temperatures are near to inflection points in Arrhenius plots of normal growth versus temperature.[239] The striking continuity in size of the delays measured over the interval 3–37° invites the suggestion that all temperature shocks affect equilibrium reactions normally involved in preparations for division, and this would be compatible with the schemes of Rasmussen and Zeuthen and in our opinion also with Cann's model.

(vii) Understanding the mechanism underlying synchronization involves appreciating the how and why of individual cells' reactions towards the synchronizing agent. Whether a cell continues to advance towards division during a temperature shock, remains "blocked" or is "set back" depends first on the shock temperature chosen, and thereafter on the cell's age. With heat shocks at temperatures not higher than 31° the age response (shown in Fig. 21, curve B_1) can be explained as follows. The newly divided cell does not carry the protein store represented by P_1, P_2 or C in the reaction schemes and the kinetic model, and thus it can not be set back. Older cells do have such a store and they may react in any of the three ways mentioned above, depending on whether at the shock temperature chosen the available store is rendered useless at a slower, equal or faster rate than the one at which it continues to grow. In these terms the meaning of "set-back" is the following: the temperature shock returns a cell which has made progress towards a division to a condition which, in certain important respects, structurally and perhaps biochemically, resembles the condition prevailing around or after division. It is significant that under the temperature conditions just specified, a single heat or cold shock never sets a normal cell back by more than the time by which the cell when first treated has advanced from the previous division: the store can be no more than emptied. In the special case of the response of heat synchronized cells to an additional temperature shock, this interpretation is also applicable, if age is taken to be the time elapsed since EH; here, however, perhaps because these cells are to some extent heat-adapted,[268] a higher temperature, e.g. 34°, is required to produce a set-back of similar duration to the one obtained in normal cells with a somewhat lower temperature.[163]

(viii) Oral structures are dynamic and subject to turnover in respect of their proteins.

In the oral primordium the number of basal bodies (kinetosomes) increases slightly but development beyond the "anarchic field" stage does not take place during the synchronizing heat treatment. After the treatment the earliest sign of continued development appears at EH + 40–45 minutes, when single and bundled microtubular and microfilamentous fibers begin to be discernible between the kinetosomes in electron micrographs of isolated oral primordia.[256] From this it has been suggested that exposing the cells to the series of heat shocks prevents fiber precursors from assembling into connections between kinetosomes, which appear to be necessary if the anarchic field is to become organized. In response to a heat shock oral primordia in stages of development prior to early stage 4 enter a course of disintegration and resorption.[256] The period during which structural development can continue at the shock temperature lasts 5–10 minutes.[80] At stage 4 stabilization towards external influences occurs. Effects on the oral primordium identical or very similar to the above can be obtained with cold, p-fluorophenylalanine, protein synthesis inhibitors and with inhibitors of RNA synthesis and energy metabolism.[78] Thus the oral primordium is a dynamic structure capable of both progressive and regressive development.

The finished oral apparatus, though structurally stable, is highly dynamic at the molecular level. Evidence for this has recently been obtained by Williams et al.,[254] whose results with exponentially multiplying cells are shown in Fig. 30. In their experiments the population was first incubated with tritiated amino acids for several generations and then transferred to fresh non-radioactive growth medium to which was added "chaser" amino acid. This transfer introduced a lag in cell multiplication. Subsequent increase in cell number was followed, and samples were removed for butanol-isolation[256] of oral apparatuses at the time of transfer, at the end of the lag phase ca. 2 hours later, and after one, two and three population doublings. Autoradiographs of the preparations showed that initially every OA was labeled in protein assumed specific for this structure. After the lag phase the number of oral apparatuses increases exponentially in step with cell number, and during this growth the label became redistributed over the increased number of oral apparatuses as shown in Figs. 30 and 31. For every time point studied the average number of grains per oral apparatus showed a spread but in most cases the frequency distribution was log-normal, i.e. monomodal. It can be suggested that the old and the new oral apparatus

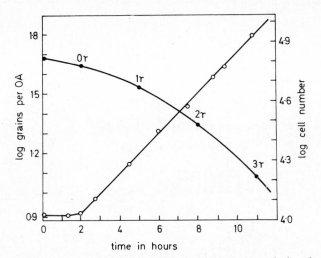

FIG. 30. Conservation of labeled protein in oral structures during the lag and logarithmic phases of growth which followed removal of tritium-labeled precursor amino acid and addition of an excess of its unlabeled counterpart. Cell multiplication (right ordinate) is indicated by the open circles, and number of grains per oral apparatus (left ordinate) by the filled circles. The scales of both ordinates are logarithmic. From Williams *et al.*[254] by courtesy of Cambridge University Press.

share proteins, and this may be by exchange to equilibrium across the cytoplasm which the two share until the two cells come apart in division. For heat-synchronized cells it can be proposed that labeled proteins deposited during the heat treatment are later taken up by the new oral apparatuses developing in phase with synchronous division in inorganic medium. Figure 32 shows the results of an experiment in which cells in growth medium were labeled with tritiated leucine during the first five heat shocks, and then washed and transferred to inorganic medium in which they were given two further heat shocks. A "chase" of non-radioactive leucine was added at EH. In the 230-minute period following EH, although the number of cells and oral apparatuses increased by a factor 3–4, the radioactivity, in grain counts per average oral apparatus did not decline much and it was monomodally distributed over the whole population of oral apparatuses at all time points studied except EH + 45 minutes.[254] Furthermore, pulse experiments with tritiated amino acids applied to growing synchronized cells showed that an oral primordium labels a bit faster than an oral apparatus but that the two become equal in this respect at the time of cell division when structural equality is also established, in the oral primordium by progressive, and in the oral apparatus by slightly regressive development.[83] The latter consists of loss of the deeper parts of a bundle of microtubular fibers which extends deep into the cell, reaching the vicinity of the macronucleus.

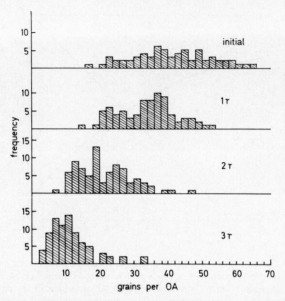

FIG. 31. Histograms of grain-count distributions from which means in Fig. 30 were calculated. Distributions found at the beginning of logarithmic growth (denoted "initial") and after 1, 2 and 3 doublings of cell number are presented. A decline in "specific activity" with time is evident. From Williams *et al.*[254] by courtesy of Cambridge University Press.

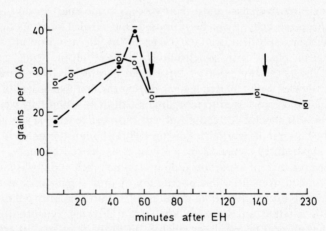

FIG. 32. Conservation of labeled protein in the oral structures of synchronized *Tetrahymena pyriformis*. See text for details. Radioactivity was determined separately for the old oral apparatus (open symbols) and the developing primordium (filled symbols) up to 60 minutes prior to the first synchronous division. The times at which the first and second synchronous divisions occurred are indicated by arrows. The limits of the standard errors of the mean value plotted are also shown. From Williams *et al.*[254] by courtesy of Cambridge University Press.

In conclusion, there is a fair chance that organelle-specific proteins shuttle back and forth between potentially identical oral structures in the same cytoplasm, continuously or at specific time intervals. If this is correct, the cytoplasm must contain proteins specified for the oral structures, and there is evidence that at least some precursors accumulate in the cytoplasm during the period with heat shocks. They are synthesized, but cannot be used, because the assembly of the new oral apparatus is blocked at the higher temperature. According to Frankel[78] in *Tetrahymena* primordial development to the stabilization point depends on concurrent synthesis of RNA and protein. The latter may be required in the assembly of the new OA which takes place after the heat shocks, at least in part from stored precursors. Whether such protein synthesized post-EH is incorporated into the structure or has catalytic function is not known.

(ix) When cells have been exposed to division-synchronizing heat shocks they need time to develop at a favourable temperature before they can divide synchronously, e.g. a period of 85 minutes at 28°. Proteins made during the period with heat shocks are insufficient to support division, either because they cannot be utilized normally, or because they do not include some specific proteins whose synthesis has been suppressed. After the heat shocks this situation is corrected by new synthesis of division-essential proteins, and at 28° these are available in critical amounts after 50–55 minutes. Consequently, up to this time, the transition point, but not later, cell division can be blocked by interference with the quality or quantity of proteins synthesized. Transition points inseparable in time are obtained on analysis of the effect of either an additional heat shock or a cold shock. Both treatments may slow protein synthesis, but a multitude of additional effects can be expected, and at this point we shall simply conclude that heat and cold applied after the standard heat shock program both make division-relevant proteins synthesized after EH useless for the forthcoming division. This effect of heat is likely to be common to all shocks in the regular program, and could thus account for the synchronization phenomenon. Even small changes from the optimal temperature of 28–29° produce detectable disturbances in the cells' progress towards division, but set-back reactions sufficient to synchronize the cells with respect to division require more drastic treatments, e.g. temporary shifts up to at least 32° or down to below 16°, a maximal effect being obtained in the case of exposure to 34° after 15 minutes.[268] Heating to 34° prevents normal structural development visible in the electron microscope. Microtubular and microfilamentous bundles connect non-ciliated kinetosomes in the oral primordia from late stage 3 (EH + 55 minutes) and onwards, but are absent in the anarchic oral primordia existing between EH and EH + 45 minutes. It can be deduced that the heat shocks interfere with the assembly of fibers and bundles of fibers. Furthermore, oral primordia at stages 2, 3 and 4, when exposed to a heat shock, suffer damage which triggers visible disorganization and later resorp-

3*

tion of the primordia. This can be seen with the light microscope and it can be suggested that heat also depolymerizes fibers not yet sufficiently bundled to be visible. Workers in this field now agree that completion of oral morphogenesis is *not* a condition for cell division in either *Stentor*,[234] *Paramecium*[100] or *Glaucoma*.[75] It is also generally held that there is a strong correlation in *Tetrahymena* between effects of external influences on cell division and oral morphogenesis respectively, suggesting a common basis (or closely related bases) for these, cf. Gavin and Frankel.[85] Williams and Zeuthen[256] have proposed that reactions involving synthesis, assembly and bundling of microtubules and microfilaments may be implicated in both oral morphogenesis and cell division. The oral primordium, and especially the finished oral apparatus, are rich in both components, and there are many demonstrations that microtubuli are present in and around the dividing macronucleus in various ciliates, including *Tetrahymena*. In the oral primordium microtubuli assist in orienting non-ciliated kinetosomes ("centrioles") relative to each other, and the plan for the oral membranelles is thereby laid down. Essentially comparable processes polarize cells for mitosis.

To account for the experimental findings accumulated to date, the hypothetical model shown schematically in Fig. 33 has been put forward. Here it is postulated that both oral morphogenesis (stomatogenesis) and cell division

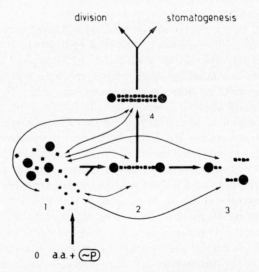

FIG. 33. Diagrammatic representation of the build-up of structures postulated to be essential for cell divisions and oral morphogenesis in *Tetrahymena*. The *heavy* arrows between reacting entities 1 and 2 and 2 and 4 indicate the path of assembly while the *thin* arrows denote that proteins already built into more complex units can be in continuous exchange with proteins in the environment. From Zeuthen and Williams.[273]

require assembly of some "structure" from a number of components as follows. At least two of the components (Fig. 33, small squares and small circles) are thought to be proteins and of these one (circles) must be continuously synthesized to promote the above assembly, indicated by heavy unidirectional arrows. (Alternatively, the limiting protein of the synchronized *Tetrahymena* system might have a catalytic function in the assembly process; in this case the polymerized structure would be built up from only two components, i.e. those represented by the small squares and large circles.) The protein components are polymerized into a strand which connects two other subunits (large circles) to form an intermediate structure which is highly labile by virtue of continuous exchange between its proteins and identical proteins in its environment. The intermediate develops into a more stable structure which is shown as two parallel strands connecting the two larger units. The thin reversibility arrows indicate that exchange is assumed to take place at all stages of assembly, but experimental evidence for this is available only at the level of the finished oral apparatus. The developing structure essential for morphogenesis and for cell division postulated in this model should be highly sensitive even to partial limitation for one of its components, and short-term limitation of precursor-supply or a temporary change in the affinity of the precursor for the structure, should therefore "trigger" full structural collapse (reaction from 2 to 3) because proteins then leave the structure faster than they enter. This model accommodates Rasmussen and Zeuthen's suggestion that an intermediary on the way to division is subject to a balance between synthesis and decay, it satisfies Cann's proposal that the decay is stimulated greatly by division-synchronizing influences, and finally, it meets Frankel's claim that partial starvation for relevant proteins triggers collapse and visible resorption of labile oral primordia. The model will also apply to stages in which there is no apparent development of the (anarchic) oral primordium if visible morphogenesis is preceded by polymerizations into labile structures not yet observable in the electron microscope. Analysis of the cells' behavior has shown that their preparation for synchronous division is continuous from EH to the transition time, and most of this time the oral primordium stays anarchic.

The fate of the primary products (denoted 3 in Fig. 33) of the structural decay which takes place when the sensitive developing oral primordia are exposed to heat, cold and inhibitors of protein synthesis, and which is postulated to occur also in cells whose oral primordia have not yet begun to differentiate visibly, is of considerable interest. These are likely themselves to be aggregates of proteins. We do not know whether these products simply dissociate and liberate proteins which can be reutilized directly in the assembly of new structures or whether they are resorbed (Frankel's expression), for example by lysosomal activity, in which case building blocks for new structures would not be made directly available. Such dissociation if rapid would permit rapid reutilization and there would be no demand for a synthesis of more protein (as indicated by reactions 0 to 1) concurrent with the reassembly of usable structures. Slow dissociation or enzymic degradation *would* require new protein synthesis, and we therefore favour one of these latter possibilities.

If heat loosens weak bonds within macromolecular aggregates, which seems possible, this would tend to shift the balance between reactions 1 to 2 and 2 to 3 in the scheme of Fig. 33 by slowing the former process, and also by stimulating the latter. This might make useful proteins less available to the cell, and assist in creating the condition observed: that more protein must be synthesized before new structures can be assembled. The closest parallel to this which we have found in the literature comes from studies of the assembly of chromosomal spindle fibers from cytoplasmic precursor protein units during mitosis. In Fig. 34 we have reproduced Inoué's schema-

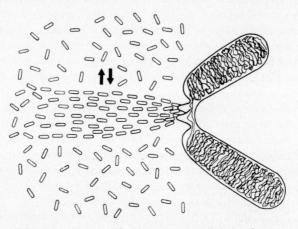

FIG. 34. Schematic diagram of orientation equilibrium of chromosomal fiber. From Inoué[116] by courtesy of Academic Press.

tic diagram of the "orientation equilibrium" of a chromosomal fiber. Fiber orientation is organized by kinetochores of chromosomes, and the oriented material is in a temperature-sensitive equilibrium with the surrounding non-oriented material. We think that this situation is comparable to that involving components essential for the morphogenesis and division of *Tetrahymena* and its nuclei, discussed in the foregoing. Here it can be mentioned that Gavin[84] has found that in *T. pyriformis* WH − 6 temperature shocks are ineffective in blocking the division of the micronucleus if they are applied after there is a space completely separating the chromosomal groups.

In the literature many instances can be found in which chilling or heating affects structures containing microtubules. Chilling disassembles the mitotic apparatus in marine eggs[116] and in the amoeba *Chaos carolinensis*[89, 201] in which reassembly occurs within a standard time after the organisms have been rewarmed. In *Actinospherium* chilling causes disappearance of the microtubular skeleton of the axopods, and sequential reorganization occurs when the cell has been warmed again.[240] In *Nephrotoma* cytoplasmic microtubules in the spermatid tails disappear in response to chilling, while microtubules running parallel in the 9 + 2 axial pattern are less sensitive.[7] In the marine eggs mentioned above heating produces an effect on the chromosomal fiber system similar to the effect of cooling.[116] Heat treatment also causes regression of the aster and spindle with reduction of the distance between centrioles at opposite poles, in eggs of the freshwater fish *Trichogaster* (extensive report of Bergan[8]). In this system there is transition to less sensitivity around metaphase. A heat shock applied 16–4 minutes before metaphase clumps the chromatin in *Physarum* and blocks mitosis.[18] It has frequently been implied that hydrogen bonding plays a role in determining the stability of cellular structures. Hydrogen bonds are known to be weakened by an increase of temperature and strengthened in the presence of heavy water. Using a micronuclear strain of *T. pyriformis*, Brown[20] observed that the presence of heavy water produced an increase in the rigidity of the mitotic apparatus and of the peripheral plasmagel, presumably by strengthening intermolecular bonding. Moner has made many experiments on the antagonistic effects of heat shocks and heavy water on the division of *Tetrahymena*. In a recent report on work along these lines[164] he concludes that some heat-sensitive process, in addition to the synthesis of RNA and protein, conditions the division, and that the presence of heavy water can stabilize this process against the damaging effect of heat. Moner does not speculate as to the nature of the process concerned, but it might well involve such an assembly of a division-relevant structure as we have described above.

As already mentioned for *Trichogaster* the relative stability of a structure is sometimes related to its stage of development. Thus Geilenkirchen[87] found that in *Limnaea* eggs the delaying effect of a heat shock on the forthcoming division was maximal when the shock was administered at the time of prophase. Moreover, at this time but not later, the mitotic apparatus could be disrupted by the weak shearing force obtained in a gentle centrifugation of the live eggs; such a centrifugation also delayed the forthcoming division maximally when carried out on prophase cells. The greater the complexity which a developing structure has attained at the time when it is damaged by some external interference, the longer may be the time required for its repair, i.e. the longer may be the age-dependent set-back obtained in response to the interference. Such responses to injurious treatments are not confined to *Tetrahymena*. They are also known for *Paramecium*,[193] for cleaving eggs of *Limnaea*[88] and *Arbacia*,[86] and we feel that if looked for carefully they would prove to be of widespread occurrence. In one interesting case (*Naegleria*) age-dependent set-backs of flagellar morphogenesis have been produced by treatment with β-mercaptoethanol.[242]

b. Relations between transcriptions and cell division

Damage done to the division machinery by one heat shock or by a series of shocks, must be repaired at a suitable temperature before the cell can divide. Repair might begin either at the level at which the primary damage is inflicted, or at preceding levels in cell metabolism. We shall first (i) describe results of experiments made with cells kept in growth medium in the period following the synchronizing heat-shock treatment, then (ii) consider comparable studies in cells developing synchronous divisions in starvation medium, and finally (iii) examine the question whether the damage itself involves the transcriptional level or only the structural levels discussed in the preceding pages.

(i) *Actinomycin D inhibits synchronous cell division.* Nachtwey and Dickinson[167] studied cells in a nutrient medium and showed that the effect of actinomycin D depends on its concentration and on the time, with respect to the occurrence of division, of its addition to the culture. These two parameters are connected by a saturation-type curve which indicates that the latest time at which this compound can be given in order to block cell division is 47 minutes before cell separation which is equivalent to EH + 48 minutes, in the case of synchronous division number 1. Before a division, food-vacuole formation ceases for a time and during this period temporary exposures to 100 μg per ml of actinomycin D produce no effect on cell division. However, since insensitivity to the drug comes before the stop for vacuole formation, it probably reflects completion of division-relevant processes rather than a block for uptake of the drug. These processes are thought to be DNA-dependent syntheses of RNA templates required for the synthesis of division-relevant proteins. Other workers have reported inhibition of division 1 by actinomycin D at concentrations of 20 and 30,[132] 15[162] and 9.6[78] μg per ml added to the cultures before EH + 30, 40 and 45 minutes respectively.

Actinomycin D also inhibits RNA synthesis. Mita[154] transferred synchronized cells at EH from a natural to a defined medium. He found no inhibition of the first synchronous division in the presence of 50 μg actinomycin D per ml, added at EH + 30 minutes, but 71 per cent inhibition of RNA synthesis with only 5 μg per ml. ³H-uracil added just after division mostly entered a macronuclear heterogenous RNA fraction with S-values higher than 24.5, and this synthesis was inhibited by 85 per cent in the presence of 5 μg actinomycin D per ml.

Actinomycin D and heat shocks have synergistic effects on cell division and on the rate of RNA synthesis in synchronized *Tetrahymena* cells. This was established by Moner[163] who used incorporation of ³H-uridine into cold TCA-precipitable material as a measure of the rate of RNA synthesis, and applied actinomycin D at concentrations of 1 or 4 μg per ml. With respect to the effects on cell division, Moner found that addition of actinomycin lowered the temperature required to block division, and increased the delay of division produced by a short heat shock. With respect to RNA synthesis, the rate curve thereof for control cells showed a maximum value at EH + 30 minutes; this maximum was considerably reduced in the presence of actinomycin. Moreover, raising the temperature in the range 28 to 34° produced a decline in the incorporation of uridine into macromolecules, and at each temperature tested the incorporation was lowered further by the presence of actinomycin. Separately, both heat and actinomycin annulled the small burst of RNA synthesis occurring at EH + 30 minutes which is considered to include mRNA synthesis relevant for the synthesis of "division proteins". When caused by heat this effect is not brought about through a decrease in the permeability of the cells, because it was shown that the soluble pool of radioactivity was larger at the higher temperature.[163]

Fluorinated pyrimidines inhibit cell division in nutrient medium.[78, 109] The effects of 5-fluoro-uracil (FU), 5-fluoro-uridine (FUR) and 5-fluoro-2-deoxyuridine (FUDR) on division of *Tetrahymena* in a Bacto-tryptone growth medium, fortified with glucose, salts and vitamins have been studied by Frankel.[78] He finds that multiplication of exponentially growing cultures is fully inhibited by FU and FUDR in 0.05 mM concentration, and by 0.02 mM FUR. Uridine (0.5 mM) releases the inhibition of FU considerably and that of FUR and FUDR only moderately, while thymidine (0.5 mM) has no releasing effect. In synchronized cells inhibition of the first synchronous division by 0.1 mM FUR added shortly after EH, is released with 1 mM uridine, but similar inhibition of later synchronous divisions is not. Oral morphogenesis in normal cells, and morphogenesis preceding the first synchronous division, can be blocked with FUR (0.2 mM) and with actinomycin D (9.6 μg per ml, i.e. 8 × 10^{-6} M). Frankel interprets his results as reflecting fluoropyrimidine contamination of the template RNA in control of syntheses of protein for oral morphogenesis and for systems used in division, and inhibition of template RNA synthesis by actinomycin D. We hold that fluoropyrimidines will contaminate not only new RNA, but also pools of preformed RNA with which new RNA becomes mixed. If this is true Frankel's results need not mean that post-EH RNA synthesis is essential for the forthcoming division.

Macronuclear function after EH is indispensable for the first synchronous division. This has been demonstrated by Nachtwey[166] who finds that surgical removal of the macronucleus before EH + 42 minutes blocks furrowing and cell separation in all synchronized cells. Emacronucleation between EH + 42 and EH + 63 minutes allows some furrowing, but no cell separation, while emacronucleation between 63 and 72 minutes after EH allows delayed cell separation; later emacronucleation does not interfere with the normal cell separation taking place at EH + 93 minutes. These results were controlled by sham operations in which only cytoplasm was removed. Possibly then, synthetic products of transcriptions and/or translations which are essential for division have to be released from the macronucleus to the cytoplasm before EH + 72 minutes, perhaps in a sequence which follows the sequence of stages in division.

From the foregoing it is clear that the first synchronous division of cells in nutrient medium can be blocked by agents assumed to interfere with production of RNA templates.

(ii) Cells transferred to inorganic starvation medium before the last heat shock perform two synchronous divisions.[28, 98, 244] In these cells post-EH synthesis of DNA[38, 39, 266], and of RNA[39] can be demonstrated autoradiographically with ^{3}H-thymidine and ^{3}H-uridine, respectively. FUDR (5 mM), deoxyadenosine (0.2 mM) and deoxyguanosine (0.2 mM) inhibit incorporation of the radioactive precursors into DNA and RNA by 45–90 per cent.[39] However, the two synchronous divisions are not affected by these compounds,[38, 39] nor are they inhibited by 5 mM FUR (Zeuthen, unpublished observations). Generally, inhibition of cell division is correlated only with inhibition of protein synthesis, see Table 2. Two analogues which do inhibit the first synchronous division in starving cells are the purines 6-methyladenine, a consistently strong inhibitor, and 8-azaguanine whose effect is somewhat variable.[109, 265] Sensitivity transition points were found with these compounds as well as with inhibitors of protein synthesis, and interference with co-factor functions of adenine and guanine was postulated.[67, 265] In sum, these results led to the conclusions (i) that starved cells can divide synchronously two times without post-EH synthesis of nucleic acids, (ii) that division of the cells depends on nucleic acids stored in excess during the period with heat shocks, (iii) that post-EH syntheses of nucleic acids are not division-relevant, and (iv) that syntheses actually occurring are not so extensive that analogue-containing products can significantly contaminate existing stores of nucleic acids.

Under starvation condition *T. pyriformis* cells lose a variety of their components to the medium. Thus it has been reported that when exponentially growing cells are transferred to an inorganic salt solution they leak hypoxanthine uracil, orthophosphate, acid labile phosphate compounds and ribose.[50, 51] Similarly, synchronized cells suspended in a non-nutrient medium lose both ninhydrin-positive and ultraviolet absorbing material; loss of the latter is stimulated by an increase of the temperature from 28° to 34°.[37, 52]

From the nature of some of the compounds leaked and the extent to which leakage occurs, it seems probable that starving cells suffer a net loss of RNA. Concurrently, however, in synchronized cells ^{3}H-uridine is readily incorporated into acid precipitable material, as reported by several authors, and together these observations suggest that some species of RNA are undergoing rapid turnover here. Byfield and Scherbaum have proposed that heat shocks synchronize *Tetrahymena* by causing destruction of mRNAs, some of which are required for the synthesis of specific division-related proteins.[22–28] Synthesis of mRNA must therefore take place after the heat shocks before the division can occur. The following is a description of one of their key experiments: working with starving cells they added ^{14}C-uracil at EH; incorporation of this into TCA precipitable material occurred linearly for 30 minutes, at which time a new heat shock (34°) was applied, and simultaneously actinomycin D (120 μg per ml) was added. In response there was a loss from macromolecules of radioactivity previously incorporated. When actinomycin D was added to cells not exposed to a heat shock at the same time, there was no further increase nor any subsequent decline in the labeling of macromolecules. The effect of heat alone in the absence of actinomycin D was not reported. Byfield and Scherbaum have repeated this experiment with the same result[26] and produced some circumstantial evidence in support of their contention above, for which they argue through a whole series of papers.[22-28] On the basis of the experimental evidence available, their interpretations are considered unconvincing, however, because they make statements about mRNA which are based solely on measurements of the total macromolecular radioactivity resulting from incubation with ^{14}C-uridine. We also point out that set-back reactions seem to be involved in the induction of division synchrony whether this is obtained with heat or with cold shocks. Since Byfield and Scherbaum found no decay of RNA with cold shocks,[26] we conclude that set-back reflects something more than destruction of RNA templates.

(iii) Throughout this review we have argued that the heat synchronized cells require post-EH synthesis of proteins for morphogenetic and other preparations of cell division. We are as yet unable to decide whether this reflects (a) that temperature shocks suppress synthesis of required proteins, (b) that such proteins are in fact synthesized but become denatured, or (c) that essential proteins synthesized are continuously lost to the cell as a result of repeated sidetrackings at the shock temperature of the ordered processes by which macromolecules meet and organize themselves into structures. We consider the last of these possibilities the more likely, but whichever one holds, some proteins required for division must be made after EH, and for this functional RNA templates must be present.

The problem under discussion is the following: is post-EH synthesis of RNA necessary for the first synchronous division? Arguments for a positive answer to this question come from work with actinomycin D, and from studies of the effects of metabolic analogues on cells in nutrient medium, while arguments to the contrary are founded in the results of comparable studies on starved cells. Let us first discuss whether or not actinomycin D as generally used on *Tetrahymena* can be considered a specific inhibitor of RNA synthesis. We tend to doubt that it can, since to produce an effect on division it has to be applied in concentrations exceeding the 2 μg per ml which Cleffmann[49] found inhibited ^{3}H-thymidine incorporation into the macronuclear TCA-insoluble material of exponentially multiplying *T. pyriformis* HSM. His results demonstrate that even at this relatively low concentration actinomycin also exerts a measurable effect on DNA synthesis. Secondly, let us weigh Frankel's[78] observation that fluorinated pyrimidines added at EH inhibited the first synchronous division in cells maintained in nutrient medium, against the finding of Holz et al.[109] that FU and FUDR had no such effect in starvation medium. As already mentioned, Frankel interpreted his results to mean that synchronous division required post-EH synthesis of RNA templates, whereas Holz et al. concluded the opposite. Frankel's interpretation is considered unsatisfactory because it does not take account of the possibility that new RNA fouled with analogue bases may foul existing pools of non-contaminated preformed RNA. In this case the inhibitory effect of these analogues could be expected to increase in proportion to the ratio between fouled RNA accumulated and pre-existing RNA. Byfield and Scherbaum[28] have shown that synthesis of RNA is much more rapid in cells in an amino acid containing medium than under starvation conditions, which could explain why the fluorinated pyrimidines inhibit synchronous division in the nutrient medium but not in the starvation medium. On the other hand, the mere observation that the analogues in question had no effect on synchronous division in a starvation medium, as reported by Holz et al., does not preclude that synchronous division requires post-EH synthesis of RNA.

Of the papers reviewed in this section, only that by Mita[154] contained any information on division-related inhibition of RNA metabolism between synchronous divisions 1 and 2. It is regrettable that other workers have not studied this, because the period from division 1 to division 2 resembles a normal cell generation much more than does the interval from EH up to division 1.

If the synchronizing heat shocks *do* damage species of RNA which have specific functions in preparing cells for division, then it is clear that post-EH transcriptional activity must be a condition for the synchronous division. However, such activity is not in itself a demonstration of a primary effect of the heat shocks on RNA, because, as stated earlier, repair of any damage inflicted may involve cellular metabolism at levels closer to the genes than the level actually affected. At present it is not possible to decide whether or not heat synchronized cells *do* need to transcribe after EH in order to be able to divide, nor is there as yet any evidence that species of RNA with specific function for division are destroyed by the heat shocks. Despite the existence of a voluminous literature on the subject a final solution of the problems discussed here must await more detailed biochemical and morphological investigations of the processes morphogenesis and cell division.

c. *Relations between DNA replication and cell division*

It appeared from earlier sections that synchronization of *Tetrahymena* with heat shocks is directed at division itself, through division-essential cytoplasmic structures, which are somehow affected by the higher temperature so that they are first brought to a common state and then recover in synchrony after the treatment. Replication of the cells' macronuclear DNA becomes synchronized only after, and presumably in response to the occurrence of division. Asynchrony of DNA synthesis until just before synchronous cell division calls for quite atypical relationships, differing from cell to cell, between the timings of DNA replication and cell division; these relationships deserve a separate study with attention paid to the fact that after division 1 most cells proceed to perform DNA synthesis in synchrony. We know practically nothing about the mechanisms by which replication within the DNA synthesis cycle is reset by impulses coupled to division, and shall confine ourselves to reporting experiments designed to provide information about the cells' capacity to carry out DNA replication and division under conditions where the time interval between these two processes becomes shorter or longer than it is normally. This can be considered a first approach towards understanding the mechanisms normally involved in separating DNA replication and cell division, and towards establishing an experimental procedure by which synchrony of cell division and of DNA synthesis will match each other fully.

The success achieved in the temperature synchronization of division in *Tetrahymena* is not yet even approached in other cell systems treated analogously, and at present we consider that this may be so because other cells are less tolerant than *Tetrahymena* with respect to acceptance of odd time intervals between DNA replication and cell division. If this is correct the studies to be described should be of help not only for the improvement of synchronization in *Tetrahymena*, but also for the development of procedures giving a useful synchronization of other cells.

We shall first describe an independent method for inducing synchrony of DNA synthesis. The method is based on starvation for thymine compounds followed by feeding of thymidine.[269] DNA synthesis can be inhibited in *Tetrahymena* by starvation for thymine compounds; to be effective, this starvation must include exogenous as well as endogenous sources when the medium is a complex mixture containing thymine compounds, as is the case with our synchronized cultures. Endogenous production of thymidylate is prevented through inhibition of the enzyme thymidylate synthetase by means of a folic acid analogue, methotrexate (M)—also known as amethopterin— and access to thymine compounds from exogenous sources is prevented by addition of uridine (U) in high concentration to the medium. Experiments show that the combination M + U added to cultures growing exponentially in the complex medium blocks DNA synthesis almost immediately and cell division after a period of 90 minutes[241]—cf. Figs. 46 and 47. This suggests that DNA replication in the S period preceding a division can be blocked with M + U and that this replication of correlated functions must take place before the division in question can occur. Removal of the inhibitors or addition of excess thymidine is immediately followed by new DNA synthesis.

Fastest increase in cell number takes place a little more than 2 hours later; this time corresponds to the duration of normal S + G_2 + D. In such experiments with exponentially multiplying cells a fair degree of synchrony of both DNA synthesis and cell division can be obtained.[241] We can conclude from these results first that M + U blocks DNA synthesis in progress at all points in the S period though only after reserves of endogenous thymine compounds have been used up, and second, that cells can progress through G_2, division and G_1 in the presence of M + U, but that they cannot progress far into a new S period. Thus as time spent in the presence of M + U is increased more and more cells are collected before an S period, and these will be released together and will synthesize DNA synchronously when thymidine is made available.

We shall now proceed to studies with the division-synchronized cells in which asynchronous DNA synthesis precedes synchronous division. In the experiments to be described (Figs. 35, 36 and 37) the division activity of the cultures has been assessed on the basis of division indices, visually estimated, and it is pointed out that the conclusions drawn will hold up only in so far as these faithfully represent cell multiplication. The conclusions stated at the

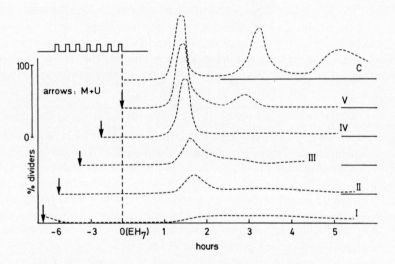

FIG. 35. Effect on the synchronized divisions of addition of 0.05 mM methotrexate (M) plus 20 mM uridine (U) at the various time points indicated by arrows prior to and during the synchronizing heat shocks. Topmost curve, C: in the absence of M + U the cells perform three synchronous divisions as shown here. Bottom curve, I: addition of M + U before the beginning of the heat shocks results in very low division activity. Curves II to V: progressively later addition of M + U results in progressively greater participation in the first synchronous division. From Zeuthen[270] by courtesy of Academic Press.

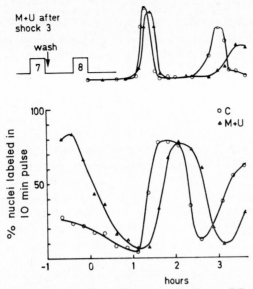

FIG. 36. Effect of the presence of 0.05 mM methotrexate (M) plus 20 mM uridine (U) during the interval from the end of the third to the end of the seventh heat shock on the ability of synchronized *Tetrahymena* cells to divide (upper curves) and to incorporate labeled thymidine into their macronuclei (lower curves). In each case the values for control cells are indicated by open circles, and those for the M + U treated population by filled triangles. The highest division index recorded had a value of about 0.9. From Zeuthen[270] by courtesy of Academic Press.

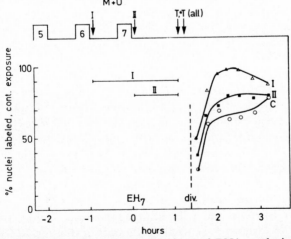

FIG. 37. Stimulating effect of prior inhibition of DNA synthesis during specific time intervals on the fraction of cells of a synchronized *Tetrahymena* population which subsequently takes part in the DNA synthesis immediately following the first synchronous division. Inhibition of DNA synthesis was achieved by addition of 0.05 mM methotrexate (M) plus 20 mM uridine (U) for the periods denoted by the horizontal bars I and II (i.e. starting at the end of heat shocks numbers 6 and 7 respectively) and released with excess thymidine at the time indicated by the arrow T. From Zeuthen[270] by courtesy of Academic Press.

end of the above paragraph apply also to cells under treatment with heat shocks in which division is blocked, but cyclic DNA synthesis continues to take place asynchronously. This section will lead up to experimental studies of the relation between synchronized DNA synthesis and synchronized division, but what interests us now is the question of the relevance of DNA replication for later divisions, and specifically of the relevance of DNA replication performed at different times in different cells (asynchronously) to a later division performed at the same time in all cells (synchronously). If the term G_2 is used to denote the interval from completion of DNA synthesis relevant to synchronous division 1 and the occurrence of this division, then G_2 must vary greatly from cell to cell in a population being prepared for synchronous division. The capacity of the individual cells to accept great variation in the length of G_2 can therefore be considered an essential feature of this system in which cell division and DNA synthesis come into phase only *after* the first synchronous division.

Causal relationships between rounds of DNA replication in the division-blocked cells and synchronous division can be studied with the reagent pair M + U, added at various times during the heat treatments, Fig. 35.[270] If M + U is added prior to the heat shocks the synchronous divisions are largely eliminated. Furthermore, the later the pair is added the more cells engage in synchronous division 1, and all cells participate in this division when M + U is added at the end of the heat shocks and also when it is added late in the shock program. In the experiments of Fig. 35 the time interval from addition of M + U to cell division lasts anything from less than 2 (curve V) to more than 7 hours (curve II). Cells blocked in or collected just before an S period later than 2 hours before the first synchronous division is due, are unaffected with regard to their capacity to divide once (curve V). On the other hand, cells thus blocked and collected from the much earlier time point indicated by the arrow on curve II, mostly fail to divide; the few which do divide in these circumstances are believed to have completed DNA replication of relevance for division 1 before the time of addition of M + U. Considered together, the curves I to V reflect that DNA synthesis is asynchronous during the period with heat shocks. It follows that cells complete DNA synthesis related to division 1 at various time points during a period which actually begins prior to the heat shocks. All cells in the culture have finished this synthesis between shock 5 and the end of shock 7, the final shock in this series of experiments. All other DNA synthesis must relate to later divisions, and it can be noted that some such synthesis, which occurs prior to division 1, is actually relevant for division 2. Thus in some cells the first synchronous division is interposed between a synthesis of DNA and the division for which it has relevance. Further experiments have shown that the division-synchronized DNA synthesis which takes place after division 1 comes to an end well before, and is a condition for, the second synchronous division.

In another series of experiments the reagent pair M + U was applied for a limited time during the period with heat shocks. Figure 36 shows the results of one such experiment in which DNA synthesis was inhibited with M + U from shock 3 through shock 7 in cells exposed to a program of eight shocks. The treatment with M + U was discontinued by washing with fresh growth medium after shock 7. Subsequently, samples of the treated cells and of untreated controls were briefly exposed to ^{3}H-thymidine at various time points, fixed and assayed for labeled macronuclei. The lower part of Fig. 36 shows percentage labeled macronuclei as a function of time, and the upper part the division indices in the treated and control populations. It can be seen that M + U collected an extra 50–60 per cent of all cells before an S period and when they were released these cells immediately entered DNA synthesis which they completed with a useful degree of synchrony. In a series of parallel experiments only the division indices were observed. The effect of M + U was counteracted by the addition of excess thymidine at various times between shock 7 and $EH_{(8)}$ + 75 minutes, and it was found that the interval between the addition of thymidine and the following division maximum remained constant throughout. This time interval was 2 hours which more or less corresponds to the duration of a normal S followed by a normal G_2 period.[32] To recapitulate, these experiments have shown that the DNA synthesis relevant for the forthcoming synchronous division, which is normally carried out asynchronously during the period with heat shocks, can be quite effectively synchronized by treatment with M + U. Furthermore, even when treatment is prolonged so that blockage of DNA synthesis is continued until well after the end of the heat shocks, the normal duration of the period $S + G_2$ is maintained.

Experimental conditions can also be arranged whereby almost all cells participate in synchronous DNA synthesis following the first synchronous division. Figure 37 shows an experiment in which DNA synthesis was stopped by addition of M + U after the last but one heat shock, and permitted to recommence in the presence of excess thymidine (0.5 mM) only from the time of the first synchronous division. At this point ^{3}H-thymidine was added to both the control and experimental cultures. In samples examined subsequently the number of labeled macronuclei rose rapidly to 100 per cent in the M + U treated cells as compared with 70 per cent in the controls. Thus in these treated cultures all cells were involved in the synchronous synthesis of DNA which took place after division 1 and which related functionally to division 2. Clearly in the control cultures some cells synthesized DNA during the last 2 hours prior to division 1. These cells cannot belong to the group which synthesizes DNA just after division 1; rather *they* are the cells which asynchronously synthesize DNA relevant for division 2, at times prior to division 1.

From the foregoing it is apparent that synchronization of DNA synthesis and cell division can be made separately in *Tetrahymena* cultures and that a functional relationship exists between DNA replication displayed early enough before the first synchronous division, and this division; a later round of DNA replication relates functionally to the second division. In heat-synchronized cells the rather confusing situation exists that synchronous division 1 splits the population in two groups with respect to the replication of DNA which has functional relationship to synchronous division 2. Cells of the first group synthesize this DNA asynchronously before division 1, and compared to normal cells the synthesis takes place unusually close to the process of division, whereas cells of the second group synthesize DNA in synchrony shortly after division 1. However, cells of the first group can be made to wait and synthesize DNA needed for the second synchronous division after division 1 and in time with the other cells. This, in a sense, normalizes the system and although the technique used is not considered fully acceptable, the results obtained supply a rational basis from which simpler procedures can be sought.

Concluding Section IIB we forward the views that synchronization of *Tetrahymena* by heat shocks requires that dynamic cellular structures essential for division are discarded at any stage of an extended process of assembly and must be remade before the cell can divide. The structures can be damaged and ultimately rendered useless by short-term heating which appears to cause limitation of essential proteins. This condition can be established by a multitude of means, including in principle any influence which can loosen bonds between proteins integrated within some structural entity, or interfere at any level between the gene and the finished product. To rebuild division-essential structures, after heat treatment the cells are compelled to synthesize protein ("translate"), they may or may not have to synthesize RNA ("transcribe") perhaps depending on environmental conditions, but they need not replicate their DNA in order to perform one synchronous division.

2. Synchrony of Cell Division Induced by Other Means

a. *Alternative temperature regimens*

In the preceding pages we have described extensive studies carried out on *T. pyriformis* GL, synchronized by the multi-heat shock method of Scherbaum and Zeuthen. Alternative temperature programs will also produce useful degrees of synchrony; thus during our preliminary work on the development of satisfactory synchronizing procedures, it was seen that a series of cold shocks was about as effective in phasing division as was a series of heat shocks. Furthermore, it is possible to keep cultures continuously cycling with respect to cell division by means of either cold (Fig. 6 of Zeuthen[263]) or heat[263, 272] shocks inflicted periodically once per cell generation. Figures 38 and 39 show some results obtained with this kind of cycling temperature program; a synchronous burst of division developed

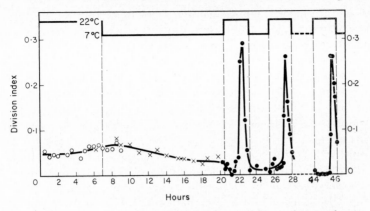

FIG. 38. Synchrony of division induced in populations of *T. pyriformis* by cold shocks applied according to the temperature regimen indicated. From Zeuthen and Scherbaum[272] by courtesy of Butterworth Scientific Publications.

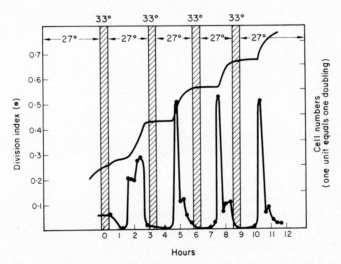

FIG. 39. Synchrony of division induced in populations of *T. pyriformis* by application of one heat shock per generation. Changes in temperature are indicated at the top of the figure. From Zeuthen[263] (based on data of Thormar and Zeuthen) by courtesy of Academic Press.

during each period at a temperature favorable for cell multiplication (here 22° and 27° respectively) and shortly after its occurrence a new cold or heat shock was initiated. In fact, the cold shock experiments of Fig. 38 provided the earliest indication that cell divisions in *T. pyriformis* populations could be synchronized by means of temperature changes. Here the cells were cultured at 22°, and then chilled to 7° for 13 hours; this resulted in first depressed then higher than normal division activity after return to 22°, the maximum division index observed being 0.3. From this point on, the periods during which the cells were allowed to remain at 22° were limited to $2\frac{1}{2}$–3 hours so that subsequent chillings were inflicted each time a burst of divisions had passed. A variation in the duration of the period at the low temperature from $2\frac{1}{2}$ to 16 hours had no effect on the division synchrony obtained. Such cultures can be kept cycling with respect to division provided the temperature treatment is continued for as long as the medium will support growth.

Padilla and Cameron have also worked out a synchrony-inducing temperature regimen involving one temperature cycle per cell generation. They used the micronucleate T. pyriformis, strain HSM, and their program consisted of $2\frac{1}{2}$-hour periods at 27° alternating with $9\frac{1}{2}$-hour periods at 12°.[173] At the beginning of each cold period the cultures were diluted with fresh medium, and in this way the population density was kept within desired limits. It is noteworthy that the two strains of T. pyriformis, GL and HSM, have different optimal growth temperatures, namely 28–29°[239] and 32.5°,[232] respectively; thus the difference between this procedure of Cameron and Padilla and that of Zeuthen and Scherbaum described in the foregoing paragraph is smaller than appears at first sight.

Although Tetrahymena populations induced to divide in synchrony by multiple heat shock and by one cold shock per generation procedures have many features in common, the two systems are also quite distinct in some respects. In the following we will compare some aspects of the results (shown in Fig. 40) obtained by Padilla and Cameron with their one cold shock per generation system with what has been described already for the multi-heat shock system. From Fig. 40 it can be seen that their cycling temperature treatment produced a burst of synchronous divisions during which a maximum of 40 per cent of the cells divided simultaneously about 100 minutes after the start of a warm period. At this time a minimum of the cells were engaged in macronuclear DNA synthesis, as shown by the dotted curve; from this curve it may also be concluded that macronuclear DNA synthesis does not become phased in this system, as pointed out by the authors themselves. Here it is recalled that in multi-heat shock synchronized T. pyriformis GL populations too very few cells synthesized DNA during the periods of synchronous division. In any synchronizing system, cells which do synthesize DNA at these times probably behave in one of the following ways: either they do not divide together with the major part of the population, or they engage in DNA synthesis soon after the macronucleus has divided, prior to the completion of cytoplasmic fission. The fact that the macronuclei of cells subjected to Padilla and Cameron's cold shock procedure do not synthesize DNA in synchrony may be correlated with the observation that structurally they resemble the macronuclei of logarithmically multiplying cells[34] more than those of cells synchronized by the

Scherbaum–Zeuthen method. Micronuclear DNA synthesis, on the other hand, occurred simultaneously in 60 per cent of the cells shortly after synchronous cell division (Fig. 40); this timing compares well with that observed in exponentially multiplying cells, whose micronuclei synthesize DNA shortly after going through mitosis. Just as in exponentially growing and heat synchronized *T. pyriformis* GL, oral morphogenesis preceded division in Padilla and Cameron's system, and both oral development and cell division were seen to be sensitive to actinomycin D added late in the cold period.[249]

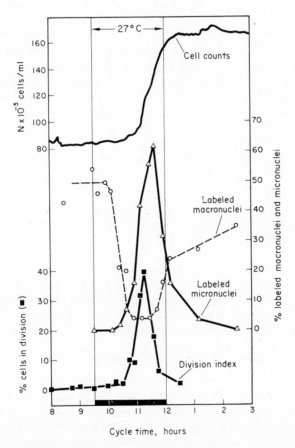

FIG. 40. Change in cell number, in the labeling of micro- and macronuclei with ³H-thymidine, and in division index in a population of *T. pyriformis* cells induced to divide in synchrony by the application of one cold shock per cell generation. The black bar indicates a "warm" period, temperature 27°. From Padilla and Cameron[173] by courtesy of the Wistar Institute.

In further work on *T. pyriformis* HSM synchronized by the continued application of one cold shock per generation, zonal centrifugation techniques were employed for the isolation of subcellular components from indolelysed cells. Thus using a low-speed rotor, Whitson *et al.*[250] were able to prepare isolates of mouth structures which, judging from the photomicrograph accompanying their text, were comparable in purity to those obtained in a differential centrifugation of butanol-lysed cells.[256] With the high-speed zonal ultracentrifuge various classes of mono- and polyribosomes could be separated from the indole lysate, and a study was made of the extent to which the presence of polyribosomes (generally associated with active protein synthesis) could be correlated with the occurrence of stomatogenesis and cell division in synchronized cells.[251] Polysomes could be detected in material harvested during the cold period, but they could be better resolved in samples taken 1 hour after initiation of the warm period, when the cells had recommenced stomatogenesis. The cells' polysome complement was reduced just after division, and also very noticeably by the presence of 10 μg per ml actinomycin D which inhibited both stomatogenesis and division. Hence it appears that a cyclic formation and breakdown of polyribosomes coupled to the processes of morphogenesis and cell division may take place both in cells synchronized with one cold shock per generation, and in those subjected to multiple heat shocks, cf. p. 41.

The division synchrony obtained with cold shocks is improved if a program involving multiple shocks is adopted in place of the one shock per cell generation procedure described above (cf. figs. 2, 5, 6d and 7d in a previous review article[268]). Here again different groups of workers have achieved good synchrony with a variety of temperature schedules, not considered to differ from one another in any fundamental way. In the following we shall refer to the work of three such groups, describing their synchronization programs and comparing their physiological and biochemical findings with one another, or with relevant data from heat synchronized systems.

The synchronizing effect of a multiple cold shock procedure worked out by Zeuthen is shown in Fig. 41, curve (a). *T. pyriformis* GL cells multiplying exponentially at 28° were subjected to 5 temperature cycles each consisting of a 2-hour cold shock at 7.5° followed by a 40-minute period at 28°. During this treatment cell division is blocked,[268] DNA synthesis is initiated in less than 20 per cent of the cells[103] and increase in dry weight is only slight.[65, 268] The cells were thereafter returned to 28° (at time EC, equal to zero minutes in Fig. 41) and at this temperature a maximum of 80 per cent showed fission simultaneously around 85–90 minutes after EC. Later on there occurred a second synchronous division burst during which a maximum division index of 0.5 was recorded. Each of the two division bursts gave rise to a doubling of the cell number. The quality of division synchrony achieved with this procedure is thus almost as good as in the multiple heat-shock synchronization (curve (b) of Fig. 41), and both systems approximate

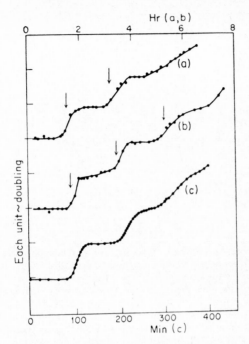

FIG. 41. Increase in cell number with time in *Tetrahymena* cultures synchronized for division in three different ways. Curve (a): synchrony induced by five cold shocks (10° for 120-minute periods alternating with 28° for 40-minute intervals). Curve (b): synchrony induced by seven heat shocks (34° for 20 minutes alternating with 28° for 40 minutes). Data of curves (a) and (b) obtained with strain GL cells; zero time represents the termination of the synchronizing treatment and the arrows indicate the times at which the division indices were maximal. Curve (c): increase in cell number in fifty clones each grown from a single cell picked from a logarithmically multiplying population. On the common time scale zero represents the point at which each isolated cell divided for the first time, and the curve is a plot of total numbers of cells in all fifty clones versus the appropriate relative time points. Data of curve (c) obtained with strain HS cells by Prescott.[187] The figure is reproduced from Zeuthen[268] by courtesy of John Wiley and Sons, Inc.

the ideal case of a naturally synchronous population, which can be established on a small scale by selecting dividing cells from an exponentially multiplying culture (curve (c) of Fig. 41). Both the first and second synchronous divisions of cells subjected to the above multiple cold shock treatment, could be inhibited with *p*-fluorophenylalanine (16 mM) added early enough in the period of preparation for a division. Later on the cells abruptly became insensitive to this compound, at times corresponding closely in position to analogous transition points in developing heat synchronized cells, i.e. at EC + 40 minutes and EC + 150 minutes (Zeuthen, unpublished).

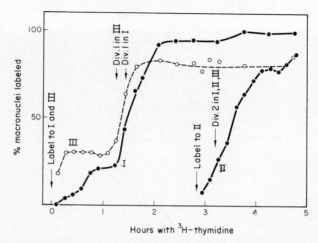

Hours with ³H-thymidine

FIG. 42. Cumulative labeling with ³H-thymidine of macronuclei in populations of *T. pyriformis* synchronized by five cold shocks. In one experiment the culture was divided into two aliquots, the first of which was incubated with ³H-thymidine from the end of the last cold shock, i.e. from time zero (curve I), and the second from shortly before division 2 (curve II). In another experiment (curve III) in which the tracer was added at zero time, the growth medium had been replaced with an inorganic medium between cold shocks 4 and 5. From Hjelm and Zeuthen[103] by courtesy of the Danish Science Press.

Synthesis of DNA has been studied also in cultures synchronized with multiple cold shocks by means of autoradiography in conjunction with continued incubation of cells with ³H-thymidine (see Fig. 42). Curve I shows that up to 20 per cent of the cells engaged asynchronously in DNA synthesis for the first 50 minutes after addition of the labeled compound at EC; during the next 25 minutes there was no increase in the proportion of labeled macronuclei and immediately after the first division a further 70 per cent of the cells began to synthesize DNA fairly synchronously. Here there was no apparent loss of previously incorporated label such as was observed in the case of heat synchronized cells[103–4] between 50 and 75 minutes after the end of the synchronizing treatment. At present we interpret this apparent loss as an artefact of increased self-absorption of sample radioactivity due to condensation of chromatin in the predivision nuclei (cf. p. 35). This difference between systems synchronized with multiple cold and multiple heat shocks, respectively, could therefore mean that prior to division the chromatin is condensed to a lesser extent in the former case. A somewhat less pronounced synchronous engagement in DNA synthesis was observed after division 2 (curve II). When transferred to an inorganic medium before the final cold shock, the cells retained the ability to participate in synchronous division 1 within standard time. In this case, however, in contrast to that of heat synchronized cells treated similarly[103] a considerable engagement in synchronous DNA synthesis occurred after division 1 (curve III).

Another effective multi-cold-shock synchronizing procedure is that of Moner and Berger,[165] in which six cold shocks at 9.5° were alternated with periods at 28°, decreasing in length from 60 to 30 minutes. During the first synchronous division burst a maximum of 70–80 per cent of the cells showed fission simultaneously at 90 minutes after EC. Moner and Berger investigated the response of this system to actinomycin D and observed a sensitivity transition point at EC + 40 minutes, i.e. coincident in time with the one found by Zeuthen with p-fluorophenylalanine, in his cold synchronized cells. Moreover, they demonstrated that the rate at which ^{14}C-uridine was incorporated into macromolecules was maximal at EC + 40 minutes, and that inhibition of incorporation at this time with actinomycin was correlated with inhibition of the subsequent division, just as in the heat-synchronized cell system studied earlier.[162] Thus many features of synchronized cell division and the foregoing period of preparation are observed consistently whether the synchrony is achieved through the use of heat shocks or cold shocks.

The third useful synchronization treatment comprising multiple cold shocks which we wish to mention here involves cycling between 28 and 10°, division being allowed to develop at either temperature.[268] In a modification of this procedure, DeBault and Ringertz[65] showed that the temperature chosen for this latter part of the treatment influenced the ratio of DNA to dry mass at division; this ratio was higher in cells dividing at 10° than in those dividing at 22°. Using Caspersssons's techniques these authors were also able to demonstrate that the division process is just as exact in cold synchronized cells as in normal cells, as far as the distribution of DNA and dry mass to the daughter cells is concerned, and that the distribution of DNA is markedly less precise than the distribution of dry mass at whichever temperature division occurs. Furthermore, DeBault and Ringertz studied the quality of the first cold synchronized division as a function of modulations of the synchronizing temperature regimen applied, altering the number of cold shocks between 3 and 5 and the temperature at which the cells were allowed to divide between 10° and 22°, as indicated in Fig. 43. The results of these experiments are depicted in Fig. 44, from which it can be seen that synchronous division was better after five (B curves) than after three (A curves) cold shocks, and better at 22° than at 10° (A_1 and B_1 better than A_2 and B_2 respectively), though remarkably good at the lower temperature after five cold shocks (curve B_2).

To bring about cold-shock-induced division synchrony in T. pyriformis GL cultures, the temperature must be shifted across the particular value of 16°,[268] at which there occurs a discontinuity in Thormar's Arrhenius curve relating cell multiplication and temperature.[239] In our experience the

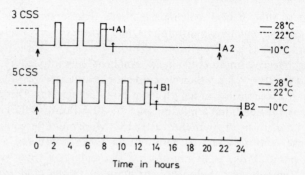

FIG. 43. Cold shock temperature regimens used to induce division synchrony in *T. pyriformis*. In the 3CSS system three cold periods of 120 minutes were alternated with 40-minute warm periods. In the 5CSS system five cold shocks were used. The cells divided either at 22° (times A1 and B1) or at 10° (times A2 and B2). From DeBault and Ringertz[65] by courtesy of Academic Press.

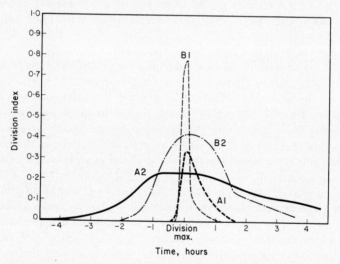

FIG. 44. Division indices during the synchronous divisions induced to take place at the times and by the treatments indicated in Fig. 43. From DeBault and Ringertz[65] by courtesy of Academic Press.

actual temperature levels between which the shifts are made do not need to differ from 16° by any critical number of degrees, although a minimal difference between the two levels may be required. Chilling is known to cause a set-back reaction in both exponentially multiplying[237] and heat synchronized[76] cells, and we feel that the set-back response is as important for the induction of synchrony with cold as with heat treatments. On this point we are at variance with Padilla *et al.*[174] who claim that the synchrony induction achieved with their cold-shock procedure,[173] developed according to the reasoning of James, is fundamentally different from that given by our multi-heat-shock treatment, though they have no telling argument to present in support of this contention.

In the foregoing we have emphasized similarities in the behaviour of multi-cold- and multi-heat-shock synchronized cell systems. The main difference between them has to do with the extent to which cell mass can increase and DNA accumulate in excess of normal values during the respective synchronizing treatments.

b. Depletion of essential nutrients

(i) *Pyrimidines in general.* When cells are deprived of one or more essential nutrients they eventually cease to divide, but they may be able to multiply anew when the compounds in question are subsequently added back to the culture medium. If resumption of division activity tends to be synchronous such a temporary differential starvation can be made the basis of a synchronizing procedure, as has been done, for example, by Cameron.[30] Building on earlier results of Lederberg and Mazia,[133] he transferred exponentially multiplying *T. pyriformis* HSM cells to a synthetic medium lacking pyrimidines, and found that cell number increased three- to four-fold while the amount of DNA doubled, after which pyrimidine reserves were exhausted, and the culture entered a non-growing state. Seventy-two hours after transfer pyrimidines were again made available to the cells, and this initiated a sequence of biosynthetic activities: synthesis of RNA was resumed soon after the addition of pyrimidines, synthesis of protein 1 hour later, and synthesis of DNA after a 2-hour lag. Figure 45

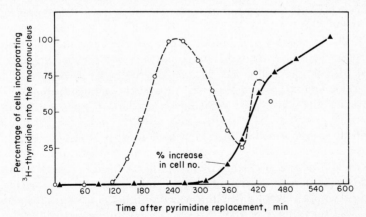

FIG. 45. Uptake of tritiated thymidine into macronuclei (open circles) and subsequent increase in cell number (solid triangles) in a culture of *T. pyriformis* HSM previously deprived of pyrimidines for 72 hours, as functions of time after the replacement of these essential nutrients. From Cameron[30] by courtesy of the Rockefeller University Press.

shows the way in which the percentage of cells synthesizing DNA and later the cell number changed with time after replacement of pyrimidines. Cell number remained constant for the first 5 hours and then increased gradually during the next 5-hour period. Participation in DNA synthesis was assessed in terms of the percentage of macronuclei which became radioactive after a brief exposure to ^{3}H-thymidine. Two hours after pyrimidine replacement some cells began to synthesizie DNA, and after a further 2 hours the whole population was engaged in this activity. About 40 per cent of the cells entered a new DNA synthesis period at the time of maximum increase in cell number. Since all cells engaged in DNA synthesis before any increase in cell number could be detected, it can be concluded that pyrimidine depletion stalled all the cells in the G_1 phase of the generation cycle; following replenishment of the pyrimidine supply these cells were then induced to take part quite synchronously first in DNA synthesis and later in the process of division.

(ii) *Thymine compounds.* In the section dealing with relations between DNA replication and synchronized cell division we have described in detail a procedure by which the cells' access to sources of thymine needed for DNA synthesis was temporarily blocked, with the result that both DNA synthesis and cell division became synchronized. In a complex medium the presence of 0.05 mM methotrexate and 20 mM uridine creates a block in the synthesis of DNA as mentioned earlier, and fluorodeoxyuridine plus uridine are known to have the same effect (Zeuthen, unpublished experiments). In a chemically defined medium the presence of either methotrexate or fluorodeoxyuridine alone is sufficient to bring about the condition of thymine starvation (Dr. Ole Westergaard, personal communication).

Biochemical analysis of *Tetrahymena* cells treated with methotrexate and uridine (M + U) yielded the results shown in Fig. 46. Immediately after the addition of the two compounds, DNA synthesis was completely inhibited for a period of about 3 hours. Effects on RNA and protein synthesis were not so dramatic initially but became progressively marked with time; measurable inhibitions were observed soon after treatment with M + U in the case of RNA synthesis, and some hours later in the case of protein synthesis.[241] Cell division was blocked only after the elapse of an interval equivalent to a normal G_2 period, which is what would be expected if completion of a round of DNA replication is a condition for completion of a cell cycle. Addition of 2.0 mM thymidine to the treated culture releases the cells from the state of thymine starvation and permits them to engage anew in DNA synthesis. When thymidine was supplied 3 hours after the addition of M + U (i.e. after a period of thymine starvation lasting a little longer than the normal generation time) DNA synthesis was immediately and synchronously resumed, and was followed by synchronous division (see Fig. 47). Percentage increases in the amount of DNA and in cell number were 78 and 93 respectively. The time interval from the addition of thymidine

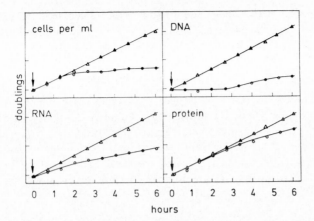

FIG. 46. Effect of methotrexate plus uridine on various growth parameters in a population of *T. pyriformis* in exponential multiplication. Triangles: untreated controls; circles: cultures exposed to the reagent pair from time zero. The ordinate scales are logarithmic, each unit representing one doubling per unit volume of culture. From Villadsen and Zeuthen[241] by courtesy of Academic Press.

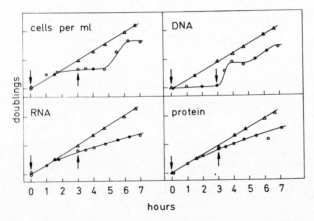

FIG. 47. As Fig. 46 except that the effects of exposure to methotrexate plus uridine from time zero were counteracted by the addition of thymidine 3 hours later. From Villadsen and Zeuthen[241] by courtesy of Academic Press.

to the earliest appearance of newly divided cells (G_2 + D periods) and to the completion of synchronous division (S + G_2 + D periods) lasted 110–120 and 170–180 minutes respectively, while synchronous DNA synthesis (S period) required 60–70 minutes. These periods are slightly extended as compared with their durations in the normal cell cycle. When added separately both methotrexate and uridine disturb and extend the cell cycle to some degree, but the particular effects on DNA replication and cell division described above are specific for their combined influence.

c. *Hypoxia and vinblastine*

Intermittent restriction of cells' access to oxygen can bring about a phasing of their divisions. Thus Rasmussen[191] obtained a fair degree of division synchrony upon exposing a population of *T. pyriformis* GL cells to two 50-minute periods of hypoxia (pO_2 less than 0.15 mm Hg), separated by an interval of 40 minutes during which the cells had access to oxygen (see Fig. 48). Following the second hypoxic period no divisions occurred for

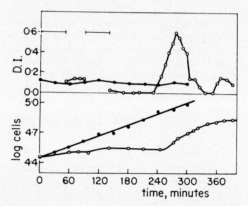

FIG. 48. Synchronization of division in *Tetrahymena* cultures in response to intermittent exposure to hypoxic conditions. The initially exponentially multiplying cells of the experimental population (open circles) were subjected to two 50-minute periods of hypoxia (horizontal bars) separated by an interval of 40 minutes, while the control population (filled circles) remained in contact with the atmosphere throughout. The upper and lower frames show division indices and population densities (logarithmic scale) respectively, as functions of time. From Rasmussen[191] by courtesy of the Danish Science Press.

some time; later the cells doubled in number during a synchronous division in which a maximum of 60 per cent divided simultaneously 2 hours after oxygen was readmitted to the culture. Rooney and Eiler[197] achieved better synchrony with six hypoxic periods (pO_2 equal to 0.7 mm Hg) spaced 40 minutes apart and each lasting 45 minutes. In a later report they described a device with which hypoxic shocks could be administered automatically.[198]

Treatment with the *Vinca* alkaloid, vinblastine, can also induce phasing of division in *Tetrahymena* cultures. A synchronizing method utilizing this compound has been developed by Stone;[231] its efficacy is illustrated in

Fig. 49. The mechanism underlying the synchronizing influence of vinblastine is hitherto unknown. Stone has discussed the possibility of a direct action on microtubular structures, but so far indirect effects via the nucleic acid metabolism cannot be excluded.[243]

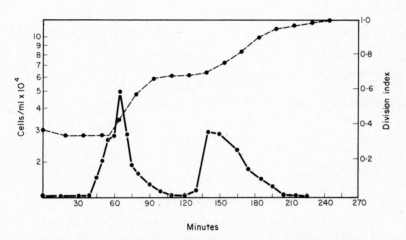

FIG. 49. Synchronized divisions in *T. pyriformis* WH6 cultures resulting from exposure to vinblastine during the exponential phase of growth. Population densities (left ordinate, dashed curve) and division indices (right ordinate, continuous curve) are plotted as functions of time after removal of the agent. From Stone[231] by courtesy of the Rockefeller University Press.

d. *Miscellaneous*

In this review we have proposed that synchronization of cell division in *T. pyriformis* with heat shocks results from disaggregation or depolymerization of structures in the making. Aggregation or polymerization requires *de novo* synthesis of protein having either enzymic or structural function. Properly timed transient interference with protein synthesis or with the polymerization process may therefore cause set-backs in development which increase with the age of the treated cell. Applied to exponentially multiplying cells, such interference may induce synchronous behavior in the culture. The heat synchronized system has proved very convenient for scanning agents for their ability to produce age-dependent set-backs in the development of *Tetrahymena* cells. Those already known to exert such effects, and which are either established or potential synchronizing influences, are listed in Table 3; though they are a heterogeneous group the effects of all of them could be accommodated by the unifying hypothesis put forward in a preceding section to account for the phenomenon of synchronization.

TABLE 3. AGENTS KNOWN TO PRODUCE SET-BACKS INCREASING WITH CELL AGE IN *Tetrahymena pyriformis*, WITH THE RESPECTIVE SOURCES OF INFORMATION

Agent	Exponentially multiplying cells	Division synchronized cells
2,4-Dinitrophenol		Hamburger and Zeuthen[98]
Hypoxia	Rasmussen[191]	Rasmussen[191]
Azide		Hamburger[97]
Sodium fluoride		Hamburger[97]
		Watanabe[244]
Fluoroacetate		Hamburger[97]
Fluorophenylalanine	Rasmussen and	Rasmussen and
	Zeuthen,[194] Frankel[78]	Zeuthen[194]
Heat shocks	Thormar[237]	Frankel[76]
		Rasmussen and
		Zeuthen[194]
Cold shocks	Thormar[237]	Frankel[76]
Mercaptoethanol		Mazia and Zeuthen[150]
High hydrostatic pressure		Lowe and Zimmerman[141]

3. Concluding Remarks

Ten years ago it was suggested that between consecutive divisions *Tetrahymena* cells build complex division-relevant structures which are held together by weak temperature-sensitive bonds, e.g. hydrogen bonds.[263] Since then the reactions of the oral primordium to temperature changes have been described and it has been shown that ultramicroscopic fibers are important components of this structure, and of other parts of the cell, especially in the stages of development preceding cell division. We tentatively regard the state of the oral primordium as an indicator of the reversible build-up from a store of macromolecular entities of a structural complex believed to be a prerequisite for the occurrence of cell division. Indeed this complex might be the equivalent of a mitotic apparatus, the existence of which has not yet been established in amicronucleate *Tetrahymena*. The development of either an oral apparatus or a mitotic spindle may typify reactions involved in general cellular morphogenesis. Such development can be visualized to take place in a stepwise manner as follows. Initially the formation of fibers or macromolecular aggregates is mediated through a weak interaction involving hydrogen and hydrophobic bonding between the side chains of the constituent macromolecules. Subsequently stronger bonds, e.g. covalent linkages, may be established and secure the stability of the evolving structural complex. We shall now proceed to consider how changes of temperature might affect structures for whose cohesion hydrogen and hydrophobic bondings are of importance.

It has been proposed earlier that the effects of heat on *Tetrahymena* cells are partly due to a disruption of hydrogen bonds. In the light of this idea the experimental results illustrated in Fig. 50 are of interest. The figure shows the extent to which the polypeptides polyglycine (G) and polyalanine (A) can exist in the ordered α-helical configuration, as opposed to the disordered random coil state, at different temperatures. Here it can be seen that the fraction of molecules having the α-helical configuration which depends on formation of hydrogen bonds along the length of the polypeptide chain, decreased with increasing temperature over a temperature range specific for each polypeptide tested. Moreover, the transition with increase in temperature from the α-helical to the random coil state was more abrupt the longer the polypeptide chain, or, in other words, the more complex the molecule. In a comparable manner, in *Tetrahymena* small increases of temperature could cause analogous changes from ordered to disordered states of developing structures unquestionably more complex than the polypeptide chains just discussed. We have described how temporary increases in temperature from 28° to 34° synchronize *Tetrahymena* populations and disrupt developing oral structures. In fact the critical events occur within a narrow temperature interval of less than one degree, and the old idea that the synchronizing heat treatment affects the cohesion of hydrogen bonded structures can still be upheld.

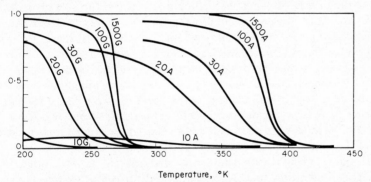

Temperature, °K

FIG. 50. Effect of temperature on the configuration of the polypeptides polyglycine (G) and poly-L-alanine (A) for the various degrees of polymerization indicated on the curves. Ordinate: computed values of the ratio of the number of molecules in the helical configuration to the number of molecules in both the helical and random coil states. From Bixon *et al.*[10] by courtesy of John Wiley and Sons, Inc.

In contrast to heat, chilling was formerly supposed to interfere with structural build-up by causing an excess formation of hydrogen bonds. This idea can now be dispensed with, and the results of chilling explained instead as an effect on hydrophobic bonding. Physico-chemical model studies have indicated that hydrophobic bonding stabilizes the α-helical protein structures,[213] and from Fig. 50 it can be seen that the temperatures needed to break down the α-helical structures of polypeptides are much higher for polyalanines, whose aliphatic side chains can take part in hydrophobic bonding, than for polyglycines consisting of the same numbers of amino acid units but lacking the aliphatic side chains. Hydrophobic bonding, however, is reduced by chilling below room temperature so that its contribution to the stability of the α-helix is lessened by this treatment.[124] In analogy, it is possible that the structural collapse of the oral primordium seen to take place when *Tetrahymena* cells are chilled, and the division synchronizing effect of a series of chillings reflect a lessening of the stabilizing influence of hydrophobic bonding within or between polypeptide chains integrated in, or to be integrated in, developing cellular structures.

It has always been puzzling that either heat or cold shocks can induce synchronous cell division in *Tetrahymena* populations, and theories on the mechanism of synchronization have been complicated by the fact that they had to account for similar effects in cells exposed to temperatures on either side of the optimum. We have just seen how earlier views have been modified slightly, so that today we formalize our interpretation of the experimental observations as follows. Binding within and/or between protein molecules specified for, but not yet organized into some structural entity, is critically reduced both by heating which breaks hydrogen bonds, and by chilling which lessens the degree of hydrophobic bonding between the side chains of the molecules. Thus heating or chilling might promote the dissociation of protein aggregates already present and/or hinder further aggregation essential for the continued development of structures required for division to take place. It may be difficult to produce conclusive experimental evidence in support of this interpretation, and at the moment we can only surmise that the broad trough in the curve showing the division-delay response of cells to various changes of temperature (from 28°) for a fixed period of time (Fig. 29) corresponds to the temperature range most conducive to the aggregation of protein subunits present in suitable conformational states, during their assembly into division-relevant structures. In any case, the observation that *T. pyriformis* GL grown at different constant temperatures assumes a minimal size at 28° and increasing size with departure to either side of this temperature[238] indicates that the balance between growth and division is temperature dependent and maximally in favour of division at temperatures in the region of the trough shown in Fig. 29.

In conclusion, it cannot be stated definitely that the *primary* effect of temperature shocks in inducing division synchrony is to disaggregate fibers of the developing oral structures. Numerous biochemical activities involve aggregation of macromolecules, or of structures and macromolecules, for example, synthesis of protein requires the association of ribosomes with messenger and transfer RNAs, and cellular control mechanisms depend on the proper aggregation and disaggregation of genes and repressor proteins, and we are aware that effects on weak bonds of importance for the cohesion of structures may be exerted at many levels of cellular metabolism.

III. ASTASIA LONGA (MASTIGOPHORA, EUGLENOIDININAE)

James' work with *Amoeba proteus*[118] provided the philosophical foundation for studies which led to the development of synchronously dividing populations of *Astasia longa*. In the original procedure[119, 176] a series of temperature cycles, each consisting of 16 hours at 15° and 8 hours at 25°, was applied. The technique was later improved by Padilla and Cook[175] and by Blum and Padilla,[11] who used a modified culture vessel (Fig. 51), secured optimal aeration, changed the temperature cycle slightly and introduced serial dilutions with sterile medium after each warm period to keep the population density within a desired range through long periods of time. This latter system is now described as one possessing continuous or repetitive synchrony; some of its growth characteristics presented in Figs. 52 and 53 are detailed below.

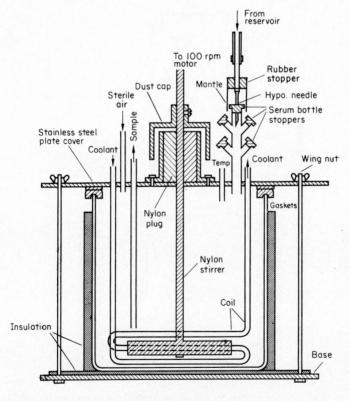

FIG. 51. An apparatus which has been used for the continuous synchronous culture of *Astasia longa*. It consists of a 15-liter pyrex glass jar, to which is clamped a stainless steel plate as cover. For further details see review article by Padilla and James.[177] Reproduced by courtesy of Academic Press.

4*

Cell division occurs only at the higher temperature and is beautifully synchronized, as much as 80 per cent of the cells taking part simultaneously in mitosis, 30 per cent in prophase (see Fig. 52). Some DNA is synthesized during the warm period, but just as much is made at 14.4°, and although conclusive evidence is not yet available it would seem that some cells begin to synthesize DNA immediately after a division, while others complete this function shortly before mitosis, i.e. DNA synthesis is not synchronized

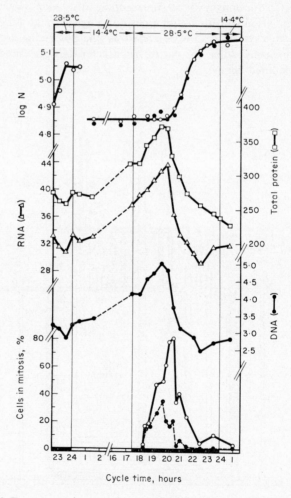

FIG. 52. From top to bottom, the curves show the changes which occurred in population density, amounts of protein, RNA and DNA (all expressed in μg per million cells) and mitotic and prophase indices, during a generation cycle in a synchronized culture of *Astasia longa*. From Blum and Padilla[11] by courtesy of Academic Press.

in this system. In an earlier investigation carried out with the original synchronizing technique the amount of DNA increased exclusively during the warm period.[121] Thus alteration of the culture conditions appears to have brought about a change in the growth pattern of the cells. In the case of RNA and protein, somewhat more than half of the amounts synthesized during one cycle are accumulated during the warm period (Fig. 53). During mitosis synthesis of the latter components ceases for a period of about

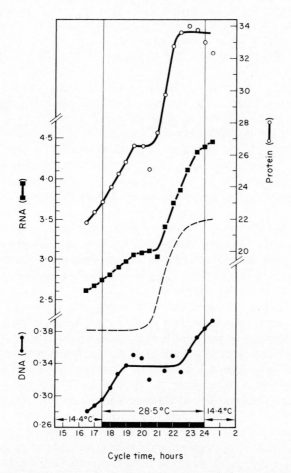

FIG. 53. From top to bottom the solid line curves represent the changes which occurred in the protein, RNA and DNA contents of an *Astasia longa* culture during a warm period of the temperature cycle shown in Fig. 52. Here the values are expressed in μg per ml culture. For orientation, the curve for change in cell number from Fig. 52 is reproduced here (dotted line). From Blum and Padilla[11] by courtesy of Academic Press.

1 hour, while synthesis of DNA is arrested for a longer interval (see Fig. 53). Such transient suspensions of macromolecular synthesizing activity are known to occur also in synchronized *Tetrahymena*.

Padilla and Blum[172] attempted to discover whether the synthesis of division-relevant species of RNA is confined to a specific interval of the cell cycle in *Astasia*. The investigation was carried out before the days of actinomycin D, with the aid of 8-azaguanine, which is known to give rise to synthesis of contaminated RNA in other cells. The azaguanine was added to synchronized *Astasia* cultures at various time points throughout the cold period, and it was observed that the later the compound was given, the larger was the fraction of cells which divided in its presence during the subsequent warm period. Thus only 20 per cent of the cells divided if azaguanine was present from the beginning of the cold period, compared to 60 per cent if it was added at the end. This shows that the synchronized *Astasia* cells do not pass simultaneously through azaguanine-sensitive stages of development, which can be interpreted to mean that they synthesize RNA relevant for division at different points in their generation cycles. Here it is quite probable that the ultimate effect on division is exerted indirectly through the production of non-functional proteins, in which connection it is of interest to note that Meeker[153] observed the same pattern of inhibition of division, on exposing synchronized *Astasia longa* to β-mercaptoethanol, an agent known to disturb the tertiary structure of proteins.

Two major investigations on the synchronized *Astasia* system deal with cell growth and respiration. The first of these was carried out by Wilson and James,[257] who determined cell number, dry weight and respiration rate by a polarographic method, in samples removed from the synchronizing apparatus. Rate of respiration per unit volume of culture measured at 14.8° was found to remain constant throughout the cold period, while during the warm period at 28.5° the rate measured at *this* temperature gradually increased 94 per cent. For purposes of comparison measurements of respiration rate were also made at 28.5° on samples removed at intervals throughout the cold period; these made it possible to show that respiration rate was unchanged throughout the cycle, if it was expressed in standard terms as the amount of oxygen taken up per cell per unit time at the higher temperature of 28.5°. Cell number also remained constant at the low temperature and doubled during the warm period. The dry weight of the synchronously dividing population increased two-fold during each complete temperature cycle; one-third of the increase occurred during the cold period, two-thirds during the warm period. From all of these observations it could be concluded that respiration per unit of cell mass doubled during the warm period, around or just before division.

These experiments were continued by James[122] who used highly specialized equipment and untraditional methods too involved to be discussed here, to relate cell density and respiration rate with the degree of synchrony obtained in these cultures. He observed that division was most synchronous in those cultures whose respiratory rates showed first a small, gradual but distinct increase during the cold period, then, immediately after the temperature was raised to 28.5°, a maximal overshoot which provided further evidence that a build-up in intrinsic respiratory capacity had occurred, and finally the greatest and most abrupt subsequent decrease. The generality of these findings is, however, questionable, since they may be mainly a reflection of the particular and limiting respiratory conditions imposed on the cells. More recently Løvlie and Farfaglio[145] studied respiration intensities throughout the cell cycle in *Euglena gracilis*, the closest known relative of *Astasia longa*. Here a considerable increase in the respiratory rate occurred between consecutive divisions of single cells maintained in darkness, and this is in striking contrast to the rather constant rates reported to characterize the cell cycle of *A. longa*.[257] This disparity in the respiratory behaviour of the two organisms, unless it be due to different environmental conditions, cannot readily be explained.

Little attempt has been made to account for synchrony induction in *Astasia*. However, it has been a basic feature of the synchronizing procedures devised that a complete temperature cycle lasts for the same length of time as one generation cycle under the conditions prevailing. When this requirement is fulfilled it is easy to see that once obtained, division synchrony can be maintained over long periods of time. What is difficult to understand is how synchronization is achieved in the first place, and discussion of this matter has hardly begun. When the synchronized *Astasia* system is allowed to run freely at constant 28.5° some phasing of cell division persists and the period of the cell cycle equals the generation time of cells multiplying logarithmically at this same temperature.[177] Studies on the effect of briefly lowering the temperature from 28.5° at different points in the cell cycle, and conversely of raising it for short periods at various stages of a cell cycle developing at the lower temperature, might tell us whether differential set-back plays a significant role in the synchronization of *Astasia*, as it does in the case of *Tetrahymena*. Cortical structures are as complex in both *Astasia longa*[120, 229] and *Euglena gracilis*[228] as they are in *Tetrahymena*, and might serve equally well as markers of set-back reactions if such can be induced in these cells. Recalling the proposal that interference with a build-up of structures rich in microtubules is responsible for the development of set-back reactions in *Tetrahymena*, we should like to draw attention to an extensive study of Sommer and Blum[230] on the tubular fibers formed in both the pellicle and nucleus of synchronized *Astasia* during premitotic and mitotic stages of development. Of course the mere occurrence of such structures does not justify immoderate extrapolations, nevertheless we think it would be worthwhile to investigate the effects on tubular fibers of *Astasia* of temperature changes causing delay in cells' preparation for division.

IV. PLASMODIUM LOPHURAE (SPOROZOA, HAEMOSPORIDIA)

In the malaria diseased homeothermic organism periodic bursts of fever is a common phenomenon. Correlated with these temperature changes the fraction of plasmodia at any particular stage of development passes through maximal and minimal values; for example, the proportion of young growth stages present is increased during the fever periods. Causal relations have been much discussed, but as far as one can gather one very basic question is yet to be answered: namely, does host periodicity control plasmodial periodicity or vice versa?

Experiments intended to throw light on this question can be made if a host-parasite system is available in which the parasites multiply in an aperiodic and asynchronous manner. *Plasmodium lophurae* grown through many passages in 10-day old chicken embryos at 38° constitute such a system,[151]

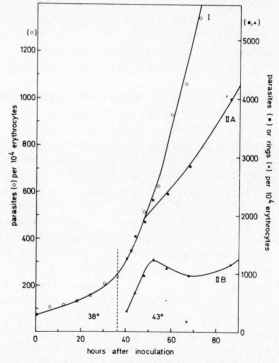

Fig. 54. Effect of temperature on multiplication of *Plasmodium lophurae* cells in the blood of chick embryos. The curves show the increase with time after inoculation in the number of parasites per 10^4 erythrocytes, under the temperature conditions indicated: Curve I: incubation at the constant temperature of 38°; curve IIA: incubation at 38° for 36 hours followed by a temperature shift up to 43°; curve IIB: same experiment as curve IIA, but values plotted represent ring forms only (see text). From Kovič and Zeuthen[129] by courtesy of the Danish Science Press.

and have found application in Kovič and Zeuthen's attempts to synchronize the development of a population of parasites within a host organism.[128-9] The results of one of their experiments are shown in Fig. 54. Here the population represented by curve I is increasing exponentially with a doubling time for cell number of 10–15 hours. Throughout the growth period the young forms of the parasite (the "rings" readily observed in the fluorescence microscope after acridine orange staining) constitute a constant fraction (27–34 per cent) of the population. This information about growth at constant temperature provides a basis for comparative studies on the effects of a temperature shock (fever period). A shock has of course two components, a shift up in temperature, later followed by a shift down, but the simplest experiment which can be made involves only a shift up. The result of such a temperature change is shown in Fig. 54, curve IIA. Here the number of parasites increases at the 38° rate for 12–15 hours after a shift up to 43°. This increase can be fully accounted for by the appearance of more ring forms (curve IIB). It is observed that "fever" and increasing numbers of rings appear together, just as in the patient. This is not because rings form faster during the "fever" but because they are removed at a slower rate. Apparently, development of rings into older trophic stages is inhibited during the first 12–15 hours at 43°. If the higher temperature is maintained for more than 12–15 hours, the number of rings ceases to increase, and most likely for two reasons: (i) from now on rings appear at a slower rate than before (the slope of the curve for population increase is deflected at this point to one-half to one-third of its previous value) and (ii) rings begin to transform into older stages. A deficiency of those division-mature cells which supply young trophozoites and rings thus appears at about the time when rings previously blocked in development by the higher temperature adapt to the new environmental conditions and progress further. At the present time it is not known whether this correlation in time is significant.

Since at least some adaptation to 43° has occurred after 12–15 hours at this temperature, it seems doubtful whether anything dramatic can be expected, when after this interval the temperature is shifted back to 38°. In the next experiment such a shift down has been made to delimit the duration of a first temperature shock to 12 hours. No response is observed that can be clearly attributed to this shift (compare Figs. 54 and 55).

In the experiment of Fig. 55 a second 12-hour heat shock has been applied 24 hours after termination of the first. The second shock is initiated at a time when division-mature cells are reappearing in great numbers, and are beginning to add to the low number of rings present at this time. The effect shown in Fig. 54, between 36 and 48 hours, is now repeated. Rapid divisions supply rings at a high rate, but all new cells formed are trapped at this stage. It is supposed that plasmodial division and infection of more erythrocytes would have occurred at the high rate observed also if the second shock had not been given. Most likely the shock temperature only has the effect of trapping the new plasmodia in the ring stage. Figure 56 shows how the percentage of ring forms can be increased from the value of about 30 per cent which is characteristic of an aperiodic population, to 55 per cent after one shock at 43° and to 70 per cent after two shocks.

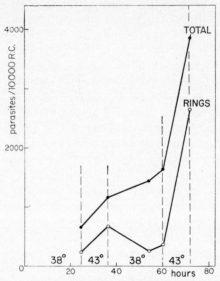

FIG. 55. Multiplication of *Plasmodium lophurae* cells in chick embryos exposed to two heat shocks. Population density expressed as number of parasites per 10^4 red cells. From Kovič and Zeuthen[129] by courtesy of the Danish Science Press.

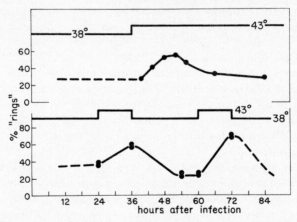

FIG. 56. Fraction of *Plasmodium lophurae* cells in the ring form in the red blood cells of chick embryos as a function of time after infection under different temperature regimes. From Kovič and Zeuthen[129] by courtesy of the Danish Science Press.

In discussing the malaria situation we want to point out the distinction between true synchrony, or degrees thereof, and periodicity. In the case of true synchrony all the cells in the population pass through a specific phase of the generation cycle as a single group and within an interval much shorter than the generation time. The term periodicity, on the other hand, can also cover the situation in which a population of cells consists of a few separate groups which operate identically except that they pass through a specific phase of development at quite different times. We expect most malaria diseases to be *periodic* in this sense, and, in addition, in attempting to adjust the temperature cycle to the plasmodial asexual generation time, proposed to last 36 hours (i.e. 12 + 24 hours) in *P. lophurae*,[102] we hope to be on the way towards establishing a *truly synchronous* development in this organism. Since the results obtained thus far are so reminiscent of the early results obtained with *Tetrahymena* (see Fig. 39) we suggest that there may be much similarity between the mechanisms of synchrony induction in *Plasmodium* and in *Tetrahymena*.

In conclusion, a partial answer can already be given to the question raised at the beginning of this section: periodic variation of host temperature can induce periodic behavior of the parasite. Whether the reverse is true we do not know.

V. AMOEBA PROTEUS AND ACANTHAMOEBA SP.
(RHIZOPODA, AMOEBOZOA)

Few published reports deal with the problem of inducing synchronized cell division an cultures of amebae. Working with *Amoeba proteus*, Dawson *et al.*[64] obtained an increase in the frequency of dividing cells in the period following a temporary cooling. In 1954 James made the accidental observation that cooling (12 hours at 18°) followed by a warm period (12 hours at 26°) produced a high proportion of prophase cells in clonal cultures of the same organism.[149] Later he showed that some degree of phasing of cell division could also be obtained in mass cultures by application of the above temperature regimen, and implied that synchronization of *A. proteus* requires a temperature shift across an inflection point in the Arrhenius plot relating cell multiplication rate and temperature.[118] In this there is a parallel with *Tetrahymena* whose synchronization by means of heat or cold shocks is known to be correlated with such shifts. Despite these early promising results, further work on *A. proteus* and the attempts of Neff and Neff[168] to synchronize the soil ameba *Acanthamoeba sp.* by means of temperature changes or shifts between aerobic and anaerobic growth conditions have been rather frustrating. It is of interest to us that the latter authors think that in *Acanthamoeba* structural disturbance of the developing mitotic apparatus may have been involved in any synchronization achieved.

4a*

VI. PHYSARUM POLYCEPHALUM (SARCODINA, MYCETOZOA)

It has been known for many years that genetically and structurally identical nuclei which share a common cytoplasm tend to perform mitoses in synchrony. Thus in many plants the nuclei of the female gametophyte, the embryo sac, the proembryo and the endosperm, which all divide without the formation of cell walls, exhibit mitotic synchrony, in some cases to a striking degree through many nuclear generations (see review by Erickson[71]). A similar situation exists in animal eggs subject to partial cleavage, as has been amply documented in a review by Agrell.[3] Noteworthy examples are afforded by discoidally cleaving fish eggs in which synchronous mitosis sometimes continues through thirteen cycles in the marginal rim of the blastoderm where no cell borders can be discerned, and by insect eggs in which the first division of the zygote nucleus is followed by a number of rounds of extremely fast synchronous mitoses (12 in *Drosophila*) prior to the onset of meroblastic cleavage. These cases serve to illustrate the point that like nuclei within a common cytoplasm can respond identically to the presumably uniform conditions prevailing in their environments at a particular time.

Work with amebae has shown that a particular type of curve can describe the growth of both mono- and multi-nucleate organisms, the species studied being *Amoeba proteus* which is almost always mononucleate, and the closely related giant ameba *Chaos chaos* (also known as *Pelomyxa carolinensis*) which contains up to several hundred nuclei depending on its size. In each of the two separate investigations involved, growth was assessed in the same way, i.e. in terms of reduced weight (RW) measured on a Cartesian Diver Balance. Prescott[184-5] studied *A. proteus* and showed that its growth rate varied periodically during the cell cycle, most rapid growth being observed immediately after, and a period of growth cessation just before, a division. Satir and Zeuthen[212] made comparable measurements on *C. chaos* and found a similar periodicity in growth rate in the vast majority of the cells investigated, and they concluded that, by and large, growth patterns are much alike in these two organisms. However, under some circumstances *C. chaos* by-passes cell division during a period of minimal growth. This happened, for instance, in an experiment in which a *C. chaos* ameba was cut into two unequal pieces, one four times bigger than the other, whose growth rates were followed until each divided once more. The resulting growth curves are reproduced in Fig. 57. The bigger fragment (represented by filled circles) showed the typical growth pattern described above for *A. proteus*, and it divided about 90 hours after cutting of the original cell (as indicated by the encircled points). In contrast, though it also grew in a cyclic manner, the smaller fragment (represented by filled squares) did not divide until somewhat later than 300 hours after cutting, by which time its reduced weight had increased some thirty fold. During the time between

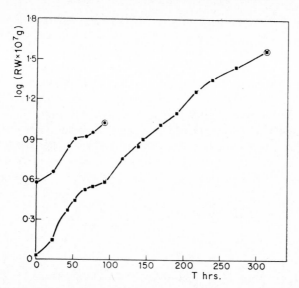

FIG. 57. Mass of *Amoeba proteus* measured as reduced weight (RW) plotted against time. At time zero one organism was cut into two fragments of differing size and the growth of each was then followed until it divided at the time corresponding to the encircled point. From Satir and Zeuthen[212] by courtesy of the Danish Science Press.

cutting and division the growth curve of the smaller fragment described three cycles, i.e. three periods with maximal, separated by intervals with minimal growth rates. Unfortunately, it was not possible to make observations of the mitotic activity, known to be fully synchronous in this organism under ordinary culture conditions, concomitantly with the growth measurements, and thus to establish whether or not the nuclei divided during the two periods of decreased growth rate occurring in the smaller cell fragment around 90 and 175 hours respectively after cutting. If they did so, the basic difference between *A. proteus* and the small *C. chaos* fragment might lie in the degree to which their nuclear and cytoplasmic divisions are coupled.

We have purposely discussed the giant ameba *C. chaos* here because we feel that it provides a natural introduction to the material on *Physarum polycephalum*, occupying as it does an intermediate position between *A. proteus* and the latter organism, with respect to the uncoupling of mitosis and cell division, as elucidated further below. In *A. proteus* mitosis and binary cell division are closely coupled and the organism is monocleate generally, binucleate only rarely.[131] In the multinucleate *C. chaos*, on the other hand, mitosis and cell division must be largely dissociated at least in small individuals, since Prescott was able to grow a clone of normal multinucleate cells from a fragment containing only one nucleus.[186] This view is also supported by the finding that cyclic growth can occur in the absence of cell division in fragments of *C. chaos* as described above. In average-sized *C. chaos* cells there *may be* a coupling between synchronous mitosis and cell division, or plasmotomy as some people prefer to call the latter process because it gives rise to a variable number (range 2 to 6, mean 3 (see review by Kudo[130]) of cells of non-uniform size. However, the only observation for which there is experimental backing here, is that cessation of growth and plasmotomy tend to occur around the same time.[212] In the giant plasmodia of *Physarum* and other myxomycetes, mitosis is not accompanied by plasmotomy, but in one case at least it has been shown that the rate of increase in the amount of plasmodial protein was slower around a mitosis than between mitosis (see Fig. 58). Both the amebae and *Physarum*

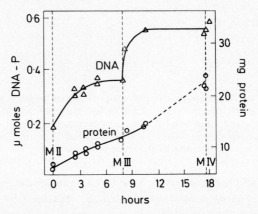

FIG. 58. Changes in the amounts of DNA and protein in plasmodia of *Physarum polycephalum* from the second (M II) to the fourth (M IV) synchronous mitosis. Redrawn from Sachsenmaier and Rusch[211] by courtesy of Academic Press.

show perhaps essentially related cytoplasmic changes around mitosis; the former round up and produce numerous small pseudopodia[40, 130] while *C. chaos*[130] and *P. polycephalum*[90] each manifests what is best described as a "freezing" of the cytoplasm.

A. DESCRIPTION OF THE SYSTEM

The slime mold *P. polycephalum* has lately received much attention, a main reason for this being that its many nuclei undergo mitosis in synchrony. It has been said that this material is "naturally synchronous" and that in it we have an organism in which synchronous events can be studied in the absence of any disturbing influences stemming from an *induction* of synchrony. The mitoses in question are indeed beautifully synchronous and in consequence many of the biochemical events occurring during interphase are likely to be synchronous too. We should like to point out, however, that the procedure used to obtain amounts of synchronous material sufficient for experiment does involve an induction period, characterized by temporary starvation, and we shall devote a section to a discussion of the factors involved in synchronization. Work on this system has already been reviewed extensively.[53, 58, 61, 91, 203-5]

P. polycephalum has a complex life cycle which includes a vegetative phase in which it grows as a plasmodium. In this phase Howard[112-13] found that mitoses occur in synchrony. Daniel and Rusch managed to grow this organism axenically[63] and they also devised a fully chemically defined medium for its culture, which is described in a review of Daniel and Baldwin.[61] Complex media were used, however, in all of the studies to be discussed here.

Reproducible axenic growth has been obtained in agitated cultures in which the organism disperses into, and multiplies as, individual microplasmodia of somewhat differing sizes suspended in the nutrient medium. Within each microplasmodium the nuclei divide in synchrony, but there is no synchrony of nuclear activity from one microplasmodium to another.[95] Such a culture appears to pass through the different phases characteristic of normal growth: namely, lag, exponential growth, negative growth acceleration, stationary and negative growth phases. The time required for a doubling of the protein content is slightly less than 24 hours at 21-22° under optimal nutritional conditions. Aeration and agitation must be kept within narrow ranges for the dual purpose of keeping the plasmodia dispersed without causing lysis, and of minimizing an agitation-stimulated secretion of a slime which can make the medium too viscous.

Plasmodia exhibiting synchronous mitoses are prepared in the following way. Microplasmodia from the agitated culture are transferred to distilled water; small samples are then placed on the central parts of round discs of filter paper or "Millipore" membrane, each supported by a layer of glass beads in a Petri dish. In the absence of growth medium these microplasmodia fuse, or coalesce, into one continuous, though branched and interdigitated, cytoplasmic system (for details see Guttes *et al.*[95] and a review by Guttes and Guttes[91]). The time required for fusion varies from 1 to 3 hours in the different reports. Fresh nutrient medium is provided only after coalescence is well advanced or complete, and then sufficient medium is added to fill the spaces between the glass beads and the filter paper. Plasmodia prepared from a vigorously growing culture of microplasmodia enter the first post-fusion mitosis (designated M I) synchronously and in approximately half the time which elapses between the two subsequent fully synchronous mitoses (designated M II and M III).

It is difficult to measure the growth rate of an individual plasmodium because it is impossible to withdraw from it respresentative samples for analysis. Some approximate evaluations can be made, however, in ways which we shall now describe. First, Rusch[204] has reported that a plasmodium which is one centimeter in diameter immediately after coalescence attains a diameter of 8 cm in 24 hours (at 24–26°), by which time it has completed three synchronous mitoses. From this it can be stated that the area is enlarged sixty-four-fold while the number of nuclei increases eight-fold, which means that a fused plasmodium spreads faster than it multiplies its nuclei. Second, curves for increase in the contents of various chemical constituents have been obtained from analyses of many, individual, fused plasmodia, all prepared identically from the same microplasmodial culture. The precision of this method depends among other things upon the re-producibility of growth from one filter to the other. Moreover, the procedure is laborious and only a few growth curves made in this way are to be found in the published literature. Figure 58 shows such curves for amounts of DNA and protein through two mitotic cycles. In each cycle both parameters nearly double in amount, and in the case of protein this result has been confirmed in a later investigation.[157] Similarly, Fig. 59 shows how the

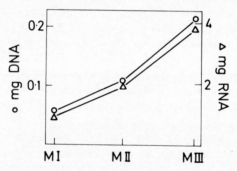

Fig. 59. Changes in the amounts of DNA and RNA in plasmodia of *Physarum polycephalum* from the first (M I) to the third (M III) synchron-ous mitosis. Redrawn from Mittermayer *et al.*[159] by courtesy of Academic Press.

plasmodial content of DNA and RNA increased in another experiment between M I and M III. Although growth is initially faster in surface cultures than in the agitated flasks,[204] it should be understood that rapid growth of a coalesced plasmodium does not necessarily continue for long—it can decline as conditions change, as, for example, when the plasmodium exceeds a certain size.

Under the conditions described so far, during synchronous mitosis in *Physarum*, locomotion, expansion and "internal shifting of material" are completely arrested.[90] In other contrasting experiments devised by An-derson,[6] the plasmodia have been made to travel continuously and vertically in an upward direction under the influence of an electric field. Under these circumstances their rates of migration are periodic, and perhaps slowed at the time of synchronous mitosis. Thus even though there is no subsequent cell division in the plasmodium, mitosis is accompanied by fundamental cytoplasmic changes, perhaps basically the same as those which occur at cell division in the organisms *Amoeba proteus* and *Chaos chaos*.

For the microscopic study of mitosis in *Physarum* small pieces of material can be removed repeatedly from one coalesced plasmodium. Such sampling does not seem to disturb the progression of the plasmodium in its cycle. The following description of nuclear events is essentially taken from Rusch.[204] The earliest signs of prophase are seen at least 1 hour before metaphase. The centrally located nucleolus begins to swell (maximal diamter: 1 μ; diameter of a nucleus: 4–5 μ) and to move towards the nuclear membrane. Shortly thereafter the chromosomes begin to condense and to move to a central position in the nucleus. The nucleolus disappears between 20 and 5 minutes before metaphase, and the rest of the mitosis runs to completion in the 10 minutes following metaphase. Whether the nuclear envelope remains intact throughout mitosis, Rusch[204] believes is still an open question; light microscopic examination suggests that it does, whereas some electron micrographs have indicated that it tears open during anaphase.[94] As soon as a nucleus has divided, many tiny, irregular bodies appear within it. Over a period of 2 hours these gradually fuse to form a complete nucleolus. We have searched the literature for photographs illustrating the above sequence of events, but found none suitable for reproduction here. In default of the same we refer readers instead to fig. 5 in the review of Guttes and Guttes[91] and figs. 9–18 in the paper by Guttes *et al.*[95] Figure 60 of the present paper shows the appearance of the nuclei during a few of the stages in synchronous mitosis. DNA synthesis commences immediately after nuclear division, and it continues at a low rate for at least an hour after reconstitution of the nucleolus, ending at somewhat different times in the different nuclei, but seldom later than 3–4 hours after metaphase. Studies of nuclear activity have lead to the conclusions: that mitosis is essentially synchronous in coalesced plasmodia with diameters smaller than 7–8 cm; that a G_1 period is absent; that the period of DNA synthesis occupies the first third of the intermitotic time and that the G_2 period covers most of the remaining interval.

The content of DNA per nucleus is rather low in the myxomycetes[236] and in the plasmodium of *P. polycephalum* it is about 1 $\mu\mu$g.[161] The chromosomes are numerous and small, so that accurate counts are difficult to obtain. Using material from Rusch's laboratory, Koevenig and Jackson[127] studied coalesced plasmodia fixed at the time of M III and reported that the mean number of chromosomes per nucleus was 56 $\pm$ 2, on the basis of counts over five selected nuclei which they assumed to be diploid. Some words of caution should be added to their statement because the published pictures show "chromosomes" at the border of resolution of the light microscope. In these experiments a small number of the nuclei were larger than the rest and were probably polyploid. Ross[200] counted a mean of

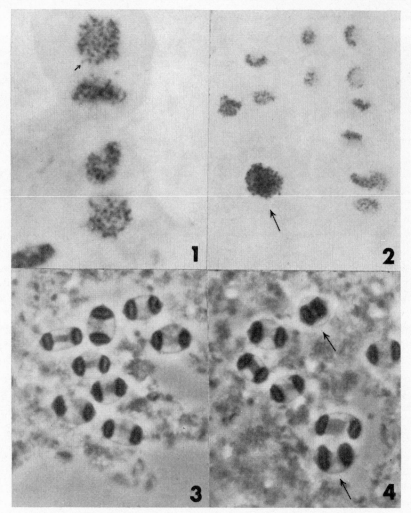

FIG. 60. Plasmodial mitosis in *Physarum polycephalum*. 1. Anaphase showing *ca.* 56 chromosomes. Arrow indicates a single chromosome ×2400. 2. Telophase nuclei. Arrow indicates a polyploid nucleus which was delayed somewhat in division ×1200. 3. Telophase nuclei showing intact nuclear membrane ×1200. 4. Anaphase and telophase nuclei. Arrows indicate nuclei which are lagging in division. From Koevenig and Jackson[127] by courtesy of Dr. Koevenig and the publishers of *Mycologia*.

fifty chromosomes per nucleus in a different strain, but observed variations from 20 to 180. The process of mitotic division in plasmodial nuclei of *P. polycephalum* involves an intranuclear spindle containing microtubules. No centrioles have been demonstrated, not even with the aid of the electron microscope.[94]

In describing the synchronous *Physarum* system we have had to take on trust some statements not actually documented in the reports accessible to us, and we have lacked information on other points: (a) though told that the mitoses of coalesced plasmodia are synchronous, in the absence of any quantitative data we cannot evaluate the degree of synchrony involved; (b) we are not informed as to whether both the plasmasol and plasmagel components of the protoplasm contribute in a uniform manner to the representative samples removed for analysis at different time points; (c) no evidence has been presented to support the statement that sampling and sectioning of the plasmodia do not cause any disturbance of the progression of the system in time. Finally we regret that the period between sampling of the microplasmodia and the occurrence of the first synchronous mitosis in the coalesced plasmodium has been so little studied both morphologically and biochemically.

B. Factors Involved in the Induction of Synchronous Mitosis

Normally a myxomycete plasmodium forms by coalescence of zygotes and small plasmodia arisen from zygotes which have already undergone nuclear division, and in *P. gyrosum*[126] as well as in many other species the coalescence is followed by synchronous nuclear divisions. The same pattern of events occurs in the laboratory when microplasmodia from agitated cultures of *P. polycephalum* are allowed to come to rest in close proximity on a mechanical support. These microplasmodia vary in size but they are all smaller than 1 mm in diameter,[91] and the number of nuclei they contain varies from just a few to several thousands.[95] In addition to the conditions mentioned above, starvation is a requirement for coalescence of the microplasmodia, supposedly in part because organisms that are well fed fail to migrate and thus to make the necessary contact with each other. Best plasmodial growth is achieved in the absence of mechanical agitation, on a wetted substrate in contact with air, as in nature; in the laboratory filter paper or "Millipore" membranes serve as suitable substrates. These factors were basic to the development of the routine procedure already described for producing material containing large numbers of synchronously dividing nuclei.

We shall now proceed to discuss factors involved in the synchronization of the first post-coalescence mitosis, M I. In the agitated cultures of microplasmodia the doubling time for protein in the phase of fastest growth in complex media ranges between 15 and 19 hours at 21–22°.[61-63] We shall assume that this also represents the average intermitotic time in the individual nuclei. On the basis of this assumption and information supplied in the review of Guttes and Guttes,[91] we have prepared Fig. 61, in which the mitotic cycles of seventeen individual microplasmodia are represented, displaced appropriately on the common time scale to illustrate the asynchrony which initially exists between them. Mitosis is of short duration and is indicated in each case by a short vertical line. On the time scale the zero point indicates the beginning of starvation and coalescence of the microplasmodia, after which an interval of $1\frac{1}{2}$ hours is allowed to elapse before nutrient medium is supplied again. Immediately after coalescence follows a period during which there are no nuclear divisions and at 6–7 hours synchronous mitosis I occurs. Figure 61 illustrates that synchronization of M I involves great variation in the length of the interval between the last mitosis in the asynchronous population of microplasmodia, and the first synchronous mitosis in the coalesced surface plasmodium.

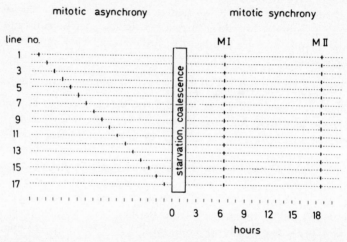

FIG. 61. A schematic diagram showing relative variations in duration of the intermitotic intervals before and after the first synchronous mitosis M I has been induced in *Physarum polycephalum* through the starvation and coalescence of microplasmodia. The behavior of seventeen individual microplasmodia representative of an exponentially growing culture is depicted. Timing of mitoses is indicated by the vertical bars. Based on data from Guttes and Guttes.[91]

In the case of microplasmodia 1 and 2 in the diagram, we do not know whether the mitosis due to take place at the beginning of or during the starvation-coalescence period did actually occur or not. Therefore we will limit ourselves to consideration of situations 3 to 17, and we observe that in these examples the time from the last mitosis in the microplasmodia to M I in the coalesced plasmodium varies between $7\frac{1}{2}$ (line 17) and $21\frac{1}{2}$ hours (line 3). Let us first compare these time intervals, during which the plasmodia were exposed to a synchronizing change of environmental conditions, with the intermitotic time of 17 hours observed under constant conditions, namely in agitated, submerged, microplasmodial cultures. The comparison shows that synchronization of mitosis is mediated through acceleration of progression in the mitotic cycle in some nuclei and through retardation in others, maximal shortening of the intermitotic period being $9\frac{1}{2}$ hours, maximal lengthening $4\frac{1}{2}$ hours. If we further compare the varying *pre*-M I intermitotic times with the value of 12 hours, which is characteristic of post-coalescence plasmodia surface-cultured at 21–22°, we find that the maximal times by which the intermitotic period is shortened and lengthened respectively are now $4\frac{1}{2}$ and $9\frac{1}{2}$ hours, i.e. numerically the reverse of the result of the first comparison. That the two comparisons give somewhat different results depends simply on the fact that the change in growth conditions is accompanied by an alteration in the duration of the intermitotic interval; thus the nuclei in surface plasmodia pass from one synchronous mitosis to the next in considerably shorter time than is required between the consecutive mitoses of microplasmodial nuclei. In other words, an improvement in growth conditions is the net result of the environmental change. At this point it may be useful to recall the several elements involved in producing this net result. These are as follows. The tearing apart of submerged plasmodia by shaking is discontinued; the microplasmodia are starved temporarily and allowed to spread over a mechanical substrate with their upper surfaces exposed to air; when coalescence has occurred the fused organisms are refed from below through the filter support.

We shall now discuss in turn the effects of starvation, coalescence and change of temperature on the *Physarum* system, although gaps in the relevant information unfortunately preclude their thorough evaluation at this time.

(i) *Effects of starvation.* We are not aware of any reports on the effects of short-term starvation on coalesced plasmodia. Of course, starvation can be expected to delay progress in the intermitotic cycle, and in fact synchronous mitosis occurs at much extended intervals when plasmodia are subjected to longer starvations during which they become reduced in size.[95] Surprisingly, this mitotic activity does not lead to an increase in the population density of the nuclei, so it has been concluded that many of them undergo degeneration under starvation conditions. In view of this we wonder whether similar nuclear degeneration takes place on a smaller scale during coalescence or when there is only a slight growth between two divisions as has occasionally been observed.[15]

(ii) *Effects of coalescence.* Intermitotic times may be shortened or extended as a result of coalescence between two plasmodia of different ages. This was shown by Rusch *et al.*[206] in fusion experiments with small pieces (*ca.* 1 × 2 cm) removed from two plasmodia B and A at specific times before and after M II respectively (see Fig. 62). Some of the pieces served as controls, while others were allowed to coalesce by partial overlapping, during (as far as we can understand) a period of only 20 minutes, after which nutrient medium was supplied. The results varied slightly depending on the intermitotic stage of the plasmodia used. In one experiment the two plasmodia differed in age by 6.3 hours, plasmodium A being in the latter

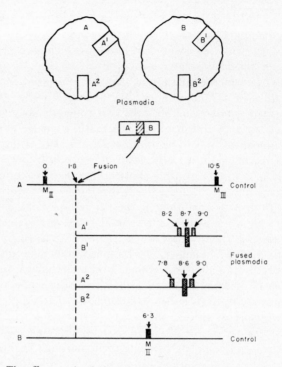

Fig. 62. The effect on the timing of mitosis due to fusion in the overlapped ends of segments cut from *Physarum polycephalum* plasmodia at different stages of the mitotic cycle. At the top of the figure the two plasmodia are sketched to show the positions from which the segments were removed. In plasmodium A mitoses M II and M III occurred at times zero and 10.5 hours respectively, whereas in plasmodium B mitosis M II occurred at 6.3 hours on the same time scale. In the fused areas resulting upon overlap of segments A^1 with B^1 and A^2 with B^2 synchronous mitoses occurred at 8.7 and 8.6 hours respectively. In the free end parts A^1 and A^2 of the overlapped segments mitosis M III took place at 9.0 hours on the common time scale in both cases, while in the opposite free ends B^1 and B^2 mitosis M II came at 8.2 and 7.8 hours respectively. From Rusch *et al.*[206] by courtesy of the Rockefeller University Press.

half of the S period following M II at coalescence, and plasmodium B in the early half of the G_2 period prior to M II. In the mid-region of the fused aggregates synchronous mitosis came 1.85 hours earlier than M III in control A and 2.35 hours later than M II in control B. Thus nuclei which find themselves in a new cytoplasmic environment are either delayed or accelerated, and in this case the delay was somewhat more pronounced than the acceleration. This has not been pointed out by the authors themselves, although delay exceeded acceleration also in another experiment in which the age difference between the original plasmodia was reduced to 4.2 hours, and the separated pieces were probably both in the G_2 phase at the time of fusion (see Rusch et al.,[206] table III). Similar retardations and accelerations were observed when whole plasmodia were allowed to fuse together.[206] In other experiments[92] small numbers of G_2 phase nuclei were taken from some plasmodia and "injected" into others, and it was found that host nuclei performing mitosis or synthesizing DNA inhibited both these functions in the injected nuclei until the time of their own subsequent mitosis. Thus it seems that the cytoplasmic environment around nuclei engaged in DNA synthesis has a strong retarding influence on neighboring nuclei more advanced in the mitotic cycle.

On the basis of these observations we can put forward in very broad terms a proposal as to how synchrony of M I is established. It will be recalled that both acceleration and retardation effects were implicated (Fig. 61). Let us suppose that S phase nuclear environments inhibit G_2 phase nuclei as described above, even when nuclei in the G_2 phase are more numerous than those in the S phase. It follows that as long as a newly coalesced plasmodium contains nuclei in the S phase, no G_2 nuclei can divide. Once this stage is passed, all the nuclei become aligned with respect to mitosis by processes which involve acceleration through the mitotic cycle of some nuclei and retardation of others. This alignment is likely to be mediated through equilibration across the cytoplasm of nuclear factors in control of nuclear progression through the G_2 phase.[204]

The aforementioned experiments of Rusch et al.[206] clearly demonstrate that in the fused aggregate formed when two plasmodia of different ages are partially overlapped, the first mitosis is not synchronous throughout the aggregate plasmodium. Rather it is expressed in the form of a mitotic wave which starts some 6 hours after overlapping and fusion, and which takes a considerable time (about 1 hour) to pass across the aggregate from its "older" to its "younger" side. In other words, the basis for a mitotic gradient persists in the aggregate for at least 6 hours, which is equivalent to about 60 per cent of the normal intermitotic time. Whether this persistence is due to poor mixing of the cytoplasm or to a strong binding to the interphase nuclei of factors in control of mitosis is not known. The preferred view of Rusch and collaborators, which is more easily harmonized with the former explanation, is that factors controlling mitosis increase in the cytoplasm throughout the G_2 phase, and are transferred to the nuclei shortly before synchronous mitosis occurs. However, the latter explanation should not be too readily dismissed, for Prescott and Goldstein[189] have shown that amebae contain proteins which are strongly retained by interphase nuclei, yet exchanged to equilibrium between nuclei of the same cytoplasm, and which are dispersed through the cytoplasm at the time of mitosis. In this connection it would be interesting to know whether or not the basis for a mitotic gradient would persist through the first mitosis in an aggregate formed between two relatively large plasmodia fused side by side.

Our understanding of how synchrony of mitosis is achieved is far from complete, but there can be no doubt that the minuteness of the micro-plasmodia and the great number in which they are used for establishing a coalesced plasmodium are important factors in the transition from the asynchronous to the synchronous state. Throughout the coalesced plas-modium each reasonably sized unit of material will contain the same standard mixture of microplasmodial constituents. Under these conditions gradients of factors in control of mitosis can only be local and in any case they disappear soon after coalescence, at the latest after the duration of an S period. The precision is impressive with which the synchronous mitoses follow each other in separate, concurrently and identically prepared coa-lesced plasmodia.[91] Apparently the progression of nuclei through the division cycle follows a very precise time course. This together with the described interactions between nuclei across a common cytoplasm must be the basis of the beautiful "natural" mitotic synchrony observed in single coalesced plasmodia.

(iii) *Effect of change of temperature.* As mentioned above, the intermitotic interval lasted 12 hours in coalesced plasmodia kept at 21–22°.[91] By raising the temperature to 26° during and after the fusion process, Rusch was able to obtain mitoses every 8–10 hours.[204] Thus temperature is another of the factors which influence the duration of the period elapsing between synchronous mitoses in a coalesced surface-grown plasmodium.

In the following we shall describe experiments of a different sort in which the response of *Physarum* to heat shocks was investigated by Brewer and Rusch.[18] As they were motivated by our studies on *Tetrahymena*, an orientating comparison of the two systems will be useful at this point. In *Tetrahymena* the synchronization of cell division with heat shocks is brought about through a differential retardation, measured as excess delay or set-back, of the cells' progress towards division, and we have proposed that this retardation is a consequence of the destruction of some protein-contain-ing structure. In *Physarum* the synchronization of mitosis which follows the fusion of microplasmodia is mediated through differential retardation or acceleration of the progress of the nuclei through the G_2 phase, as explained above, both types of response being in some way due to an exchange be-tween nuclei across the cytoplasm of factors in control of mitosis.

In their studies on the effects of heat shocks on synchronous mitosis in *Physarum*, Brewer and Rusch began by testing the response to different shock temperatures, exposing plasmodia cultured at 26° to single 30-minute shocks, beginning 2 hours before the metaphase of M III. Here they observed delays of mitosis which were proportional to the shock temperature in the range 32° (no delay) to 47° (3 hours delay). They then showed that exposure to the fixed shock temperature of 37° for periods of time varying from 15 to 90 minutes produced almost constant excess delays (set-backs). This tem-perature does not affect the viability of *P. polycephalum*, but after 3 days at

37° the organisms transform into dormant spherules capable of reforming plasmodia upon return to 26°.

In additional experiments sensitivity to a single heat shock (exposure to 37° for 30 minutes) was investigated throughout the synchronous mitotic cycle (see Fig. 63). Plasmodia exposed to heat during the S period were only slightly delayed in performing the subsequent synchronous mitosis; those exposed during the G_2 period 2 hours before a mitosis was due, i.e. prior to visible chromosome condensation, were most delayed. A detailed analysis of the 40-minute period immediately preceding metaphase showed that 10-minute shocks, begun between 16 and 4 minutes before the metaphase was due, interfered with the polarization of the mitotic figure and blocked normal nuclear division. The affected nuclei assumed larger than normal size, and by-passed one division, but took part in the next synchronous mitosis. Heat shocks applied immediately before metaphase had no effect. The similarities between these responses and those of *Tetrahymena* to comparable treatments are striking.

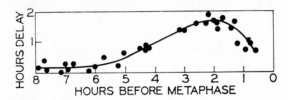

Fig. 63. Delays of onset of metaphase (ordinate) resulting from one 30-minute heat shock (37°) applied at various times (abscissa) in the mitotic cycle of *Physarum polycephalum*. From Brewer and Rusch[18] by courtesy of Academic Press.

Lastly, Brewer and Rusch investigated the effect of a series of heat shocks. Starting 2 hours before M III was due, they exposed the plasmodia to three 30-minute heat shocks at 37°, separated by intervals of 1 hour. The total time spent at 37° was thus 1.5 hours and the observed delay of M III was 3.1 hours, which corresponds to a set-back or delay in excess of the treatment of 1.6 hours. Furthermore, they showed that whether two or three heat shocks were applied, a constant interval of time (1.6 hours) elapsed between termination of the last shock and the occurrence of the subsequent synchronous mitosis. In this we see a strong resemblance to the response of *Tetrahymena* to multiple heat shocks. Thus each new shock annulled progress made towards the synchronous event, i.e. mitosis in the case of *P. polycephalum*. Moreover, the experiments showed that a great part of the time "lost" in response to the heat shocks in the generation M II to M III was regained in the next generation (M III to M IV) which lasted only slightly more than 5 hours. Another similarity between the *Tetrahymena* and

Physarum systems is that, in contradiction of a statement made by Brewer and Rusch,[18] DNA synthesis begins in both of them immediately after the occurrence of the heat-delayed events, cell division and mitosis respectively. On the basis of their results the latter authors propose that heat-sensitive factors having to do with mitosis increase during the first hours of the G_2 interval and begin to disappear or to become more heat resistant in the last 2 hours before mitosis.

In the case of *Tetrahymena* the heat-sensitive factors are thought to be structures in the making, assembled from macromolecules. Similar heat-sensitive structures could exist in *Physarum*, where the structures might be nuclear (e.g. the developing spindle) and the constituent precursor macromolecules of cytoplasmic origin. At any rate, the above comparisons indicate some points of similarity between *Tetrahymena* cells preparing to divide synchronously after a series of heat shocks, and coalesced *Physarum* plasmodia developing synchronous mitoses.

C. MACROMOLECULAR SYNTHESES

Plasmodia of *P. polycephalum* showing synchronous mitosis have been used extensively for studies in the biochemistry of the mitotic cycle. As seen from a technical point of view this material is possessed of both advantages and disadvantages. On the one hand, samples representing known fractions of a growing plasmodium cannot be obtained from it. Consequently, if progression through the mitotic cycle is to be followed with direct chemical analyses, these must be made on identically prepared, parallel plasmodia sacrificed at known times, and the reliability of the results then depends strongly upon the reproducibility of growth from one plasmodium to the other. On the other hand, the removal of small parts from single plasmodia does not disturb their behavior, and analysis of such samples can give useful results, particularly if these are expressed as *ratios* rather than *absolute amounts*, and, for example, as specific activities in experiments with radioactive tracers. It is procedures of this latter type which have been used most often in the investigation of *P. polycephalum's* biochemistry.

Plasmodia of this organism contain the macromolecular components DNA, RNA and protein in the proportions $1 : 5 : 94$ respectively.

DNA. During the S period, which began immediately after mitosis and ended when roughly one-third of a full mitotic cycle had been completed,[208, 211] the amount of DNA increased to nearly double the original quantity, as seen in Fig. 58. Tracer experiments with the DNA precursors 6-^{14}C-orotic acid[171] or ^{3}H-thymidine[13, 56, 207] indicated the same location within the mitotic cycle of the DNA synthesis period.

The replication of DNA seems to take place according to a strict program.[13] Evidence for this is furnished by experiments with radioactive thymidine and the heavy DNA precursor–analogue bromodeoxyuridine (BUdR). These showed that DNA which replicated early in one S period also replicated early in the next, and conversely. Furthermore, the BUdR incorporation patterns were consistent with a semiconservative replication of nuclear DNA.[207]

Only a few per cent of the total DNA is present in the plasmodial mitochondria. This DNA replicates throughout the cycle and asynchronously. We shall not deal further with it here, but it has been extensively studied.[13, 19, 59, 72, 93, 96]

The occurrence of mitosis appears to depend on the completion of nuclear DNA synthesis. Thus Sachsenmaier and Rusch[208, 211] using FUdR and uridine in combination to block DNA synthesis specifically for a limited period of time, showed that this treatment delayed the onset of the subsequent mitosis for an equivalent interval. Moreover, Devi *et al.*[66] and Sachsenmaier[208] examined the mitosis-delaying effect of irradiating the plasmodia with ultraviolet light at various times during the preceding intermitotic period. The results they obtained were compatible with the suggestion[208] that not only the replication of DNA but also its function, which extends into the G_2 period, is indispensable for the forthcoming mitosis. Rusch[204] has proposed that full replication of the DNA complement is required before a plasmodium can engage in G_2 phase activities; however, his own results, presented in Fig. 58, suggest that somewhat less than a doubling of DNA may suffice.

The physical properties of DNA from *P. polycephalum* have been investigated in some detail. Single-stranded DNA obtained from nuclei in the S phase was found to have a molecular weight of about 1.5×10^7, as determined by alkaline gradient centrifugation, while that harvested from nuclei in the G_2 phase was larger (about 4×10^7). Assuming that there are fifty chromosomes and 1 $\mu\mu$g DNA per nucleus, McGrath and Williams[152] calculated that the DNA from an average S phase chromosome dissociates in alkali into 300 single-stranded pieces. A satellite DNA comprising about 1 per cent of the total DNA has been detected; this is distinct from mitochondrial DNA[13, 204] and appears to be synthesized throughout the mitotic cycle, possibly with the exception of the period of nuclear DNA synthesis.

The control of DNA synthesis has been studied in isolated nuclei, prepared according to the methods of Mittermayer *et al.*[160] and Mohberg and Rusch.[161] Such nuclei when harvested from plasmodia in the S phase[16] readily incorporated the four deoxynucleotide triphosphates into acid-precipitable material in the presence of ATP and magnesium ions. Added singly, native salmon sperm DNA and spermine both stimulated this incorporation, while in combination their effect was much enhanced[17]—see Fig. 64. This incorporation activity was minimal in nuclei harvested 2 hours prior to a mitosis, maximal in those collected $1\frac{1}{2}$ hours after mitosis, i.e. before the end of the S period. This result may reflect that the activity of DNA polymerase fluctuates through the cell cycle, but it might also be due to changes with time in the ease with which the added template DNA penetrates the isolated nuclei. Unfortunately, no information is given about the state of preservation of the nuclear envelopes in these preparations, but it should be kept in mind that published electron micrographs show rifts in them around mitosis. Moreover, Rusch and colleagues[13, 16, 17] have submitted repeatedly that under *in vivo* conditions the replication of DNA may be regulated by the supply of templates rather than by the activity of DNA polymerase. Obviously there are many obscure points to be clarified before the mechanisms controlling *in vivo* DNA synthesis can be understood.

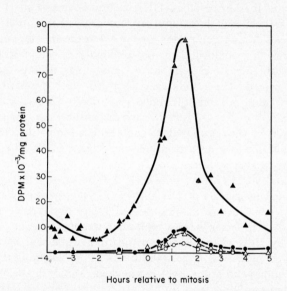

Hours relative to mitosis

FIG. 64. Incorporation of ³H-dATP by *Physarum polycephalum* nuclei isolated at various times during the mitotic cycle, as indicated on the abscissa. All incubations were carried out in a standard medium supplemented as follows: open circles: no addition; filled circles: spermine added; open triangles: DNA added; filled triangles: both spermine and DNA added. From Brewer and Rusch[17] by courtesy of Academic Press.

Relations between syntheses of proteins and of DNA are dealt with in the section on proteins.

RNA. Amounts of RNA essentially double between successive synchronous mitoses.[159] Just how total RNA increases through a complete mitotic cycle remains unknown to us, as we have not come across any detailed investigation of this parameter with quantitative chemical methods, and information pieced together from studies with pulses of radioactive precursors does not form a consistent picture. Thus intact plasmodia incorporated (a) ¹⁴C-orotic acid at a rate which was more or less constant throughout the mitotic cycle (171), (b) ¹⁴C-cytidine at a rate which was minimal around mitosis, and then increasing during the first half and decreasing during the second half of the cycle,[207-8] and (c) ^{3}H-uridine at a rate which fluctuated with time so as to present two minima per mitotic cycle, one at mitosis and the other at the end of the S period.[158] The differences between these rate curves cannot be accounted for at the moment, but they might be due to some extent to a variation in size between the several precursor pools concerned.

Major efforts have been exerted to characterize the types of RNA synthesized at various times through the mitotic cycle. Results obtained from three separate experimental approaches to this problem are summarized below:

(i) The simplest procedure involved testing the sensitivity of RNA synthesis towards actinomycin D. Here it was found that the two maxima in the rate curve for ^{3}H-uridine incorporation could be distinguished on this basis; on pretreatment with the drug in the very high concentration 175 μg per ml[158] for a period of 2 hours, the post-mitotic peak was reduced by half while the pre-mitotic peak was eliminated. Throughout the mitotic cycle there was some incorporation of uridine which remained unaffected by actinomycin.

(ii) In another approach, the RNA was pulse-labeled *in vivo* with ^{3}H-uridine during periods of 3 or 10 minutes at different points in the mitotic cycle, and then analysed by the technique of sucrose gradient centrifugation after extraction with phenol. Figure 65 shows some of the UV absorbance

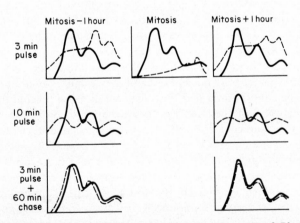

FIG. 65. Characteristics of RNA extracted from plasmodia of *Physarum polycephalum* at three different time points in the mitotic cycle. Solid line curves show sedimentation profiles (S values of the three peaks from left to right: 27, 17 and 4 S, respectively) and curves traced by a broken line the corresponding radioactivity values (in counts per minute) after incubation with ^{3}H-uridine as indicated on the left-hand side of the figure. From Braun *et al.*[14] by courtesy of the Elsevier Publishing Co.

and radioactivity distribution patterns obtained.[14] From these it can be seen that plasmodia in the late G_2 and S phases gave identical distribution patterns, whereas plasmodia at the stage of mitosis incorporated the precursor at a reduced rate and preferentially into the lighter fractions of RNA (those to the right in the diagram). Turning briefly to consider the RNA synthesized *in vitro* by isolated nuclei, we would point out that the size of this product never exceeded that of the tRNAs;[160] this may indicate that isolated nuclei have some functional characteristics in common with nuclei performing mitosis.[14] It was also demonstrated (Fig. 65) that the rapidly labeled RNA existing immediately after short pulses with ³H-uridine, was distributed over the entire size range of RNA species (except at mitosis), whereas the stable labeled RNA remaining 1 hour after the addition of a chase was concentrated mainly in the ribosomal fractions, as can be seen from the lowest row of sedimentation patterns. Both the 27S and 17S ribosomal RNAs (left and middle peaks) appeared to be synthesized in the same proportions during the late G_2 and S phases.

(iii) Finally, attempts have been made to demonstrate a sequential transcription of DNA in the course of the mitotic cycle. This is a very difficult task, which in our opinion has not yet been satisfactorily concluded. Experimentally the aim has been to show that RNA synthesized in different parts of the mitotic cycle contains adenylic and guanylic acids in different proportions; this it was proposed would constitute evidence for a sequential reading of the genome. First Cummins *et al.*[60] published the results of a study in which they employed *in vivo* labeling of RNA with a pulse of radioactive phosphate, followed by quantitative isolation of the RNA nucleotides and measurement of their relative radioactivities. They found that the A/G radioactivity ratio did indeed vary, as it was greater than 1.0 for RNA synthesized near the beginning of the mitotic cycle, i.e. shortly after a mitosis, but less than 1.0 for RNA synthesized late in the cycle. However, they pointed out that the nucleotide pool may require from 2 to 5 hours to reach isotope equilibration and that this complication precludes unambiguous interpretation of their results. In later work Cummins and Rusch[57] attempted to circumvent this difficulty by using isolated nuclei to synthesize the RNA, and the nucleotides ATP, GTP, CTP and UTP labeled with ³²P in the phosphate group at the alpha-position as the radioactive precursors. The method of "nearest neighbor" nucleotide analysis was then applied to the RNA obtained, and from the results of this Cummins and Rusch concluded that the relative frequencies of dinucleotides beginning with A (adenylic acid) or U (uridylic acid) were higher at the beginning than at the end of the mitotic cycle, whereas the frequencies of dinucleotides beginning with G (guanylic acid) or C (cytidylic acid) followed the reverse order. Their evidence for this statement, as depicted in Fig. 66, is hardly convincing, however. While pool equilibration is likely to be less of a problem here than in the previous study, this experimental system has other shortcomings

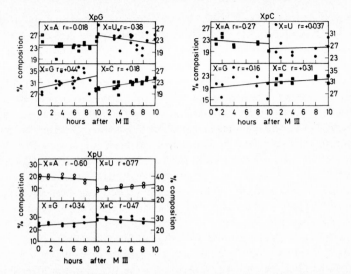

FIG. 66. The results of "nearest neighbor" nucleotide analysis on RNA synthesized *in vitro* by nuclei of *Physarum polycephalum* isolated from plasmodia in different stages of the mitotic cycle following synchronous mitosis M III. The ordinates express the relative frequencies of occurrence of each of the four nucleotides adenylic acid (A), uridylic acid (U), guanylic acid (G) and cytidylic acid (C) in combination with G (upper left diagram), C (upper right diagram) and U (lower left diagram). *r* is the coefficient of correlation between the two variables frequency of a particular dinucleotide and time. A value of *r* of ±1 would demonstrate perfect correlation between the two variables, whereas a value of zero would indicate no detectable relationship between them. From Cummins and Rusch[57] by courtesy of the Elsevier Publishing Co.

which should not be ignored. Firstly, since none of the RNA synthesized by isolated nuclei is larger than tRNA, it cannot be considered comparable to that synthesized *in vivo*. Secondly, it is not stated that the amount of RNA did indeed double during the cycle in question (M III to M IV); instances have been described in which RNA increased to a lesser extent,[15] and in such cases differences in the composition of the RNA synthesized at different times could be a reflection of transition from the exponential to the stationary growth phase, and thus of preparations for the sclerotizations or sporulations which are later differentiations in the life cycle of *P. polycephalum*. In addition, misleading results due to the action of nuclear enzymes, e.g. phosphatases, on the radioactive substrates might be more prevalent under *in vitro* conditions. Finally, we feel that investigations of the plasmodial mitotic cycle should always be extended over more than *one* cycle and that studies based on incorporation of labeled precursors should be complemented with chemical analyses demonstrating the extent to which synthesis of the relevant macromolecules has taken place.

In this field some workers have also tried to pinpoint the timing of synthesis of the RNA required for a particular mitosis, using actinomycin. Sachsenmaier and colleagues[208] were able to show that actinomycin D could greatly delay the onset of mitosis, provided it was added more than 2 hours in advance of this event. They also found some correlation between this delay of mitosis and the degree to which the actinomycin inhibited RNA synthesis as measured by ^{3}H-cytidine incorporation. Previously, Mittermayer et al.[159] had reported that in order to produce an effect on mitosis, this drug had to be added much earlier—at least one and a half mitotic periods before the mitosis was expected. This discrepancy between the findings of these two groups is probably accountable to the circumstance that besides using a higher concentration of actinomycin, Sachsenmaier et al. had purposely selected for experiment a number of sublines of P. polycephalum which were more sensitive to the drug than those usually employed. In fact the plasmodia are relatively impermeable to actinomycin D. Thus it was observed[159] that with an external concentration of 100 μg per ml, only 17 μg per ml was retained by the plasmodium after 5 hours, and the possibility remains that some of this was adsorbed to the plasmodial surface. Consequently very high concentrations (100–250 μg per ml) have been used consistently for in vivo studies with this drug.[159, 208-9] Moreover, the in vitro nuclear synthesis of low molecular weight species of RNA could be inhibited 50 per cent with only 0.1 μg actinomycin per ml.[160] In conclusion, no very precise statement can yet be made concerning the timing of the synthesis of RNA in control of mitosis, although the work of Sachsenmaier et al. indicates that it is only completed within the 2 hours immediately prior to mitosis.

Protein. Under optimal growth conditions plasmodial protein doubles in amount between two consecutive mitoses. As assessed by chemical determination of total protein content, the growth rate was rather uniform through the mitotic cycle, except for a period of slow growth just prior to mitosis (see Fig. 58). However, from measurements of plasmodial protein radioactivities after pulse exposures to ^{3}H-lysine a different, more detailed growth pattern was obtained which was characterized by an increase in the rate of protein synthesis about 1 hour after mitosis.[157] As could have been expected, this increase was preceded by a rise in the ratio of polysomes to monoribosomes, the value of which had decreased during mitosis to reach a minimum 40 minutes after telophase.[157]

The ratio of histone to DNA appears to be constant throughout the mitotic cycle[161] and to the extent that this is correct it means that the synthesis of histones parallels that of DNA.

The relation between protein synthesis and mitosis has been studied with the aid of the metabolic inhibitor actidione (cycloheximide). This drug strongly inhibited the incorporation of ^{14}C-labeled amino acids by intact plasmodia supplied with a medium containing only citric acid, glucose and salts, while having little influence on nucleoside incorporation.[55] The effect on mitosis depended on the stage of the mitotic cycle at which actidione was added.[54, 55] Addition during interphase or early prophase completely prevented mitosis.[55] Later additions revealed a transition from full sensitivity of the plasmodia (immediate block in the progression of mitosis) to complete insensitivity, which occurred between prophase and prometaphase, i.e. between 13 and 5–3 minutes prior to metaphase. When the actidione was added during this transitional interval mitosis proceeded slowly and the early telophase stage was reached 1 to 2 hours later.[54] On the basis of these results it was suggested that structural proteins essential for mitosis and nuclear reconstitution are completed before the dissolution of the nucleolus in prophase, while proteins which determine the duration of the interval from metaphase to nuclear reconstitution are synthesized from late prophase to prometaphase.[54]

This work was afterwards extended to cover the resultant effects on the subsequent DNA synthesis commencing at telophase.[56] When actidione was added during early prophase, mitosis was blocked as described above and DNA synthesis completely inhibited. When the drug was introduced at late prophase so that mitosis occurred slowly in the absence of protein synthesis, synthesis of DNA was initiated at telophase, but not carried to completion. On addition of actidione just before or within the S period protein synthesis was severely inhibited and the incorporation of ^{3}H-thymidine was progressively retarded until it almost stopped within 1 to $1\frac{1}{2}$ hours, when 20 to 30 per cent of the DNA had replicated. Evidence was presented which indicated that DNA replicates semiconservatively also in the presence of actidione. The interpretation put on these findings was that nuclear DNA synthesis is maintained by newly synthesized protein.

When attempting to evaluate these results and conclusions the reader would do well to bear in mind that the drug was used in the concentration range 10–50 µg per ml and mostly at the higher concentration.[54–56, 209] It has been reported that 5 µg actidione per ml inhibits growth, and that use of concentrations greater than 20 µg per ml leads to breakdown of the plasmodia within a few hours.[55]

Concluding remarks. At the present time most cell biologists adhere to the view that progression of the cell in the mitotic cycle results from a sequential gene transcription followed by translation into proteins, with subsequent organization of macromolecules into structural aggregates required to function at particular stages of the cycle. In the following we shall give some examples of points in which this general description fits the organism *Physarum polycephalum*, in so far as this can be done on the basis of the limited experimental evidence available to date.

Let us regard the mitotic cycle as beginning at the telophase stage of mitosis. This is soon followed, in *Physarum* as in other systems, by the initiation of DNA synthesis. It has been suggested that the unfolding of deoxynucleo-protein fibers plays a role in making the DNA templates accessible to a DNA polymerase already present and active, or capable of being activated. It is postulated that the initiation and maintenance of DNA synthesis in *Physarum* require previous and continued synthesis of protein. It is not known why synthesis of DNA stops when the DNA complement has been duplicated, nor is it clear why the following G_2 phase during which mitosis is prepared is so long relative to the S phase in *Physarum*. Some of the material reviewed seems to indicate that the G_2 period can actually be shortened when "young" G_2 protoplasm is mixed with "old". If this is so, we believe that the *Physarum* system could be exploited advantageously in the future to elucidate causal relationships between events of the G_2 period.

Conclusive evidence for a sequential transcription of genetic information during the mitotic cycle in *Physarum* has not yet been produced, but from two sources comes support for the notion that at least some of the RNA is synthesized differentially with time. Thus it has been claimed that changes occurring during the mitotic cycle in the relative frequencies of particular RNA dinucleotides corroborate this idea. So does the observed rise in thymidine kinase activity with time (see Fig. 67), provided that it is closely preceded by synthesis of the mRNA which codes for this enzyme.[209] Actually thymidine kinase activity begins to increase in the late G_2 phase and reaches a maximum shortly after mitosis,[210] and this is probably one of the many conditions which must be fulfilled before DNA replication can follow mitosis and run to completion. It has been pointed out that the increase in thymidine kinase activity coincides with an increase in the activity of DNA

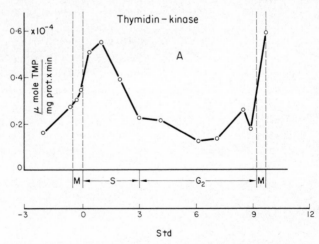

FIG. 67. Activity of thymidine kinase at different stages between two synchronous mitoses in plasmodia of *Physarum polycephalum*. From Sachsenmaier and Ives[210] by courtesy of Springer Verlag.

polymerase,[17] but whether this is fortuitous, or whether, as has been proposed,[17, 210] it reflects that messengers for enzymes involved with DNA replication are transcribed from the same operon, is so far unknown. Ultimately, whatever the underlying mechanisms G_2 processes lead to a new mitosis thus closing the cycle.

VII. GENERAL CONSIDERATIONS

Let us as a starting point for this discussion accept current views according to which progression of a cell from one division to the next is the result of a programmed gene activation which is repeated cyclically, once in each cell generation. The event of DNA replication and the occurrence of cell division can be considered useful markers of this cyclic activity, itself dependent on specific physical and chemical environmental conditions. In our view it is a feature of the normal cell cycle, as exemplified by exponentially multiplying cells that the gene activation program always runs its full course during the interval between successive cell divisions. Lacking much information on this point in the case of *Tetrahymena*, we shall here consider its cell cycle normal only when cell division is followed rather closely by a new replication of DNA in the usual way.

In accordance with these ideas the synchronization produced in *Tetrahymena* populations by *multiple heat shocks* can be considered to consist of three separate stages.

(i) The first stage is the period with heat shocks, during which cell division is blocked due to prevention of the assembly of protein precursors into structures which normally arise towards the end of the cell cycle, i.e. close to the occurrence of division. Since it has also been shown that the individual cells can engage repeatedly in cycles of DNA replication during this stage, it seems that the blockage of cell division does not phase or synchronize the programmed reading of the genome.

(ii) The second stage begins when the heat-shock treatment is discontinued, and it lasts until the cells divide synchronously after a standard time interval. During this stage the structures required for division become assembled and we have evidence that this process depends on concomitant protein synthesis, but so far it is uncertain whether it also requires immediately preceding transcription or whether already transcribed genetic information is translated.

(iii) The third stage begins just after the first synchronous division. At this point two-thirds of the cells begin to synthesize DNA anew, and in synchrony. This must mean that somehow—perhaps by the mediation of feed-backs from structures having functions in cellular or nuclear division— these cells have been induced to discontinue the programmed reading of the genome and to resume reading together from the point at which newly divided cells usually start, or close to this point. The remaining one-third of the cells do not synthesize DNA at any time between the first and second synchronous divisions, so at this point they can not be considered to behave normally according to the criterion put forward above. However, during the later synchronous division cycles an increasingly greater fraction of the population participates in DNA synthesis shortly after division. Thus with time an increasing fraction of the population becomes normal with respect to the timing of gene activation, though at the price of loss of synchrony of cell activities.

It is likely that the mechanisms described above operate also in both *Tetrahymena* and *Astasia* synchronized by the application of *one temperature shock per cell generation*. With this technique we know that cell divisions can become beautifully synchronized, but whether or not the gene activation cycle comes into phase with the division cycle may depend on the particular combination of environmental factors chosen to produce the division synchrony. In other words, "repetitive" synchronous division activity does not in itself guarantee that gene activations are also synchronized. In fact, with the one shock per generation temperature regimens published to date, we know that cell division is uncoupled from its normal position relative to DNA synthesis, so here cell division and gene activation may not be phased. It is generally agreed by workers in this field that thus far the price for repetitive division synchrony has been acceptance of lack of synchrony of a number of biochemical events which in normal cells occur at set times between divisions.

A somewhat different situation may exist in *Physarum* where a heterogeneous distribution of cytoplasmic factors in control of mitosis in a population of microplasmodia is changed after coalescence of the latter to a homogeneous distribution, with the result that all the nuclei of the fused plasmodium are assigned the same position in the mitotic cycle. When this has happened the nuclei continue to advance towards mitosis together and they divide in synchrony. The price paid for this mitotic synchrony is acceptance of the absence of cell division.

VIII. ACKNOWLEDGMENTS

The following have worked with us in the laboratory and materially helped to promote studies of synchronized protozoa: Drs. H. A. Andersen, J. Barton, E. Bernstein, C. F. Brunk, J. R. Cann, R. E. Cerroni, C. Chapman-Andresen, R. L. Crockett, P. B. Dunham, J. Frankel, A. Forer, K. Hamburger, K. K. Hjelm, E. Kay Hoffmann, G. G. Holz, Jr., M. Ikeda, M. Kovič†, B. Kramhøft, V. Leick, B. A. Lowy, A. Løvlie, O. Michelsen, J. G. Moner, D. S. Nachtwey, R. H. Neff, R. J. Neff, J. R. Nilsson, R. E. Pearlman, P. Plesner, B. Satir, P. Satir, O. Scherbaum†, H. Thormar, I. S. Villadsen, M. Wand, O. Westergaard, N. E. Williams.

We are greatly indebted to Dr. Rosemary Rasmussen for the revision of our text. Being a scientist herself and having worked with synchronized *Tetrahymena* (under the name R. L. Crockett) she has insisted not only on proper English, but also on understanding and correctly expressing the opinions we have wished to put forward. Therefore, if this text is clearer than those we have produced before, either separately or jointly, the honour is hers.

The number of corrections made has put increased demands on the patience of Mrs. A. Hvillum and Miss Eugénie Rosher who have typed the various drafts of the manuscript and in many other ways skilfully helped us to complete this report. We thank them heartily.

IX. REFERENCES

1. ADAMS, M. H. (1954) Synchronization of cell division in microorganisms. *Science* **120**, 793.
2. AGRELL, I. (1961) Cytochemical demonstration of a varied linkage of nuclear RNA during growth and division. *Pathologie-Biologie* **9**, 775–8.
3. AGRELL, I. (1964) Natural division synchrony and mitotic gradients in metazoan tissues. In ZEUTHEN, E. (ed.) *Synchrony in Cell Division and Growth*, pp. 39–67. Interscience Publishers, New York.
4. ALLEN, R. D. (1968) Basal body replication in *Tetrahymena pyriformis*. *J. Protozool.* **15** (Suppl.), 7.
5. ANDERSEN, H. A., BRUNK, C. F., and ZEUTHEN, E. (1970) Studies on the DNA replication in heat synchronized *Tetrahymena pyriformis*. *C.R. Trav. Lab. Carlsberg* **38**, 123–31.
6. ANDERSON, J. D. (1964) Regional differences in ion concentration in migrating plasmodia. In ALLEN, R. D., and KAMIYA, N. (eds.), *Primitive Motile Systems in Cell Biology*, pp. 125–34, Academic Press, New York.
7. BEHNKE, O., and FORER, A. (1967) Evidence for four classes of microtubules in individual cells. *J. Cell. Sci.* **2**, 169–92.
8. BERGAN, P. (1960) On the blocking of mitosis by heat shock applied at different mitotic stages in the cleavage divisions of *Trichogaster trichopterus* var. *sumatranus*. *Nytt Magasin for Zoologi* **9**, 37–121.
9. BERNSTEIN, E., and ZEUTHEN, E. (1966) The relationship of RNA-synthesis to temperature as an inducer of synchronous division. *C.R. Trav. Lab. Carlsberg* **35**, 501–17.
10. BIXON, M., SCHERAGA, H. A., and LIFSON, S. (1963) Effect of hydrophobic bonding on the stability of poly-L-alanine helices in water. *Biopolymers* **1**, 419–29.
11. BLUM, J. J., and PADILLA, G. M. (1962) Studies on synchronized cells: the time course of DNA, RNA and protein synthesis in *Astasia longa*. *Exptl. Cell Res.* **28**, 512–23.
12. BRADBURY, P., and PITELKA, D. R. (1965) Observations on kinetosome formation in an apostome ciliate. *J. Microscopie* **4**, 805–10.
13. BRAUN, R., MITTERMAYER, C., and RUSCH, H. P. (1965) Sequential temporal replication of DNA in *Physarum polycephalum*. *Proc. Natl. Assoc. Sci.* **53**, 924–31.
14. BRAUN, R., MITTERMAYER, C., and RUSCH, H. P. (1966) Sedimentation patterns of pulse-labeled RNA in the mitotic cycle of *Physarum polycephalum*. *Biochim. Biophys. Acta* **114**, 27–35.
15. BRAUN, R., MITTERMAYER, C., and RUSCH, H. P. (1966) Ribonucleic acid synthesis *in vivo* in the synchronously dividing *Physarum polycephalum* studied by cell fractionation. *Biochim. Biophys. Acta* **114**, 527–35.
16. BREWER, E. N., and RUSCH, H. P. (1965) DNA synthesis by isolated nuclei of *Physarum polycephalum*. *Biochem. Biophys. Res. Comm.* **21**, 235–41.

17. BREWER, E. N., and RUSCH, H. P. (1966) Control of DNA replication: effect of spermine on DNA polymerase activity in nuclei isolated from *Physarum polycephalum*. *Biochem. Biophys. Res. Comm.* **25**, 579–84.

18. BREWER, E. N., and RUSCH, H. P. (1968) Effect of elevated temperature shocks on mitosis and on the initiation of DNA replication in *Physarum polycephalum*. *Exptl. Cell Res.* **49**, 79–86.

19. BREWER, E. N., DE VRIES, A., and RUSCH, H. P. (1967) DNA synthesis by isolated mitochondria of *Physarum polycephalum*. *Biochim. Biophys. Acta* **145**, 686–92.

20. BROWN, M. J. (1966) The nature of the mitotic arrest by deuterium oxide in *Tetrahymena pyriformis*. *J. Protozool.* **13** (Suppl.), 25.

21. BURTON, K. (1956) A study of the conditions and mechanism of the diphenylamine reaction for the colorimetric estimation of deoxyribonucleic acid. *Biochem. J.* **62**, 315–23.

22. BYFIELD, J. E., and SCHERBAUM, O. H. (1966) Suppression of RNA and protein accumulation by temperature shifts in a heat-synchronized protozoan. *Life Science* **5**, 2263–9.

23. BYFIELD, J. E., and SCHERBAUM, O. H. (1966) Temperature dependent RNA decay in *Tetrahymena*. *J. Cell. Physiol.* **68**, 203–6.

24. BYFIELD, J. E., and SCHERBAUM, O. H. (1967) Stability of division-related protein and nucleic acid fractions in synchronized *Tetrahymena*. *J. Cell. Physiol.* **70**, 265–74.

25. BYFIELD, J. E., and SCHERBAUM, O. H. (1967) Temperature effect on protein synthesis in a heat-synchronized protozoan treated with actinomycin D. *Science* **156**, 1504–5.

26. BYFIELD, J. E., and SCHERBAUM, O. H. (1967) Temperature dependent decay of RNA and of protein synthesis in a heat-synchronized protozoan. *Proc. Natl. Acad. Sci.* **57**, 602–6.

27. BYFIELD, J. E., and SCHERBAUM, O. H. (1968) Non-coordinate synthesis of division-related m-RNA. *Nature* **218**, 1271–3.

28. BYFIELD, J. E., and SCHERBAUM, O. H. (1968) Amino acid control of RNA synthesis in an animal cell (*Tetrahymena*) and its relation to some aspects of gene expression. *Exptl. Cell. Res.* **49**, 202–6.

29. BYFIELD, J. E., SERAYDARIAN, K. H., and SCHERBAUM, O. H. (1960) Enzymatic determination of ATP during growth and cell division of *Tetrahymena*. *Physiologist* **3**, 35.

30. CAMERON, I. L. (1965) Macromolecular events leading to cell division in *Tetrahymena pyriformis* after removal and replacement of required pyrimidines. *J. Cell Biol.* **25**, part 2, 9–18.

31. CAMERON, I. L., and GUILE, E. E., Jr. (1965) Nucleolar and biochemical changes during unbalanced growth of *Tetrahymena pyriformis*. *J. Cell Biol.* **26**, 845–55.

32. CAMERON, I. L., and NACHTWEY, D. S. (1967) DNA synthesis in relation to cell division in *Tetrahymena pyriformis*. *Exptl. Cell Res.* **46**, 385–95.

33. CAMERON, I. L., and PADILLA, G. M. (eds.) (1966) *Cell Synchrony*; *Studies in Biosynthetic Regulation*. Academic Press, New York.

34. CAMERON, I. L., PADILLA, G. M., and MILLER, O. L., Jr. (1966) Macronuclear cytology of synchronized *Tetrahymena pyriformis*. *J. Protozool.* **13**, 336–41.

35. CAMPBELL, A. (1964) The theoretical basis of synchronization by shifts in environmental conditions. In ZEUTHEN, E. (ed.) *Synchrony in Cell Division and Growth*, pp. 469–84. Interscience Publishers, New York.

36. CANN, J. R. (1963) A kinetic model of induced division-synchrony in *Tetrahymena pyriformis*. *C.R. Trav. Lab. Carlsberg* **35**, 431–53.

37. CANN, J. R. (1968) On the leakage of ultraviolet absorbing materials and alanine from synchronized *Tetrahymena pyriformis*. *C.R. Trav. Lab. Carlsberg* **36**, 319–25.

38. CERRONI, R. E., and ZEUTHEN, E. (1962) Asynchrony of nuclear incorporation of tritiated thymidine into *Tetrahymena* cells synchronized for division. *C.R. Trav. Lab. Carlsberg* **32**, 499–511.

39. CERRONI, R. E., and ZEUTHEN, E. (1962) Inhibition of macromolecular synthesis and of cell division in synchronized *Tetrahymena*. *Exptl. Cell Res.* **26**, 604–5.

40. CHAKLEY, H. W. (1951) Control of fission in *Amoeba proteus* as related to the mechanism of cell division. *Ann. N.Y. Acad. Sci.* **51**, 1303–10.

41. CHAPMAN-ANDERSEN, C., and NILSSON, J. R. (1968) On vacuole formation in *Tetrahymena pyriformis* GL. *C.R. Trav. Lab. Carlsberg* **36**, 405–32.

42. CHOU, S. C., and SCHERBAUM, O. H. (1963) Temperature-induced changes in phosphorus metabolism in synchronized *Tetrahymena*. *Biochim. Biophys. Acta* **71**, 221–4.

43. CHOU, S. C., and SCHERBAUM, O. H. (1965) Isolation and preliminary characteristics of two phosphorus-containing deoxysugars accumulating in division blocked *Tetrahymena*. *Exptl. Cell Res.* **39**, 346–54.

44. CHOU, S. C., and SCHERBAUM, O. H. (1965) Occurrence of acid soluble phosphorylated deoxysugars in division synchronized cells. *Exptl. Cell Res.* **40**, 217–23.

45. CHOU, S. C., and SCHERBAUM, O. H. (1966) Ethanolamine and phosphonic acid complexes in heat-treated *Tetrahymena* cells. *Exptl. Cell Res.* **45**, 31–38.

46. CHRISTENSSON, E. (1959) Changes in free amino acids and protein during cell growth and synchronous division in mass cultures of *Tetrahymena pyriformis*. *Acta Physiol. Scand.* **45**, 339–49.

47. CHRISTENSSON, E. (1962) Different RNA-fractions during cell growth and synchronous division in mass cultures of *Tetrahymena pyriformis*. *Acta Physiol. Scand.* **54**, 1–8.

48. CHRISTENSSON, E. G. (1967) Histones and basic proteins during cell division in heat-synchronized mass cultures of *Tetrahymena pyriformis* GL. *Arkiv för Zoologi, Serie 2,* **19**, 297–308.

49. CLEFFMANN, G. (1965) Die Schwellen der Hemmung der Nukleinsäuresynthese und der Teilung durch Actinomycin bei *Tetrahymena pyriformis*. *Zeitschrift für Zellforschung* **67**, 343–50.

50. CLINE, S. G. (1966) Influence of ions and tonicity on RNA metabolism in *Tetrahymena pyriformis*. *J. Cell. Physiol.* **68**, 157–64.

51. CLINE, S. G., and CONNER, R. L. (1966) Orthophosphate excretion as related to RNA metabolism in *Tetrahymena*. *J. Cell. Physiol.* **68**, 149–56.

52. CROCKETT, R. L., DUNHAM, P. B., and RASMUSSEN, L. (1965) Protein metabolism in *Tetrahymena pyriformis* cells dividing synchronously under starvation conditions. *C.R. Trav. Lab. Carlsberg* **34**, 451–86.

53. CUMMINS, J. E. (1969) Nuclear DNA replication and transcription during the cell cycle of *Physarum*. In PADILLA, G. M., WHITSON, G. L., and CAMERON, I. L. (eds.) *The Cell Cycle: Gene–Enzyme Interactions*, pp. 141–58. Academic Press, New York.

54. CUMMINS, J. E., BLOMQUIST, J. C., and RUSCH, H. P. (1966) Anaphase delay after inhibition of protein synthesis between late prophase and prometaphase. *Science* **154**, 1343–4.

55. CUMMINS, J. E., BREWER, E. N., and RUSCH, H. P. (1965) The effect of actidione on mitosis in the slime mold *Physarum polycephalum*. *J. Cell. Biol.* **27**, 337–41.

56. CUMMINS, J. E., and RUSCH, H. P. (1966) Limited DNA synthesis in the absence of protein synthesis in *Physarum polycephalum*. *J. Cell Biol.* **31**, 577–83.

57. CUMMINS, J. E., and RUSCH, H. P. (1967) Transcription of nuclear DNA in nuclei isolated from plasmodia at different stages of the cell cycle of *Physarum polycephalum*. *Biochim. Biophys. Acta* **138**, 124–32.

58. CUMMINS, J. E., and RUSCH, H. P. (1968) Natural synchrony in the slime mould *Physarum polycephalum*. *Endeavour* **27**, 124–9.

59. CUMMINS, J. E., RUSCH, H. P., and EVANS, T. E. (1967) Nearest neighbor frequencies and the phylogenetic origin of mitochondrial DNA in *Physarum polycephalum*. *J. Mol. Biol.* **23**, 281–4.

60. CUMMINS, J. E., WEISFELD, G. E., and RUSCH, H. P. (1966) Fluctuations of ^{32}P distribution in rapidly labeled RNA during the cell cycle of *Physarum polycephalum. Biochim. Biophys. Acta* **129**, 240–8.

61. DANIEL, J. W., and BALDWIN, H. H. (1964) Methods of culture for plasmoidal myxomycetes. In PRESCOTT, D. M. (ed.) *Methods in Cell Physiology* 1, pp. 9–41. Academic Press, New York.

62. DANIEL, J. W., KELLEY, J., and RUSCH, H. P. (1962) Hematin-requiring plasmodial myxomycete. *J. Bact.* **84**, 1104–10.

63. DANIEL, J. W., and RUSCH, H. P. (1961) The pure culture of *Physarum polycephalum* on a partially defined soluble medium. *J. Gen. Microbiol.* **25**, 47–59.

64. DAWSON, J. A., KESSLER, W. R., and SILBERSTEIN, J. K. (1937) Mitosis in *Amoeba proteus. Biol. Bull.* **72**, 125–44.

65. DeBAULT, L. E., and RINGERTZ, N. R. (1967) A comparison of normal and cold synchronized cell divisions in *Tetrahymena. Exptl. Cell Res.* **45**, 509–18.

66. DEVI, V. R., GUTTES, E., and GUTTES, S. (1968) Effect of ultraviolet light on mitosis in *Physarum polycephalum. Exptl. Cell Res.* **50**, 589–98.

67. DEWEY, V. C., HEINRICH, M. R., MARKEES, D. C., and KIDDER, G. W. (1960) Multiple inhibition by 6-methylpurine. *Biochem. Pharmacol.* **3**, 173–80.

68. ELLIOTT, A. M. (1963) The fine structure of *Tetrahymena pyriformis* during mitosis. In LEVINE, L. (ed.) *The Cell in Mitosis*, pp. 107–24. Academic Press, New York.

69. ELLIOTT, A. M., KENNEDY, J. R., and BAK, I. L. (1962) Macronuclear events in synchronously dividing *Tetrahymena pyriformis. J. Cell. Biol.* **12**, 515–31.

70. ENGELBERG, J., and HIRSCH, H. R. (1966) On the theory of synchronous cultures. In CAMERON, I. L., and PADILLA, G. M. (eds.) *Cell Synchrony*, pp. 14–37. Academic Press, New York.

71. ERICKSON, R. O. (1964) Synchronous cell and nuclear division in tissues of the higher plants. In ZEUTHEN, E. (ed.) *Synchrony in Cell Division and Growth*, pp. 11–37. Interscience, New York.

72. EVANS, T. E. (1966) Synthesis of a cytoplasmic DNA during the G_2 interphase of *Physarum polycephalum. Biochem. Biophys. Res. Comm.* **22**, 678–83.

73. FLICKINGER, C. J. (1965) The fine structure of the nuclei of *Tetrahymena pyriformis* throughout the cell cycle. *J. Cell Biol.* **27**, 519–29.

74. FORER, A., NILSSON, J. R., and ZEUTHEN, E. (1970) Studies of the oral apparatus of *Tetrahymena pyriformis* GL. *C.R. Trav. Lab. Carlsberg*, **38**, 67–86.

75. FRANKEL, J. (1960) Effects of localized damage on morphogenesis and cell division in a ciliate, *Glaucoma chattoni. J. Exptl. Zool.* **143**, 175–91.

76. FRANKEL, J. (1962) The effects of heat, cold and *p*-fluorophenylalanine on morphogenesis in synchronized *Tetrahymena pyriformis* GL. *C.R. Trav. Lab. Carlsberg* 33, 1–52.

77. FRANKEL, J. (1964) Cortical morphogenesis and synchronization in *Tetrahymena pyriformis* GL. *Exptl. Cell. Res.* **35**, 349–60.

78. FRANKEL, J. (1965) The effect of nucleic acid antagonists on cell division and oral organelle development in *Tetrahymena pyriformis. J. Exptl. Zool.* **159**, 113–48.

79. FRANKEL, J. (1967) Studies on the maintenance of oral developments in *Tetrahymena pyriformis* GL-C. II. The relationship of protein synthesis to cell division and oral organelle development. *J. Cell. Biol.* **34**, 841–58.

80. FRANKEL, J. (1967) Studies on the maintenance of development in *Tetrahymena pyriformis*, GL-C. I. An analysis of the mechanism of resorption of developing oral structures. *J. Exptl. Zool.* **164**, 435–60.

81. FRANKEL, J., and WILLIAMS, N. E. Cortical development in *Tetrahymena*. To appear in ELLIOTT, A.M.(ed.) *The Biology of Tetrahymena*. Appleton, Century, Crofts, New York.

82. FURGASON, W. H. (1940) The significant cytostomal pattern of the "Glaucoma-Colpidium group", and a proposed new genus and species, *Tetrahymena geleii. Arch. für Protistenk.* **94**, 224–66.

83. Furgason, W. H., Small, E. B., and Loefer, J. (1960) New ideas regarding the morphogenesis of the buccal apparatus in hymenostome ciliates. *Xth Intern. Congr. Soc. Cell Biol.*, p. 230. L'Expansion Scientifique Française.

84. Gavin, R. H. (1965) The effects of heat and cold on cellular development in synchronized *Tetrahymena pyriformis* WH-6. *J. Protozool.* **12**, 307–18.

85. Gavin, R. H., and Frankel, J. (1966) The effects of mercaptoethanol on cellular development in *Tetrahymena pyriformis*. *J. Exptl. Zool.* **161**, 63–82.

86. Geilenkirchen, W. L. M. (1964) The cleavage schedule and the development of *Arbacia* eggs as separately influenced by heat shocks. *Biol. Bull.* **127**, 370.

87. Geilenkirchen, W. L. M. (1914) Periodic sensitivity of mechanisms of cytodifferentiation in cleaving eggs of *Limnaea stagnalis*. *J. Embryol. Exptl. Morphol.* **12**, 183–95.

88. Geilenkirchen, W. L. M. (1966) Cell division and morphogenesis of *Limnaea* eggs after treatment with heat pulses at successive stages in early division cycles. *J. Embryol. Exptl. Morphol.* **16**, 321–37.

89. Goode, M. D. (1967) Kinetics of microtubule assembly after cold disaggregation of the mitotic apparatus. *J. Cell Biol.* **35**, 47A–48A.

90. Guttes, E., and Guttes, S. (1963) Arrest of plasmodial motility during mitosis in *Physarum polycephalum*. *Exptl. Cell Res.* **30**, 242–4.

91. Guttes, E., and Guttes, S. (1964) Mitotic synchrony in the plasmodia of *Physarum polycephalum* and mitotic synchronization by coalescence of microplasmodia. In Prescott, D. M. (ed.) *Methods in Cell Physiology* **1**, pp. 43–54, Academic Press, New York.

92. Guttes, E., and Guttes, S. (1967) Transplantation of nuclei and mitochondria of *Physarum polycephalum* by plasmodial coalescence. *Experientia* **23**, 713–15.

93. Guttes, S., and Guttes, E. (1968) Regulation of DNA replication in the nuclei of the slime mold *Physarum polycephalum*. *J. Cell Biol.* **37**, 761–72.

94. Guttes, S., Guttes, E., and Ellis, R. A. (1968) Electron microscope study of mitosis in *Physarum polycephalum*. *J. Ultrastruct. Res.* **22**, 508–29.

95. Guttes, E., Guttes, S., and Rusch, H. P. (1961) Morphological observations on growth and differentiation of *Physarum polycephalum* grown in pure culture. *Develop. Biol.* **3**, 588–614.

96. Guttes, E. W., Hanawalt, P. C., and Guttes, S. (1967) Mitochondrial DNA synthesis and the mitotic cycle in *Physarum polycephalum*. *Biochim. Biophys. Acta* **142**, 181–94.

97. Hamburger, K. (1962) Division delays induced by metabolic inhibitors in synchronized cells of *Tetrahymena pyriformis*. *C.R. Trav. Lab. Carlsberg* **32**, 359–70.

98. Hamburger, K., and Zeuthen, E. (1957) Synchronous divisions in *Tetrahymena pyriformis* as studied in an inorganic medium. The effect of 2,4-dinitrophenol. *Exptl. Cell Res.* **13**, 443–53.

99. Hamburger, K., and Zeuthen, E. (1960) Some characteristics of growth in normal and synchronized populations of *Tetrahymena pyriformis*. *C.R. Trav. Lab. Carlsberg* **32**, 1–18.

100. Hanson, E. D. (1962) Morphogenesis and regeneration of oral structures in *Paramecium aurelia*: An analysis of intracellular development. *J. Exptl. Zool.* **150**, 45–65.

101. Hardin, J. A., Einem, G. E., and Lindsay, D. T. (1967) Simultaneous synthesis of histone and DNA in synchronously dividing *Tetrahymena pyriformis*. *J. Cell Biol.* **32**, 709–17

102. Hewitt, R. (1942) Studies on the host–parasite relationships of untreated infections with *Plasmodium lophurae* in ducks. *Amer. J. Hygiene* **36**, 6–42.

103. Hjelm, K. K., and Zeuthen, E. (1967) Synchronous DNA synthesis induced by synchronous cell division in *Tetrahymena*. *C.R. Trav. Lab. Carlsberg* **36**, 127–60.

104. Hjelm, K. K., and Zeuthen, E. (1967) Synchronous DNA synthesis following heat-synchronized cell division in *Tetrahymena*. *Exptl. Cell Res.* **48**, 231–2.

105. HOLM, B. (1966) Stability of DNA in synchronized cultures of *Tetrahymena pyriformis. Biochim. Biophys. Acta* **119**, 647–9.
106. HOLM, B. (1968) Changes in the amount of DNA in *Tetrahymena. J. Protozool.* **15** (Suppl.), 35–36.
107. HOLM, B. (1968) Changes in the amount of DNA in synchronized cultures of *Tetrahymena* cells. *Exptl. Cell Res.* **53**, 18–36.
108. HOLZ, G. G., Jr. (1960) Structural and functional changes in a generation in *Tetrahymena. Biol. Bull.* **118**, 84–95.
109. HOLZ, G. G., JR., RASMUSSEN, L., and ZEUTHEN, E. (1963) Normal versus synchronized division in *Tetrahymena pyriformis.* A study with metabolic analogs. *C.R. Trav. Lab. Carlsberg* **33**, 289–300.
110. HOLZ, G. G., SCHERBAUM, O. H., and WILLIAMS, N. E. (1957) The arrest of mitosis and stomatogenesis during temperature-induction of synchronous division in *Tetrahymena pyriformis*, mating type I, variety 1. *Exptl. Cell Res.* **13**, 618–21.
111. HOTCHKISS, R. D. (1954) Cyclical behaviour in pneumococcal growth and transformability occasioned by environmental changes. *Proc. Natl. Acad. Sci. U.S.* **40**, 49–55.
112. HOWARD, F. L. (1931) The life story of *Physarum polycephalum. Am. J. Bot.* **18**, 116–33.
113. HOWARD, F. L. (1932) Nuclear division in plasmodia of *Physarum polycephalum. Ann. Bot.* (London) **46**, 461–77.
114. IKEDA, M., NILSSON, J. R., WAND, M., and ZEUTHEN, E. Changes in state of condensation of chromatin during heat synchronization of *Tetrahymena pyriformis.* In preparation.
115. IKEDA, M., and WATANABE, Y. (1965) Studies on protein-bound sulfhydryl groups in synchronous cell division of *Tetrahymena pyriformis. Exptl. Cell Res.* **39**, 584–90.
116. INOUÉ, S. (1964) Organization and function of the mitotic spindle. In ALLEN, R. D., and KAMIYA, N. (eds.) *Primitive Motile Systems in Cell Biology*, pp. 549–94. Academic Press, New York.
117. IWAI, K., MITA, T., SENSHU, T., SHIOMI-YABUKI, H., OCHIAI, H., ISHIKAWA, K., and HAYASHI, H. (1967) Chemical aspects of histones from calf thymus and from synchronously growing *Tetrahymena. Abstracts 7th Internatl. Congress of Biochemistry* (Tokyo) **3**, 511.
118. JAMES, T. W. (1959) Synchronization of cell division in *Amoeba. Ann. N.Y. Acad. Sci.* **78**, 501–14.
119. JAMES, T. W. (1960) Controlled division synchrony and growth in protozoan microorganisms. *Ann. N.Y. Acad. Sci.* **90**, 550–8.
120. JAMES, T. W. (1963) Cell division and growth studies on synchronized flagellates. In HARRIS, R. J. C. (ed.) *Cell Growth and Cell Division*, pp. 9–26.
121. JAMES, T. W. (1964) Induced division synchrony in the flagellates. In ZEUTHEN, E. (ed.) *Synchrony in Cell Division and Growth*, pp. 323–49. Interscience, New York.
122. JAMES, T. W. (1965) Dynamic respirometry of division synchronized *Astasia longa. Exptl. Cell Res.* **38**, 439–53.
123. KAMIYA, T. (1959) Carbohydrate changes in the acid-soluble fractions of the ciliate protozoon *Tetrahymena geleii*, W, during the course of synchronous culture. *J. Biochem.* **46**, 1187–92.
124. KAUZMAN, W. (1959) Some factors in the interpretation of protein denaturation. In ANFINSEN, C. B., JR., ANSON, M. L., BAILEY, K., and EDSALL, J. T. (eds.) *Advances in Protein Chemistry* **14**, 1–63. Academic Press, New York.
125. KIRBY, K. S. (1964) A new method for the isolation of ribonucleic acids from mammalian tissues. *Biochem. J.* **56**, 405–8.
126. KOEVENIG, J. L. (1964) Studies on life cycle of *Physarum gyrosum* and other myxomycetes. *Mycologia* **56**, 170–84.

127. KOEVENIG, J. L., and JACKSON, R. C. (1966) Plasmodial mitosis and polyploidy in the myxomycete *Physarum polycephalum*. *Mycologia* **58**, 662–7.

128. KOVIČ, M., and ZEUTHEN, E. (1966) Malarial periodicity and body temperature. An experimental study of *Plasmodium* in chicken embryos. In CORRADETTI, A. (ed.) *Proc. First Intern. Congr. Parasitology, Roma*, 1964, **1**, pp. 241–4. Pergamon Press.

129. KOVIČ, M., and ZEUTHEN, E. (1967) Malarial periodicity and body temperature. An experimental study of *Plasmodium lophurae* in chicken embryos. *C.R. Trav. Lab. Carlsberg* **36**, 209–23.

130. KUDO, R. R. (1947). *Pelomyxa carolinensis* Wilson. II. Nuclear division and plasmotomy. *J. Morphol.* **80**, 93–144.

131. KUDO, R. R. (1959) *Pelomyxa* and related organisms. *Ann. N.Y. Acad. Sci.* **78**, 474–86.

132. LAZARUS, L. H., LEVY, M. R., and SCHERBAUM, O. H. (1964) Inhibition of synchronized cell division in *Tetrahymena* by actinomycin D. *Exptl. Cell Res.* **36**, 672–6.

133. LEDERBERG, S., and MAZIA, D. (1960) Protein synthesis and cell division in a pyrimidine-starved protozoan. *Exptl. Cell Res.* **21**, 590–5.

134. LEE, K. H., YUZIRIKA, Y. O., and EILER, J. J. (1959) Studies on cell growth and cell division. II. Selective activity of chloramphenicol and azaserine on cell growth and cell division. *J. Am. Pharm. Assoc. Sci. Ed.* **48**, 470–3.

135. LEICK, V. (1967) Growth rate dependency of protein and nucleic acid composition of *Tetrahymena pyriformis* and the control of synthesis of ribosomal and transfer RNA. *C.R. Trav. Lab. Carlsberg* **36**, 113–26.

136. LEICK, V. (1969) Effect of actinomycin D and D,L-*p*-fluorophenylalanine on ribosome formation in *Tetrahymena pyriformis*. *Eur. J. Biochem.* **8**, 215–20.

137. LEICK, V. (1969) Formation of subribosomal particles in the macronuclei of *Tetrahymena pyriformis*. *Eur. J. Biochem.* **8**, 221–8.

138. LEICK, V., and PLESNER, P. (1968) Formation of ribosomes in *Tetrahymena pyriformis*. *Biochim. Biophys. Acta* **169**, 398–408.

139. LEICK, V., and PLESNER, P. (1968) Precursor of ribosome sub-units in *Tetrahymena pyriformis*. *Biochim. Biophys. Acta* **169**, 409–15.

140. LINDH, N. O., and CHRISTENSSON, E. (1962) The carbohydrate metabolism during growth and division of *Tetrahymena pyriformis* GL. *Arkiv för Zoologi, Ser. 2*, **15**, 163–80.

141. LOWE, L., and ZIMMERMAN, A. M. (1967) Hydrostatic pressure effects on the macromolecular synthesis in *Tetrahymena pyriformis*. *J. Protozool.* **14** (Suppl.), 9.

142. LOWRY, O. H., ROSEBROUGH, N. J., FARR, A. L., and RANDALL, R. J. (1951) Protein measurement with the folin phenol reagent. *J. Biol. Chem.* **193**, 265–75.

143. LOWY, B., and LEICK, V. (1969) The synthesis of DNA in synchronized cultures of *Tetrahymena pyriformis* GL. *Exptl. Cell Res.* **57**, 277–88.

144. LØVLIE, A. (1963) Growth in mass and respiration rate during the cell cycle of *Tetrahymena pyriformis*. *C.R. Trav. Lab. Carlsberg* **33**, 377–413.

145. LØVLIE, A., and FARFAGLIO, G. (1965) Increase in photosynthesis during the cell cycle of *Euglena gracilis*. *Exptl. Cell Res.* **39**, 418–34.

146. MACDONALD, A. G. (1967) Effect of high pressure on the cell cycle of *Tetrahymena pyriformis*. *J. Protozool.* **14** (Suppl.), 42–43.

147. MACDONALD, A. G. (1967) Delay in the cleavage of *Tetrahymena pyriformis* exposed to high hydrostatic pressures. *J. Cell. Physiol.* **70**, 127–30.

148. MACDONALD, A. G. (1967) The effect of high hydrostatic pressure on the cell division and growth of *Tetrahymena pyriformis*. *Exptl. Cell Res.* **47**, 569–80.

149. MAZIA, D., and PRESCOTT, D. M. (1954) Nuclear function and mitosis. *Science* **120**, 120–2.

150. MAZIA, D., and ZEUTHEN, E. (1966) Blockage and delay of cell division in synchronized populations of *Tetrahymena* by mercaptoethanol (monothioethylene glycol). *C.R. Trav. Lab. Carlsberg* **35**, 341–61.

151. McGHEE, B. R. (1949) The course of infection of *Plasmodium lophurae* in chick. *J. Parasit.* **35**, 411–16.

152. McGRATH, R. A., and WILLIAMS, R. W. (1967) Interruptions in single strands of the DNA in slime mold and other organisms. *Biophys. J.* **7**, 309–17.

153. MEEKER, G. L. (1964) The effect of β-mercaptoethanol on synchronously dividing populations of *Astasia longa. Exptl. Cell Res.* **35**, 1–8.

154. MITA, T. (1965) Effects of actinomycin D on the RNA synthesis and the synchronous cell division of *Tetrahymena pyriformis* GL. *Biochim. Biophys. Acta* **113**, 182–5.

155. MITA, T., ONO, T., and SUGIMURA, T. (1959) Cited in SCHERBAUM, O. H. (1964) Biochemical studies on synchronized *Tetrahymena*. In ZEUTHEN, E. (ed.) *Synchrony in Cell Division and Growth*, pp. 177–95. Interscience, New York.

156. MITA, T., and SCHERBAUM, O. H. (1965) Chromatographic fractionation of DNA isolated from normally and synchronously dividing *Tetrahymena pyriformis* GL. *J. Biochem.* **58**, 130–6.

157. MITTERMAYER, C., BRAUN, R., CHAYKA, T. G., and RUSCH, H. P. (1966) Polysome patterns and protein synthesis during the mitotic cycle of *Physarum polycephalum. Nature* **210**, 1133–7.

158. MITTERMAYER, C., BRAUN, R., and RUSCH, H. P. (1964) RNA synthesis in the mitotic cycle of *Physarum polycephalum. Biochim. Biophys. Acta* **91**, 399–405.

159. MITTERMAYER, C., BRAUN, R., and RUSCH, H. P. (1965) The effect of actinomycin D on the timing of mitosis in *Physarum polycephalum. Exptl. Cell Res.* **38**, 33–41.

160. MITTERMAYER, C., BRAUN, R., and RUSCH, H. P. (1966) Ribonucleic acid synthesis *in vitro* in nuclei isolated from the synchronously dividing *Physarum polycephalum. Biochim. Biophys. Acta* **114**, 536–46.

161. MOHBERG, J., and RUSCH, H. P. (1964) Isolation of nuclei and histones from the plasmodium of *Physarum polycephalum. J. Cell Biol.* **23**, 61 A.

162. MONER, J. G. (1965) RNA synthesis and cell division in heat synchronized populations of *Tetrahymena pyriformis. J. Protozool.* **12**, 505–9.

163. MONER, J. G. (1967) Temperature, RNA synthesis and cell division in heat-synchronized cells of *Tetrahymena. Exptl. Cell Res.* **45**, 618–30.

164. MONER, J. G. (1968) The absence of cell division in cells of heat-synchronized *Tetrahymena* completing division-related RNA and protein synthesis. *J. Protozool.* **15** (Suppl.), 30.

165. MONER, J. G., and BERGER, R. O. (1966) RNA synthesis and cell division in cold-synchronized cells of *Tetrahymena pyriformis. J. Cell. Physiol.* **67**, 217–23.

166. NACHTWEY, D. S. (1965) Division of synchronized *Tetrahymena pyriformis* after emacronucleation. *C.R. Trav. Lab. Carlsberg* **35**, 25–35.

167. NACHTWEY, D. S., and DICKINSON, W. J. (1967) Actinomycin D: Blockage of cell division of synchronized *Tetrahymena pyriformis. Exptl. Cell Res.* **47**, 581–95.

168. NEFF, R. J., and NEFF, R. H. (1964) Induction of synchronous division in amoebae. In ZEUTHEN, E. (ed.) *Synchrony in Cell Division and Growth*, pp. 213–46. Interscience, New York.

169. NILSSON, J. R., and WILLIAMS, N. E. (1966) An electron microscope study of the oral apparatus of *Tetrahymena pyriformis. C.R. Trav. Lab. Carlsberg* **35**, 119–41.

170. NISHI, A., and SCHERBAUM, O. H. (1962) Levels of pyrimidine nucleotides and DPNH oxidase activity in growing cultures. *Biochim. Biophys. Acta* **65**, 411–18.

171. NYGAARD, O. F., GUTTES, S., and RUSCH, H. P. (1960) Nucleic acid metabolism in a slime mold with synchronous mitosis. *Biochim. Biophys. Acta* **38**, 298–306.

172. PADILLA, G. M., and BLUM, J. J. (1963) Studies on synchronized cells. Inhibition of cell division of *Astasia longa* by 8-azaguanine. *Exptl. Cell Res.* **32**, 289–304.

173. PADILLA, G. M., and CAMERON, I. L. (1964) Synchronization of cell division in *Tetrahymena pyriformis* by a repetitive temperature cycle. *J. Cell. Comp. Physiol.* **64**, 303–8.

174. PADILLA, G. M., CAMERON, I. L., and ELROD, L. H. (1966) The physiology of repetitively synchronized *Tetrahymena*. In CAMERON, I. L., and PADILLA, G. M. (eds.) *Cell Synchrony*, pp. 269–88. Academic Press, New York.

175. PADILLA, G. M., and COOK, J. R. (1964) The development of techniques of synchronizing flagellates. In ZEUTHEN, E. (ed.) *Synchrony in Cell Division and Growth*, pp. 521–35. Interscience, New York.

176. PADILLA, G. M., and JAMES, T. W. (1960) Synchronization of cell division in *Astasia longa* on a chemically defined medium. *Exptl. Cell Res.* **20**, 401–15.

177. PADILLA, G. M., and JAMES, T. W. (1964) Continuous synchronous cultures of protozoa. In PRESCOTT, D. M. (ed.) *Methods in Cell Physiology* **1**, 141–57. Academic Press, New York.

178. PLESNER, P. (1958) The nucleoside triphosphate content of *Tetrahymena pyriformis* during the cell division cycle in synchronously dividing mass cultures. *Biochim. Biophys. Acta* **29**, 462–3.

179. PLESNER, P. (1961) Changes in ribosome structure and function during synchronized cell division. *Cold Spring Harbor Symp. Quant. Biol.* **26**, 159–62.

180. PLESNER, P. (1963) Nucleotide metabolism and ribosomal activity during synchronized cell division. In HARRIS, R. J. C. (ed.) *Cell Growth and Cell Division*, pp. 77–91. Academic Press, New York.

181. PLESNER, P. (1964) Nucleotide metabolism during synchronized cell division in *Tetryhymena pyriformis*. *C.R. Trav. Lab. Carlsberg* **34**, 1–76.

182. PLESNER, P. (1964) Nucleotide metabolism during synchronized cell division in *Tetrahymena pyriformis*. In ZEUTHEN, E. (ed.) *Synchrony in Cell Division and Growth*, pp. 197–212. Interscience, New York.

183. PLESNER, P., RASMUSSEN, L., and ZEUTHEN, E. (1964) Techniques used in the study of synchronous *Tetrahymena*. In ZEUTHEN, E. (ed.) *Synchrony in Cell Division and Growth*, pp. 543–63. Interscience, New York.

184. PRESCOTT, D. M. (1955) Relations between cell growth and cell division. I. Reduced weight, cell volume, protein content, and nuclear volume of *Amoeba proteus* from division to division. *Exptl. Cell Res.* **9**, 328–37.

185. PRESCOTT, D. M. (1956) Relations between cell growth and cell division. II. The effect of cell size on cell growth rate and generation time in *Amoeba proteus*. *Exptl. Cell Res.* **11**, 86–94.

186. PRESCOTT, D. M. (1956) Mass and clone culturing of *Amoeba proteus* and *Chaos chaos*. *C.R. Trav. Lab. Carlsberg, Serie Chimique* **30**, 1–12.

187. PRESCOTT, D. M. (1959) Variations in the individual generation times of *Tetrahymena pyriformis*. *Exptl. Cell Res.* **16**, 279–84.

188. PRESCOTT, D. M. (1964) The normal cell cycle. In ZEUTHEN, E. (ed.) *Synchrony in Cell Division and Growth*, pp. 71–97. Interscience, New York.

189. PRESCOTT, D. M., and GOLDSTEIN, L. (1967) Nuclear-cytoplasmic interaction in DNA synthesis. *Science* **155**, 469–70.

190. RANDALL, J., and DISBREY, C. (1965) Evidence for the presence of DNA at basal body sites in *Tetrahymena pyriformis*. *Proceedings Royal Society*, B, **162**, 473–91.

191. RASMUSSEN, L. (1963) Delayed divisions in *Tetrahymena* as induced by short-time exposures to anaerobiosis. *C.R. Trav. Lab. Carlsberg* **33**, 53–71.

192. RASMUSSEN, L. (1967) Effects of DL-*p*-fluorophenylalanine on *Paramecium aurelia* during the cell generation cycle. *Exptl. Cell Res.* **45**, 501–4.

193. RASMUSSEN, L. (1967) Effects of metabolic inhibition on *Paramecium aurelia* during the cell generation cycle. *Exptl. Cell Res.* **48**, 132–9.

142   ERIK ZEUTHEN AND LEIF RASMUSSEN

194. RASMUSSEN, L., and ZEUTHEN, E. (1962) Cell division and protein synthesis in *Tetrahymena* as studied with *p*-fluorophenylalanine. *C.R. Trav. Lab. Carlsberg* **32**, 333–58.
195. RASMUSSEN, L., and ZEUTHEN, E. (1966) Amino acid antagonisms in *Tetrahymena*. *C.R. Trav. Lab. Carlsberg* **35**, 85–100.
196. RASMUSSEN, L., and ZEUTHEN, E. (1966) Cell division in *Tetrahymena* adapting to DL-*p*-fluorophenylalanine. *Exptl. Cell Res.* **41**, 462–5.
197. ROONEY, D. W., and EILER, J. J. (1967) Synchronization of *Tetrahymena* cell division by multiple hypoxic shocks. *Exptl. Cell Res.* **48**, 649–52.
198. ROONEY, D. W., and EILER, J. J. (1969) Hypoxic synchronization of *Tetrahymena* cell division with an automatic apparatus. *Exptl. Cell Res.* **54**, 49–52.
199. ROSENBAUM, N., ERWIN, J., BEACH, D., and HOLZ, G. G., JR. (1966) The induction of a phospholipid requirement and morphological abnormalities in *Tetrahymena pyriformis* by growth at supra-optimal temperatures. *J. Protozool.* **13**, 535–46.
200. ROSS, I. K. (1966) Chromosome numbers in pure and gross cultures of myxomycetes. *Am. J. Bot.* **53**, 712–18.
201. ROTH, L. E. (1967) Electron microscopy of mitosis in amebae. III. Cold and urea treatments of mitotic inhibitors on microtubuli formation. *J. Cell Biol.* **34**, 47–59.
202. ROTH, L. E., and MINICK, O. T. (1961) Electron microscopy of nuclear and cytoplasmic events during division in *Tetrahymena pyriformis*, strains W and HAM 3. *J. Protozool.* **8**, 12–21.
203. RUSCH, H. P. (1962) The use of *Physarum polycephalum* for studies on growth and differentiation. In BRENNAN, M. J., and SIMPSON, W. L. (eds.). *Biological Interaction in Normal and Neoplastic Growth*, pp. 21–24. Churchill, London.
204. RUSCH, H. P. (1970) Some biochemical events in the life cycle of *Physarum polycephalum*. In PRESCOTT, D. M., GOLDSTEIN, L., and McCONKEY, E. H. (eds.) *Advances in Cell Biology* 1, pp. 297–327. Appleton, Century, Crofts.
205. RUSCH, H. P., BRAUN, R., DANIEL, J. W., MITTERMAYER, C., and SACHSENMAIER, W. (1964) The role of DNA and RNA in mitosis and differentiation in *Physarum polycephalum*. In EMMELOT, P., and MÜHLBOCK, O. (eds.) *Cellular Control Mechanisms and Cancer*, pp. 80–85. Elsevier, Amsterdam.
206. RUSCH, H. P., SACHSENMAIER, W., BEHRENS, K., and GRUTER, V. (1966) Synchronization of mitosis by the fusion of the plasmodia of *Physarum polycephalum*. *J. Cell. Biol.* **31**, 204–9.
207. SACHSENMAIER, W. (1964) Zur DNS- und RNS-synthese im Teilungscyklus synkroner Plasmodium von *Physarum polycephalum*. *Biochem. Zeitschrift* **340**, 541–7.
208. SACHSENMAIER, W. (1966) Analyse des Zellcyklus durch Eingriffe in die Makromolekylbiosynthese. In SITTE, P. (ed.) *Probleme der Biologischen Reduplikation*, pp. 139–60. Springer Verlag, Berlin.
209. SACHSENMAIER, W., von FOURNIER, D., and GÜRTLER, K. F. (1967) Periodic thymidine-kinase production in synchronous plasmodia of *Physarum polycephalum*: Inhibition by actinomycin and actidione. *Biochem. Biophys. Res. Comm.* **27**, 655–60.
210. SACHSENMAIER, W., and IVES, D. H. (1965) Periodische Änderungen der Thymidinkinase-Aktivität in synkronen Mitosecyklus von *Physarum polycephalum*. *Biochem. Zeitschrift* **343**, 399–406.
211. SACHSENMAIER, W., and RUSCH, H. P. (1964) The effect of 5-fluoro-2′-deoxyuridine on synchronous mitosis in *Physarum polycephalum*. *Exptl. Cell Res.* **36**, 124–33.
212. SATIR, P., and ZEUTHEN, E. (1961) Cell cycle and the relationship between growth rate to reduced weight (RW) in the giant ameba *Chaos chaos*. *C.R. Trav. Lab. Carlsberg* **32**, 241–63.
213. SCHERAGA, H. A. (1966) Principles of protein structure. In HAYASHI, T., and SZENT-GYÖRGYI, A. G. (eds.) *Molecular Architecture in Cell Physiology*, pp. 39–61. Prentice-Hall.

214. SCHERBAUM, O. (1957) The content and composition of nucleic acids in normal and synchronously dividing mass culture of Tetrahymena pyriformis. Exptl. Cell Res. **13**, 24–30.

215. SCHERBAUM, O. H. (1960) Synchronous division of microorganisms. Ann. Rev. Microbiol. **14**, 283–310.

216. SCHERBAUM, O. H. (1960) Possible sites of metabolic control during induction of synchronous cell division. Ann. N.Y. Acad. Sci. **90**, 565–79.

217. SCHERBAUM, O. H. (1963) Chemical prerequisites for cell division. In LEVINE, L. (ed.) The Cell in Mitosis, pp. 125–57. Academic Press, New York.

218. SCHERBAUM, O. H., CHOU, S. C., SERAYDARIAN, H., and BYFIELD, J. E. (1962) The effects of temperature shifts on the intracellular level of nucleoside triphosphates in Tetrahymena pyriformis. Canad. J. Microbiol. **8**, 753–60.

219. SCHERBAUM, O. H., JAMES, T. W., and JAHN, T. L. (1959) The amino acid composition in relation to cell growth and cell division in synchronized cultures of Tetrahymena pyriformis. J. Cell Comp. Physiol. **53**, 119–37.

220. SCHERBAUM, O. H., and LEE, Y. C. (1966) Nucleohistone composition in stationary and division synchronized Tetrahymena cultures. Biochemistry **5**, 2067–75.

221. SCHERBAUM, O. H., and LEVY, M. (1961) Some aspects of the carbohydrate metabolism in relation to cell growth and cell division. Pathologie-Biologie **9**, 514–17.

222. SCHERBAUM, O. H., and LOEFER, J. B. (1964) Environmentally induced growth oscillations in protozoa. In HUTNER, S. H. (ed.) Biochemistry and Physiology of Protozoa **3**, 9–59. Academic Press, New York.

223. SCHERBAUM, O. H., LOUDERBACK, A. L., and JAHN, T. L. (1958) The formation of subnuclear aggregates in normal and synchronized protozoan cells. Biol. Bull. **115**, 269–75.

224. SCHERBAUM, O. H., LOUDERBACK, A. L., and JAHN, T. L. (1959) DNA-synthesis, phosphate content and growth in mass and volume in synchronously dividing cells. Exptl. Cell Res. **18**, 150–66.

225. SCHERBAUM, O., and ZEUTHEN, E. (1954) Induction of synchronous cell division in mass cultures of Tetrahymena pyriformis. Exptl. Cell Res. **6**, 221–7.

226. SCHNEIDER, W. C. (1957) Determination of nucleic acids in tissues by pentose analysis. In COLOWICK, S. P., and KAPLAN, N. O. (eds.) Methods in Enzymology **3**, pp. 680–4. Academic Press, New York.

227. SINGER, W., and EILER, J. J. (1960) The biological action of cellular depressants and stimulants. V. The effect of phenylurethane on cellular synthesis by Tetrahymena pyriformis GL. J. Am. Pharm. Assoc., Sci. Ed. **49**, 669–73.

228. SOMMER, J. R. (1965) The ultrastructure of the pellicle complex of Euglena gracilis. J. Cell Biol. **24**, 253–7.

229. SOMMER, J. R., and BLUM, J. J. (1964) Pellicular changes during division in Astasia longa. Exptl. Cell Res. **35**, 423–5.

230. SOMMER, J. R., and BLUM, J. J. (1965) Cell division in Astasia longa. Exptl. Cell Res. **39**, 504–27.

231. STONE, G. E. (1968) Synchronized cell division in Tetrahymena pyriformis following inhibition with vinblastine. J. Cell Biol. **39**, 556–63.

232. STONE, G. E., and CAMERON, I. L. (1964) Methods for using Tetrahymena in studies of the normal cell cycle. In PRESCOTT, D. M. (ed.) Methods in Cell Physiology **1**, 127–40. Academic Press, New York.

233. TAMIYA, H., SHIBATA, K., SASA, T., IWAMURA, T., and MORIMURA, Y. (1953) Effect of diurnally intermittent illumination on the growth and some cellular characteristics of Chlorella. Carnegie Institute, Washington, Publ. **600**, 76–84.

234. TARTAR, V. (1966) Fission after division primordium removal in the ciliate Stentor coerulens and comparable experiments on reorganizers. Exptl. Cell Res. **42**, 357–70.

235. DE TERRA, N. (1967) Macronuclear DNA synthesis in *Stentor*. Regulation by a cytoplasmic initiator. *Proc. Natl. Acad. Sci.* **57**, 607–14.

236. THERRIEN, C. D. (1966) Microspectrophotometric measurement of nuclear deoxyribonucleic acid content in two myxomycetes. *Can. J. Bot.* **44**, 1667–76.

237. THORMAR, H. (1959) Delayed division in *Tetrahymena pyriformis* induced by temperature changes. *C.R. Trav. Lab. Carlsberg* **31**, 207–25.

238. THORMAR, H. (1962) Cell size of *Tetrahymena pyriformis* incubated at various temperatures. *Exptl. Cell Res.* **27**, 585–6.

239. THORMAR, H. (1962) Effect of temperature on the reproduction rate of *Tetrahymena pyriformis*. *Exptl. Cell Res.* **28**, 269–79.

240. TILNEY, L. G., and PORTER, K. R. (1967) Studies on the microtubules in heliozoa. II. The effect of low temperature on these structures in the formation and maintenance of axopodia. *J. Cell Biol.* **34**, 327–43.

241. VILLADSEN, I., and ZEUTHEN, E. (1970) Synchronization of DNA synthesis in *Tetrahymena* populations by temporary limitation of access to thymine compounds. *Exptl. Cell Res.* **61**, 302–10.

242. WADE, J., and SATIR, P. (1968) The effect of mercaptoethanol on flagellar morphogenesis in the amoebo-flagellate *Naegleria gruberi* (Schardinger). *Exptl. Cell Res.* **50**, 81–92.

243. WAGNER, E. K., and ROIZMAN, B. (1968) Effect of the *Vinca* alkaloids on RNA synthesis in human cells *in vitro*. *Science* **162**, 569–70.

244. WATANABE, Y. (1963) Some factors necessary to produce division conditions in *Tetrahymena pyriformis*. *Jap. J. Sci. Med. Biol.* **16**, 107–24.

245. WATANABE, Y., and IKEDA, M. (1965) Evidence for the synthesis of the "division protein" in *Tetrahymena pyriformis*. *Exptl. Cell Res.* **38**, 432–4.

246. WATANABE, Y., and IKEDA, M. (1965) Isolation and characterization of the division protein in *Tetrahymena pyriformis*. *Exptl. Cell Res.* **39**, 443–52.

247. WATANABE, Y., and IKEDA, M. (1965) Further confirmation of the "division protein" fraction in *Tetrahymena pyriformis*. *Exptl. Cell Res.* **39**, 464–9.

248. WELLS, C. (1960) Identification of free and bound amino acids in three strains of *Tetrahymena pyriformis* using paper chromatography. *J. Protozool.* **7**, 7–10.

249. WHITSON, G. L., and PADILLA, G. M. (1964) The effects of actinomycin D on stomatogenesis and cell division in temperature-synchronized *Tetrahymena pyriformis*. *Exptl. Cell Res.* **36**, 667–71.

250. WHITSON, G. L., PADILLA, G. M., CANNING, R. E., CAMERON, I. L., ANDERSON, N.G., and ELROD, L. H. (1966) The isolation of oral structures from *Tetrahymena pyriformis* by low-speed zonal centrifugation. *Natl. Cancer Inst. Monogr.* **21**, 317–21.

251. WHITSON, G. L., PADILLA, G. M., and FISHER, W. D. (1966) Cyclic changes in polysomes of synchronized *Tetrahymena pyriformis*. *Exptl. Cell Res.* **42**, 438–46.

252. WILLIAMS, N. E. (1964) Induced division synchrony in *Tetrahymena vorax*. *J. Protozool.* **11**, 230–6.

253. WILLIAMS, N. E., and LUFT, J. H. (1968) Use of a nitrogen mustard derivative in fixation for electron microscopy and observations on the ultrastructure of *Tetrahymena*. *J. Ultrastruct. Res.* **25**, 271–92.

254. WILLIAMS, N. E., MICHELSEN, O., and ZEUTHEN, E. (1969) Synthesis of cortical proteins in *Tetrahymena*. *J. Cell Sci.* **5**, 143–62.

255. WILLIAMS, N. E., and SCHERBAUM, O. H. (1959) Morphogenetic events in normal and synchronously dividing *Tetrahymena*. *J. Embryol. Expt. Morphol.* **7**, 241–56.

256. WILLIAMS, N. E., and ZEUTHEN, E. (1966) The development of oral fibers in relation to oral morphogenesis and induced division synchrony in *Tetrahymena*. *C.R. Trav. Lab. Carlsberg* **35**, 101–20.

257. WILSON, B. W., and JAMES, T. W. (1963) The respiration and growth of synchronized populations of the cell *Astasia longa*. *Exptl. Cell Res.* **32**, 305–19.

258. WOLFE, J. (1967) Structural aspects of amitosis: a light and electron microscope study in the isolated macronuclei of *Paramecium aurelia* and *Tetrahymena pyriformis*. *Chromosoma (Berl.)* **23**, 59–79.

259. WU, C., and HOGG, J. F. (1956) Free and non-protein amino acids of *Tetrahymena pyriformis*. *Arch. Biochem. Biophys.* **62**, 70–77.

260. WU, C., and HOGG, J. F. (1962) The amino acid composition and nitrogen metabolism of *Tetrahymena geleii*. *J. Biol. Chem.* **196**, 753–64.

261. WUNDERLICH, F., and FRANKE, W. W. (1968) Structure of macronuclear envelopes of *Tetrahymena pyriformis* in the stationary phase of growth. *J. Cell Biol.* **38**, 458–62.

262. ZEUTHEN, E. (1953) Growth as related to the cell cycle in the single-cell cultures of *Tetrahymena pyriformis*. *J. Embryol. Exptl. Morphol.* **1**, 239–49.

263. ZEUTHEN, E. (1958) Artificial and induced periodicity in living cells. *Adv. Biol. Med. Phys.* **6**, 37–73.

264. ZEUTHEN, E. (1961) Cell division and protein synthesis. In GOODWIN, T. W., and LINDBERG, O. (eds.) *Biological Structures and Function* **2**, 537–48. Academic Press, London.

265. ZEUTHEN, E. (1961) Synchronized growth in Tetrahymena cells. In ZARROW, M. X. (ed.) *Growth in Living Systems*, pp. 135–58. Basic Books, New York.

266. ZEUTHEN, E. (1963) Independent cycles of cell division and of DNA synthesis in Tetrahymena. In HARRIS, R. J. C. (ed.) *Cell Growth and Cell Division*, pp. 1–7. Academic Press, New York.

267. ZEUTHEN, E. (ed.) (1964) *Synchrony in Cell Division and Growth*. Interscience, New York.

268. ZEUTHEN, E. (1964) The temperature induced division synchrony in *Tetrahymena*. In ZEUTHEN, E. (ed.) *Synchrony in Cell Division and Growth*, pp. 99–158. Interscience, New York.

269. ZEUTHEN, E. (1968) Thymine starvation by inhibition of uptake and synthesis of thymine-compounds in Tetrahymena. *Exptl. Cell Res.* **50**, 37–46.

270. ZEUTHEN, E. (1970) Independent synchronization of DNA synthesis and of cell division in same culture of *Tetrahymena* cells. *Exptl. Cell Res.* **61**, 311–25.

271. ZEUTHEN, E., and RASMUSSEN, L. (1966) Incorporation of DL-*p*-fluorophenylalanine into proteins of Tetrahymena. *J. Protozool.* **13**, (Suppl.), 29.

272. ZEUTHEN, E., and SCHERBAUM, O. H. (1954) Synchronous divisions in mass cultures of the ciliate protozoon *Tetrahymena pyriformis*, as induced by temperature changes. In KITCHING, J. A. (ed.) *Recent Developments in Cell Physiology*, pp. 141–56. Butterworth, London.

273. ZEUTHEN, E., and WILLIAMS, N. E. (1969) Division-limiting morphogenetic processes in Tetrahymena. In COWDRY, E. V., and SENO, S. (eds.) *Nucleic Acid Metabolism, Cell Differentiation and Cancer Growth*, pp. 203–16. Pergamon Press, Oxford.

274. ZIMMERMAN, A. M. (1969) Effects of high pressure on macromolecular synthesis in synchronized Tetrahymena. In PADILLA, G. M., WHITSON, G. L., and CAMERON, I. L. (eds.) *The Cell Cycle, Gene-Enzyme Interactions*, pp. 203–25. Academic Press, New York.

NUCLEAR PHENOMENA DURING CONJUGATION AND AUTOGAMY IN CILIATES

I. B. Raikov

Institute of Cytology of the Academy of Sciences,
Leningrad F-121, USSR

CONTENTS

I. INTRODUCTION

Conjugation is a special type of nuclear reorganization process peculiar to ciliates; in most cases it is characterized by only temporary mating of two individuals. As a rule, conjugation is not directly connected with reproduction phenomena: the number of individuals does not increase after its completion. Conjugation does not lead to production of progeny but "only" to genetic recombination and replacement of the old macronucleus by a new one (see also Vol. 3, Chapter 1).

Comparing conjugation with other types of nuclear reorganization processes in Protozoa, Grell (Vol. 2, Chapter 3, p. 199) shows it to be "a special type of gamontogamy† not found in other Protozoa". This type is characterized by (1) formation by the gamonts of only gamete nuclei, not of gametes, and (2) monoecy (hermaphroditism) of the gamonts resulting in their mutual fertilization.

The external phenomena of conjugation, i.e. temporary pairing of infusorians, has for a long time attracted the attention of investigators; an excellent critical review of observations and theories of the older authors was given by Maupas[165] at the beginning of his fundamental work of 1889. This allows me to limit myself to a brief summary of the main steps in the development of the concept of conjugation in ciliates.

Until 1858, couples of ciliates were almost universally considered to be stages of longitudinal fission and not the result of mating. Balbiani[4] first showed the "syzigia" of ciliates to be a result of mating. But unicellularity of the Protozoa was not yet known, and Balbiani erroneously took the macronucleus for an "ovary", the division products of the micronucleus—for "seminal capsules", and the anlagen of new macronuclei—for "fertilized ova" developing inside the infusorian body. At first Balbiani[4] thought that "ova" develop into "embryos" leaving the infusorian body. These "embryos" were in reality parasitic suctorians. Later Balbiani[5] abandoned the idea of "embryos", but Stein[247] maintained it even in 1867.

A revolution in the concept of conjugation was made in 1876 by the work of Bütschli,[14] who demonstrated that both the macronucleus and the micronucleus are true cell nuclei, that micronuclear spindles are not "seminal capsules" but mitotic figures, that transparent "germinative spheres" are not ova but anlagen of new macronuclei, and that these anlagen develop from products of micronuclear divisions. Thus, Bütschli first showed the old macronucleus to be replaced during conjugation by a new macronucleus

† Gamontogamy is "mating of gamonts which form gametes or gamete nuclei fusing later on" (Vol. 2, Chapter III by Grell, p. 151).

which has micronuclear origin. But the exact sequence of micronuclear events still escaped to Bütschli.

The concept of nuclear phenomena during conjugation of ciliates took its final form in 1889, as a result of the work of R. Hertwig[106] and especially of Maupas.[165] Simultaneously and independently, these investigators gained a correct understanding of the full sequence of maturation divisions, discovered the exchange of migratory pronuclei and the karyogamy, and made clear the sequence of synkaryon divisions which were previously confused with maturation divisions. Hertwig studied these phenomena in *Paramecium aurelia*, and Maupas in many holotrichs, heterotrichs, hypotrichs, and peritrichs. These classical investigations still retain their significance and in some aspects are even unsurpassed.

Several types of conjugation may be distinguished among the ciliates. In most cases, fusion of the gamonts (or conjugants) is partial and temporary: two exconjugants result from separation of the mates. This type of conjugation is called *temporary* (or partial). In other cases, the gamonts fuse completely, forming only one exconjugant, frequently called a synconjugant; this is called *total* conjugation.

Differences in body size and other characters may exist between the two conjugants (gamonts). In temporary conjugation these differences are usually absent, or are accidental. This is the so-called *temporary isogamontic conjugation*, the most frequent among ciliates (e.g. in *Paramecium*, Fig. 5). But in rare cases there are regular differences between the two partners undergoing temporary conjugation; this is the *temporary anisogamontic conjugation* (e.g. in *Opisthotrichum*, Fig. 1, *b*). In total conjugation, there are cases when the mates are at first alike (*total isogamontic conjugation*, e.g. in *Metopus*, Fig. 14). But far more frequently the partners undergoing total conjugation are so different that two categories of conjugants may be distinguished, the macroconjugants and the microconjugants. This is *total anisogamontic conjugation* (e.g. in *Carchesium*, Fig. 18).

The terminology describing the cytological phenomena of conjugation lacks uniformity. For example, the term "progamic division" has been used in two different senses: for cell division preceding conjugation and giving rise to two preconjugants (ref. 76, and others), and for micronuclear divisions preceding formation of the pronuclei (many American authors). The term "metagamic division" has also been used, on one hand, to designate the divisions of the synkaryon (without cytokinesis), and, on the other hand, to designate the divisions of the exconjugant already having anlagen of new macronuclei. Therefore it seems useful to give up both these terms. The following terminology is used in this chapter:

1. *Preconjugation (cell) divisions*: special divisions of the ciliate preceding conjugation (synonym: progamic cell divisions).

2. *Maturation divisions* (of the micronucleus): two meiotic and one equational (postmeiotic) micronuclear division preceding formation of pronuclei

(synonyms: progamic, pregamic, pregametic, or prezygotic micronuclear divisions).

3. *Pronuclei* (the migratory and the stationary): haploid nuclei, derivatives of the maturation divisions, fusing in karyogamy (synonym: gamete nuclei).

4. *Synkaryon*: diploid product of fusion of the pronuclei (synonym: amphinucleus).

5. *Synkaryon divisions*: mitoses of the synkaryon without cell division (synonyms: metagamic, postgamic, or postzygotic nuclear divisions).

6. *Exconjugant divisions*: cell divisions segregating the macronuclear anlagen (synonym: metagamic cell divisions).

Autogamy, a somewhat similar process of nuclear reorganization, occurs at times among the ciliates, usually without mating. But the micronucleus undergoes the maturation divisions leading, as in conjugation, to formation of two pronuclei. These sister pronuclei fuse with each other in their own cell (the autogamont), forming there a synkaryon. This is the pattern of *autogamy in singles*. *Autogamy in pairs*, or autogamy during conjugation, usually facultative, may occur in some ciliates. All the nuclear phenomena here resemble those during the normal (reciprocal) conjugation, except there is no exchange of migratory pronuclei, the synkarya in both mates being formed by autogamy. This type of "double autogamy" has been called "cytogamy",[267] but the use of this term meets serious objections.[61, 35]

Reviews of the nuclear phenomena during conjugation have been published repeatedly.[76, 254, 88, 264] However, most of them are limited in scope, specialized in certain respects, and do not include the most recent contributions.

Two problems related to ciliate conjugation have already been considered by myself in the chapter "Macronucleus of ciliates" (Vol. 3, Chapter 1): (1) development of the macronuclear anlagen in the exconjugant, and (2) degeneration of the old macronucleus. No discussion of these topics will be repeated in the present chapter.

The author is sincerely grateful to Dr. L. Schneider (Würzburg, West Germany) and Dr. E. Vivier (Lille, France) for providing originals of their electron micrographs published elsewhere, and to Dr. Ruth Stocking Lynch (Los Angeles, U.S.A.) for correcting the English translation of the chapter.

II. PLACE OF CONJUGATION AND AUTOGAMY IN THE LIFE CYCLE OF THE SPECIES

1. Conjugation and Periods of Clonal Life

Extensive consideration of the so-called "life cycle" of the ciliates is outside the scope of this chapter. It may only be mentioned that in some ciliates the ability of the individuals to conjugate changes definitely during cultivation of a clone or of a caryonide. Such cycles were recorded in some ciliates by Maupas[164-5] and were extensively studied by Jennings[130-1] in *Paramecium bursaria*, and by Sonneborn[245-6] in *P. aurelia*.

During its first period, the clone appears to be immature; neither conjugation nor autogamy occur. The length of the period of immaturity may differ from species to species as well as in various syngens of one species. In *P. bursaria* it usually lasts for 3 to 5 months,[130] in *P. aurelia* it is generally shorter than a month, and even totally absent in syngens 4, 8, 10, and 14.[246] No immaturity period occurs in *P. aurelia* after autogamy.

Later on, the clone enters the period when the animals are capable of normal conjugation. Conjugation has an appreciable "rejuvenating" effect only during this period. This maturity period lasts in different syngens of *P. aurelia* from 3 days to a month or more; in *P. bursaria* it may last even several years.

Then comes a period of senility. In *P. aurelia*, the animals do not conjugate, but still undergo autogamy. However, the proportion of abnormalities that occur during autogamy steadily grows, until autogamy becomes lethal.[245] In *P. bursaria*, the period of senility is characterized by increased mortality of exconjugants (there is no autogamy in this species). The period of senility terminates with death of the clone.

Thus at least in some ciliates mating occurs only after a number of generations has passed, following the previous conjugation. But a clone (more exactly, a caryonide) usually does not enter conjugation spontaneously even during the maturity period (except the selfing caryonides). Mixing of two clones of complementary mating types is needed to produce conjugation. Differentiation of the mating types may be considered an evolutionary adaptation which favors interbreeding of animals of different origin and prevents intraclonal conjugation (inbreeding).

2. Seasonal Conjugations

In wild ciliate populations, conjugation often occurs only during some definite season. For example, near Leningrad (latitude 60°N.) *Loxodes striatus* conjugates in August,[11] *Zoothamnium arbuscula* in late June and early July,[91] *Bursaria truncatella* in September and October,[199] and *Nassula ornata* in May (Raikov, unpublished). Seasonal conjugations are probably conditioned by some unknown combination of ecological factors and the concurrent end of the period of immaturity following the conjugation of the preceding year.

The clearest picture of seasonal conjugations is found in some parasitic ciliates, where the nuclear reorganization cycle is correlated to certain seasonal phenomena in the life cycle of the host. For example, in *Nyctotherus cordiformis* conjugation occurs only in the gut of tadpoles of definite age which have lately ingested cysts of *Nyctotherus*, and continues during metamorphosis of the tadpoles.[10, 93, 265] The start of conjugation is probably conditioned here by some hormonal changes in the host organism.

3. Preconjugants

It has been shown for a long time that ciliate individuals uniting in con-jugation are often smaller than vegetative animals. Maupas[165] showed that such small individuals arise in *Tetrahymena* (= *Leucophrys*) *patula*, *Prorodon teres*, *Didinium nasutum*, and some other ciliates by two, three, or more rapid cell divisions. Such ciliates have no time to grow to the usual size during the intervals between these divisions. Since conjugation usually occurred in moderately starved cultures, Hertwig[107] tried to relate these rapid divisions to starvation (in *Dileptus*) and called them "hunger divis-ions" (Hungerteilungen). Later, rapid cell divisions preceding conjugation were described in *Didinium*,[204] *Dileptus*,[255] *Fabrea*,[84] and other species (see Table 1, p. 178). But the majority of free-living ciliates seem to have no special preconjugation divisions. In those species of *Paramecium*, *Tetrahy-mena*, *Euplotes*, *Stylonychia*, etc., where mating types are known, no special cell divisions occur during the interval between mixing of the complementary cultures and start of conjugation. However, in *Stentor* the mating reaction is followed by appearance of morphologically distinct preconjugants which have modified peristomal regions facilitating pairing.[261a] In species which have no preconjugation divisions, the early conjugants are either not smaller than vegetative cells [*Paramecium bursaria*, *Chilodonella uncinata*, *Sty-lonychia pustulata*, *Euplotes eurystomus*;[165] *Bursaria truncatella*[199]], or they are smaller simply because preferential mating of the smaller individuals occurs.[189, 129, 261]

A preconjugation cell division is obvious in many parasitic ciliates. It is usually single. Dogiel[76] thoroughly studied the preconjugation division (he called it "progamic" division) in some Entodiniomorphida (*Diplodi-nium*, *Cycloposthium*, *Opisthotrichum*) and in *Isotricha ruminantium*. He showed that the preconjugation division is morphologically different from the usual vegetative cell division, and gives rise to a special generation of individuals called preconjugants. Dogiel considered the life cycle of the ciliates he studied to be a long series of agamic generations, in which a generation of preconjugants may appear.

Preconjugation divisions and preconjugants were later found in other entodiniomorphids,[201, 183] in parasitic holotrichs, e.g. *Bütschlia*, *Paraiso-tricha*,[77, 78] *Balantidium*, [119, 179] *Cryptochilum*, [43] and in the heterotrich *Nyctotherus*.[265, 93]

In Entodiniomorphida and parasitic Holotricha, the preconjugation division differs from the usual cell division by the much larger size of the micronuclear mitotic spindle and by the absence of a long connecting strand between the daughter micronuclei in telophase. The micronucleus of a preconjugant does not condense after mitosis but remains swollen and spindle-like (Fig. 1, *a*); it often acquires the form characteristic of the meiotic prophase. For example, in *Cryptochilum*[43] the micronucleus of a preconjugant enters the "crescent stage" even before mating (Fig. 7, *a*).

The preconjugants may differ from vegetative cells also in the form of their macronucleus. The hook-shaped macronucleus of *Cycloposthium bipalmatum* condenses (rounds up) before a usual division, but does not before the preconjugation division. The two preconjugants therefore receive macronuclei of different shape: the anterior cell has a short hook-shaped macronucleus, and the posterior cell, an elongate macronucleus.[76] In *Bütschlia*, the macronuclei of the preconjugants have pointed ends while macronuclei of vegetative cells have rounded ends.[77]

The preconjugants are significantly smaller than vegetative animals in all species studied. This is related to the absence of growth of the preconjugants after the preconjugation division. Some cell organelles do not form at all after this division: e.g. in *Opisthotrichum* and *Diplodinium* vegetative animals have two contractile vacuoles while preconjugants have only one (the second does not form). In *Cycloposthium*, which has four or five contractile vacuoles, the preconjugants show only two or three. The preconjugants of *Diplodinium*, *Opistotrichum*, and *Isotricha* ingest no food.[76]

Of special interest are the preconjugants of *Opisthotrichum janus* (Fig. 1), studied by Dogiel.[76] The preconjugation division is sharply unequal in this species (Fig. 1, *a*); it gives rise to a large anterior preconjugant (or

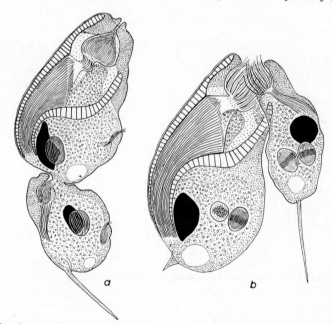

Fig. 1. *Opisthotrichum janus*: *a*—late stage of unequal preconjugation division; *b*—conjugation between a macroconjugant and a microconjugant. Iron hematoxylin staining. After Dogiel.[76]

macroconjugant) and a small posterior preconjugant (or microconjugant). The skeleton of the mother cell goes entirely to the large anterior preconjugant, and the long caudal spine of the mother cell to the small posterior preconjugant. No anlage of a new skeletal plate appears in the posterior preconjugant, which is entirely devoid of skeletal elements (Fig. 1, a, b).

The differentiation of preconjugants of *Opisthotrichum* into macroconjugants and microconjugants seems to be the initial stage in development of secondary gonochorism (dioecy) in hermaphroditic (monoecious) organisms. Here, the conjugation becomes anisogamontic (Fig. 1, b). But anisogamonty is not yet obligatory in *Opisthotrichum*: in 83 per cent of pairs, a macroconjugant mates with a microconjugant, but in 17 per cent of pairs conjugation occurs between two macroconjugants. No conjugation between two microconjugants was observed. Unlike the ciliates with obligatory anisogamonty (e.g. Peritricha), the conjugation of *Opisthotrichum* is still temporary (not total): after separation, viable exconjugants originate from both macroconjugant and microconjugant.

Another special type of preconjugation divisions is characteristic of the apostomatid family Foettingeriidae (Fig. 2). These ciliates have a complicated life cycle including a stage of hypertrophic growth followed by a stage of repeated palintomic divisions giving rise to small tomites.[22] Conjugation appears to be superimposed on such a palintomic cycle. Large vegetative animals (the trophonts) start mating but the maturation divisions of their micronuclei are delayed while both of the united partners divide transversely several times in succession (linear palintomy). The resulting daughter cells remain attached to each other, forming a double chain of small individuals. In *Polyspira delagei* this process goes on without a preceding encystment (Fig. 2, a), while in *Gymnodinioides* and some other genera it occurs within a cyst (Fig. 2, b). Only after the end of palintomy do the micronuclei start the maturation divisions and the double chain of individuals breaks into many small conjugating pairs in which the nuclear phenomena are concluded (Fig. 2).

This process was discovered by Minkiewicz[168] in *Polyspira* and called syndesmogamy. It has been more extensively studied by Chatton and Lwoff[22] who gave it another name, zygopalintomy. The appearance of zygopalintomy seems to be the result of a precocious mating of two trophonts before preconjugation divisions; the palintomic divisions leading to the double chain formation are obviously homologous to repeated preconjugation divisions.

Finally, preconjugants are characteristic of all peritrichs and some suctorians. In Peritricha, the conjugation is total and obligatorily anisogamontic: it always occurs between a large sessile macroconjugant and a small free-swimming microconjugant. It is difficult to distinguish macroconjugants from vegetative individuals; the latter seem to be able to transform directly into macroconjugants, at least in some species. As to the micro-

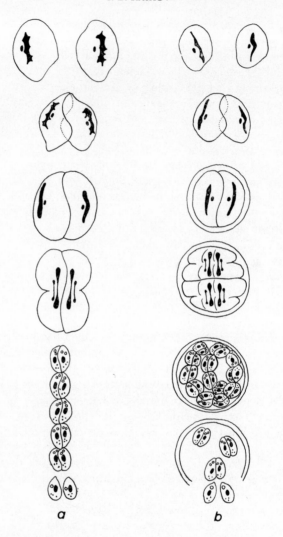

FIG. 2. Diagram of zygopalintomy in *Polyspira delagei* (*a*) and *Gymno-dinioides inkystans* (*b*). After Chatton and Lwoff.[22]

conjugants, three methods of their formation exist (for reviews see Finley,[87, 88] and Dass[47]).

The first method is unequal preconjugation division of a vegetative cell, separating a small swimming microconjugant from the much larger part which becomes a macroconjugant (Fig. 3). In *Vorticella microstoma* (Fig. 3, *a*) the microconjugant is much smaller than the macroconjugant, while in other species, e.g. in *V. campanula*, this difference may not be so sharp

(Fig. 3, *b*). This method of microconjugant formation has been described in a number of species of *Vorticella*, in *Opercularia coarctata*, *Lagenophrys labiata*, *Telotrochidium henneguyi*, etc. (see Table 3, p. 188).

The mode of formation of microconjugants in *Lagenophrys tattersalli*[274] and *Rhabdostyla vernalis*[89, 90] may be considered a variant of the first method. Here the unequal division separates a small individual called a protoconjugant which divides once (*Lagenophrys*) or twice (*Rhabdostyla*), giving rise, respectively, to two or four microconjugants.

The second method consists of one, two, or three rapidly successive equal preconjugation divisions of some vegetative animals (or of some zooids in colonial peritrichs). A "rosette" of two, four, or eight small cells on a common stalk is formed (Fig. 4, *a*). All these cells detach and become swimming microconjugants. The macroconjugants are usually similar to vegetative individuals or zooids (Fig. 4, *b*). This method of microconjugant formation

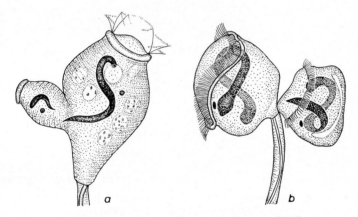

FIG. 3. Unequal preconjugation division in *Vorticella*: *a*—in *V. microstoma* (after Finley[87]); *b*—in *V. campanula* (after Mügge[173]).

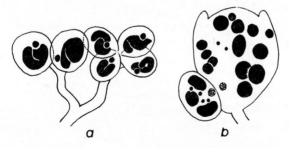

FIG. 4. *Carchesium spectabile*: *a*—formation of "rosettes" of microconjugants; *b*—conjugation between a macroconjugant and a microconjugant (a single pronucleus seen in each conjugant). Feulgen reaction, ×700. After Dass.[47]

is known in *Vorticella monilata, V. nebulifera, V. cucullus,* and in most colonial peritrichs—*Epistylis, Carchesium, Zoothamnium* (except *Z. alternans*), and *Ophrydium* (for references see Table 3, p. 188).

An interesting differentiation of zooids of a colony exists in some colonial Peritricha. For example, in *Zoothamnium arbuscula*[91] the colonies are potentially monoecious, but macroconjugants arise only on relatively young colonies, while microconjugants are formed only on mature colonies. This prevents conjugation of a macroconjugant with a microconjugant formed on the same colony. In a colony only two macroconjugants differentiate, each at the top of one of its two main branches. The microconjugants are formed by one equal division of small vegetative zooids (microzooids) in any region of a colony.

The third method of microconjugant formation is known only in *Zoothamnium alternans*.[248] As in *Z. arbuscula,* the zooids of a colony are differentiated. Only one macroconjugant arises in a colony, formed by the terminal macrozooid of the trunk. The microconjugants arise without cell division from terminal or subterminal microzooids of the branches; the microzooids detach from the colony and directly transform into swimming microconjugants.

No special preconjugants exist in the majority of Suctoria (see Table 2, p. 186). However, a sharply anisogamontic conjugation has been described in *Discophrya* (= *Prodiscophrya*) *collini, D. endogama,* and *D. buckei* by Kormos and Kormos,[151, 153-4] and also in *Discophrya collini* by Canella.[19a] The macroconjugants resemble vegetative individuals, while the microconjugants are formed by budding of the latter. The microconjugants resemble vegetative swarmers in both structure and mode of formation, but are much smaller than the swarmers. A microconjugant passes entirely into a macroconjugant (the so-called "inner conjugation"). It is obvious that these suctorians have two types of budding, vegetative budding and preconjugation budding.

III. NORMAL PATTERNS OF NUCLEAR PHENOMENA IN VARIOUS TYPES OF CONJUGATION

1. Temporary (Partial) Conjugation

Temporary conjugation is characteristic of the vast majority of ciliates (see Tables 1 and 2, p. 178). This type of conjugation also shows the widest diversity of nuclear phenomena. Let us consider several representative patterns of nuclear behavior during temporary conjugation.

Isogamontic conjugation of *Paramecium caudatum* (Figs. 5, 6), a classical example of conjugation in general, is now known in considerable detail.[19, 54, 59, 148-9, 165, 191, 194, 267] Soon after mating, the micronucleus of each conjugant enters the characteristic crescentic form of the first maturation prophase (Fig. 5, *a*) and later completes the division (b). Both daughter nuclei undergo the second maturation division (*c, d*). As will be shown below (p. 196), both these divisions are meiotic. Of the four derivatives of these maturation divisions, three become pycnotic and are later resorbed, and one, lying in the oral region of the cell, undergoes a third maturation division (Figs. 5, *e*; 6). This division is equational and gives rise to two spindle-shaped pronuclei (Fig. 5, *f*). The pronucleus lying deeper in the cytoplasm is stationary, and the pronucleus located near the surface of contact of the conjugants, in a small protrusion of the body (the paroral cone), is migratory. The migratory pronucleus penetrates the joined pellicles of the conjugants passing into the partner, so that exchange of migratory pronuclei occurs between the mates (Figs. 5, *f*; 6). After that, the migratory pronucleus approaches the stationary pronucleus of the partner, and in parallel positions the two fuse to form the synkaryon (Fig. 5, *g*).

The synkaryon divides mitotically three times, the first division (Fig. 5, *h*) immediately following pronuclear fusion. The conjugants usually separate after this division, so that the second and the third synkaryon divisions take place in exconjugants (Fig. 5, *i, j*). Eight synkaryon derivatives are formed (four in the anterior and four in the posterior end of the body). The four anterior nuclei increase in size and become macronuclear anlagen, while the four posterior nuclei condense (Fig. 5, *k*). Later on, this characteristic position of the nuclei is lost (1); of the four small nuclei, three degenerate, and the only one that remains is the new micronucleus (Figs. 5, *m*; 6).

Two exconjugant divisions follow, during which the micronucleus mitotically divides (Fig. 6, *a, b*), and the macronuclear anlagen segregate among the daughter cells at first by twos (Fig. 5, *n*), then by singles (Fig. 6).

The old macronucleus becomes lobed during the maturation divisions (Fig. 5, *a–i*). During the synkaryon divisions, it transforms into a dense skein of strands (*j*), which loosens a little later (*k*). Finally, the strands disintegrate into rounded fragments (*l, m*), which segregate at the exconjugant divisions (*n*) and gradually become resorbed.

Another pattern of the temporary isogamontic conjugation has been studied by Dain[43] in *Cryptochilum echini* (Figs. 7, 8). Mating occurs between preconjugants whose micronuclei are already in meiotic prophase (Fig. 7, *a*). In each mate the first (Fig. 7, *b, c*) and the second (*d*) maturation divisions yield four identical nuclei (*e*). One nucleus, which is near the contact region of the conjugants, undergoes the third maturation division, and the other three nuclei degenerate (*f, g*), as in *Paramecium caudatum*.

The mode of transfer of the migratory pronuclei differs in *Cryptochilum* from that in *P. caudatum*: the anterior derivative of the third maturation division, which is to become the migratory pronucleus, appears to be pushed into the partner by the very long telophase spindle (Fig. 7, *g*). The stationary

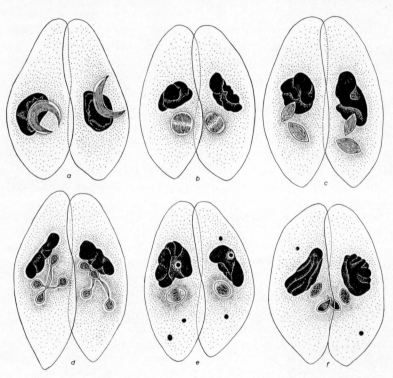

FIG. 5. Conjugation in *Paramecium caudatum*: *a*—meiotic prophase ("crescent stage"); *b*—anaphase of the first maturation division; *c, d*—late prophase and telophase of the second maturation division; *e*—metaphase of the third maturation division; *f*—exchange of migratory pronuclei;

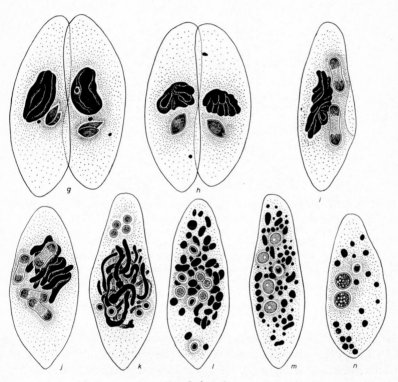

FIG. 5. (*cont.*).

g—fusion of pronuclei; *h*—first, *i*—second, *j*—third synkaryon division; *k*—differentiation of synkaryon derivatives into four macronuclear anlagen (anterior) and four micronuclei (posterior); *l*—macronuclear fragmentation; *m*—growth of the four macronuclear anlagen, only one of the four micronuclei persists; *n*—first exconjugant division offspring with two macronuclear anlagen and a single micronucleus. Feulgen reaction. Original.

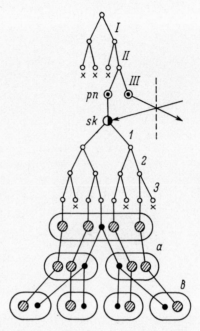

FIG. 6. Diagram of nuclear behavior during conjugation in *Paramecium caudatum* (only one of the two partners shown). *I, II, III*—respective maturation divisions of the micronucleus, *pn*—pronuclei, *sk*—synkaryon, *1, 2, 3*—synkaryon divisions, *a, b*—two exconjugant divisions. In the exconjugant and its offspring, hatched circles are macronuclear anlagen, small black circles are new micronuclei. Crosses mark degenerating nuclei.

pronucleus passes into the posterior body region (*h*). The pronuclei fuse in parallel spindles (*i*).

After separation of the mates, the synkaryon divides twice (Fig. 7, *j, k*); of its four derivatives, one differentiates into a micronucleus, and three into macronuclear anlagen (*l*). There are two exconjugant divisions (Fig. 8, *a, b*). During the first division, two macronuclear anlagen pass into the anterior, one into the posterior daughter cell (Fig. 7, *m, n*). After the second division of the anterior daughter cell, all the exconjugant offspring receive the normal number of nuclei (Fig. 8).

The old macronucleus does not fragment in *Cryptochilum* but becomes pycnotic as a whole during development of the anlagen of new macronuclei (Fig. 7, *l*).

More simple is the conjugation process (Figs. 9, 10) in species of *Chilodonella*.[86, 118, 160-1, 165, 207a] No special preconjugants exist in *Chilodonella*; the conjugation is isogamontic at early stages but often becomes slightly heterogamontic at late stages of the process due to a decrease in the size of the left partner.[86] Only one of the derivatives of the first maturation

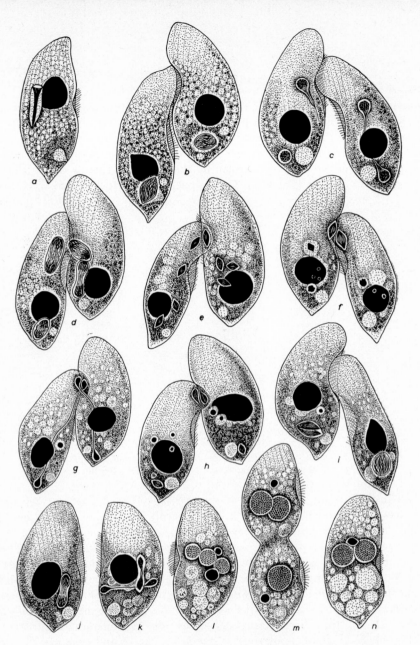

FIG. 7. Conjugation in *Cryptochilum echini*: *a*—preconjugant; *b, c*—metaphase and telophase of the first maturation division; *d*—second maturation
division; *e*—its four derivatives in each conjugant; *f*—third maturation
division; *g*—late telophase of the third maturation division (exchange of
the future migratory pronuclei); *h*—pronuclei; *i*—synkaryon formation
(pronuclei are fused in the right mate but still separate in the left mate);
j—exconjugant showing first synkaryon division; *k*—second synkaryon
division; *l*—exconjugant with three macronuclear anlagen, one micronucleus, and pycnotic old macronucleus; *m*—first exconjugant division;
n—its anterior offspring with two macronuclear anlagen and a single
micronucleus. Iron hematoxylin staining. After Dain.[43]

division of the micronucleus (Fig. 9, *a*) undergoes the second maturation division; the other nucleus becomes pycnotic (Figs. 9, *b*; 10). Of the two derivatives of the second maturation division, again one degenerates, and the other enters the third division. As in *Cryptochilum*, the migratory pronucleus is pushed into the partner by the telophase spindle of the third maturation division (Fig. 9, *c*). The pronuclei fuse to form a synkaryon (*d*)

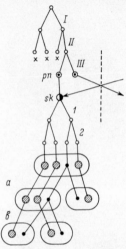

FIG. 8. Diagram of nuclear behavior during conjugation in *Cryptochilum echini* (only one partner shown). Abbreviations as in Fig. 6.

which divides only once (*e*, *f*). Of the two synkaryon derivatives, one becomes the micronucleus, while the other sharply increases in size and becomes the anlage of the new macronucleus (Fig. 9, *g*). The reconstruction of the nuclear apparatus is accomplished without special exconjugant divisions. The old macronucleus degenerates by pycnosis without fragmentation (Fig. 9, *g*).

A peculiar example of nuclear behavior is given by the conjugation of the apostomatid *Collinia* (syn. "*Anoplophrya*"), studied by Collin[38] and other authors.[13, 249] The conjugation is isogamontic. After the meiotic prophase (Fig. 11, *a*), the micronucleus undergoes the first maturation division (Fig. 11, *b*, *c*). Both its derivatives divide for the second time (*d*); of the four products of the second maturation division (*e*), three degenerate while one enters the third maturation division (*f*). The spindle of the third division yields two pronuclei located near the junction of the partners (Fig. 11, *g*). After the exchange of the migratory pronuclei (*h*), a synkaryon forms, the pronuclei fusing in interphase state (*h*, *i*). There are two synkaryon divisions (*j*, *k*) giving rise to four, at first identical, nuclei (*l*). The conjugants separate at this stage; of the four nuclei, one becomes a macronuclear anlage, one, a micronucleus, and two degenerate (Figs. 11, *m*, *n*; 12). There are no exconjugant divisions.

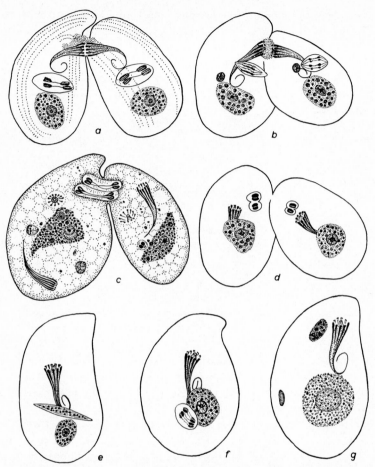

FIG. 9. Conjugation in *Chilodonella uncinata*: *a*—first maturation division; *b*—second maturation division (a pycnotic first division derivative seen in each partner); *c*—telophase of the third maturation division and exchange of the future migratory pronuclei (old pharyngeal baskets resorbing, anlagen of new baskets appearing); *d*—fusion of pronuclei, growth of new pharyngeal baskets; *e, f*—exconjugants showing first synkaryon division; *g*—exconjugant with a macronuclear anlage (at lower right), a micronucleus (at lower left), and the pycnotic old macronucleus (at top). Iron hematoxylin staining. After MacDougall.[160]

Most peculiar is the behavior of the old macronucleus in *Collinia*. The macronuclei of both partners elongate during the third maturation division (Fig. 11, *f*). After synkaryon formation, the macronuclei reciprocally penetrate the partner (*i, j*) and advance into it approximately up to half of their lengths (*k*). When the partners separate, both macronuclei constrict (*l*), each exconjugant receiving two macronuclear halves: a half of its own macronucleus and a half of the partner's macronucleus (m, n). Both macronuclear halves round up (o) and become resorbed without fusion or fragmentation (p, q).

Temporary conjugation occurs in some sessile ciliates as well, mainly among the Suctoria (see Table 2, p. 186). Conjugation of *Dendrocometes paradoxus*, studied by Hickson and Wadsworth,[109] may serve as our example (Fig. 13).

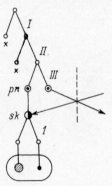

FIG. 10. Diagram of nuclear behavior during conjugation in *Chilodonella uncinata* (only one partner shown). Abbreviations as in Fig. 6.

As in most suctorians, conjugation occurs in *Dendrocometes* between two neighboring sessile animals. They form special cytoplasmic protrusions, the so-called conjugative processes (Fig. 13, *a*), which meet and ensure contact between the mates (*b*).

Hickson and Wadsworth[109] describe only two maturation divisions in *D. paradoxus*. All the micronuclei are said to undergo the first division (Fig. 13, *b*), with only one nucleus taking part in the second division, which occurs inside the conjugative processes, near the boundary of the mates (*c*). However, there are grounds for believing this description erroneous and that *D. paradoxus* has three maturation divisions, as other suctorians have (see Table 2). The last micronuclear division (Fig. 13, *c*) is surely homologous to the third (not the second) maturation division in suctorians like *Cyclophrya katharinae*.[155]

The last maturation division gives rise to two pronuclei (Fig. 13, *d*); the migratory pronucleus passes through the pellicle separating the conjugants and fuses with the stationary pronucleus of the partner (*e*). The synkaryon divides twice (*f*, *g*); of its four products, one becomes the macronuclear anlage while three become micronuclei (Fig. 13, *h*).

As in *Collinia*, the old macronucleus of *Dendrocometes* strongly elongates during conjugation (Fig. 13, *b*, *f*, *g*, *h*). But it does not penetrate the partner. The old macronucleus fragments and becomes resorbed after separation of the mates. Both exconjugants are viable in *Dendrocometes*.

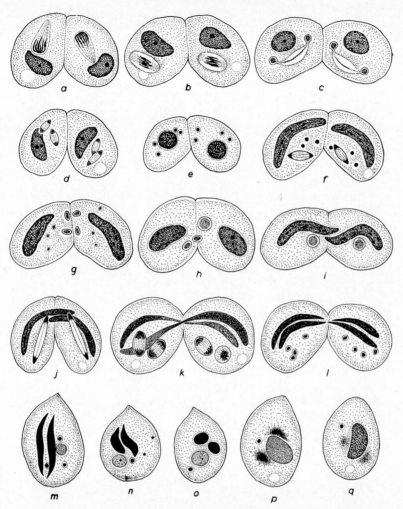

FIG. 11. Conjugation in *Collinia branchiarum*: *a*—meiotic prophase; *b*, *c*—metaphase and telophase of the first maturation division; *d*—anaphase of the second maturation division; *e*—its four derivatives; *f*—anaphase of the third maturation division (three pycnotic nuclei in each partner, stretching of the macronuclei); *g*—pronuclei; *h*—passage of a migratory pronucleus (from right to left), synkaryon already formed in the right mate; *i*—synkarya in both mates (the macronuclei begin to penetrate into the partner); *j*—telophase of the first synkaryon division; *k*—anaphase of the second synkaryon division, exchange of macronuclear halves; *l*—derivatives of the second synkaryon division, constriction of both macronuclei; *m*—exconjugant containing a macronuclear anlage, three micronuclei, and two macronuclear halves; *n*—pycnosis of two micronuclei; *o*—the macronuclear halves round up; *p*, *q*—resorption of both halves of the old macronuclei, development of the macronuclear anlage. Iron hematoxylin staining, ×1400. After Collin.[38]

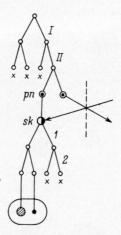

FIG. 12. Diagram of nuclear behavior during conjugation in *Collinia branchiarum* (only one partner shown). Abbreviations as in Fig. 6.

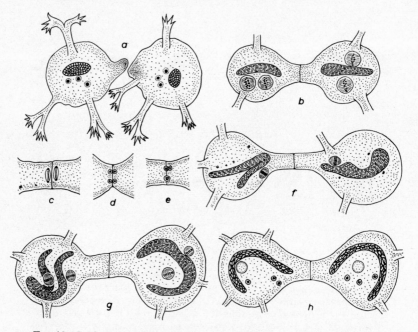

FIG. 13. Conjugation of *Dendrocometes paradoxus*: *a*—two neighbouring individuals with conjugative processes; *b*—first (?) maturation division, beginning of macronuclear stretching; *c*—second (third?) maturation division inside the conjugative processes; *d*—pronuclei; *e*—pronuclear fusion; *f, g*—first and second synkaryon divisions (the old macronucleus shows strong stretching); *h*—a single macronuclear anlage and three micronuclei in each partner. Iron hematoxylin staining. After Hickson and Wadsworth.[109]

2. Total Isogamontic Conjugation

Among free-swimming ciliates, the total isogamontic conjugation has been described in *Metopus sigmoides* by Noland,[184] in *Urostyla hologama* by Heckmann,[105] and in *U. polymicronucleata* by Moldenhauer.[172] Beyond any doubt, the "copulation" of *Urostyla flavicans*, previously recorded by Ilowaisky,[113] should also be interpreted as total conjugation.

Let us consider the case of *Metopus* (Figs. 14, 15). At early stages of conjugation, the partners are alike and connected with a broad cytoplasmic bridge (Fig. 14, *a*). The micronucleus of each conjugant undergoes the first maturation division (*b–e*); then both daughter nuclei take part in the second maturation division (*f*). At first, the four derivatives of the second division are alike and there is still no appreciable difference between the two conjugants (Fig. 14, *g*). Three of the four nuclei soon degenerate, and the fourth nucleus undergoes the third maturation division (*h*). A peculiar process of transfer of the cytoplasm of one conjugant into the other conjugant starts at the same time. Both pronuclei of the "donor" conjugant, formed by that time as a result of the third maturation division, pass into the partner together with the cytoplasm (Figs. 14, *i*; 15), leaving only the old macronucleus in the shrunken residue of the "donor" conjugant. Of the four pronuclei thus present in the cytoplasm of the "recipient" conjugant, two degenerate by pycnosis, while the other two swell, approach each other (Fig. 14, *i*, *j*), and finally fuse into the synkaryon (*l*).

The synkaryon divides only once (Fig. 14, *m*). One of its derivatives becomes the micronucleus, the other forms the macronuclear anlage (*n*). The old macronucleus of the "recipient" partner degenerates by pycnosis, without fragmentation (*n*). The residue of the "donor" conjugant does not fuse totally with the "recipient" but separates from it, usually during the synkaryon division. This small exconjugant contains only the old macronucleus (Fig. 14, *o*) and soon perishes.

The total conjugation in *Urostyla* goes on in a more or less similar way. In *U. hologama*[105] and *U. polymicronucleata*,[172] it is always the left conjugant which becomes the "donor" (though it does not differ at first from the right conjugant). In these species, the cytoplasm of the "donor" totally flows over into the "recipient" (no small non-viable exconjugant is formed). The synconjugant encysts in both *Urostyla* species.

The total isogamontic (or slightly anisogamontic) conjugation is frequent among sessile ciliates. The conjugation of the suctorian *Ephelota gemmipara* (Figs. 16, 17), studied by Collin[39] and more thoroughly by Grell,[96] may be taken as an example.

Two neighboring stalked individuals of *Ephelota* unite in conjugation. They may be of equal (Fig. 16, *a*) or of unequal size (but these differences, if present, are accidental). Soon after mating, one of the two partners, usually the smaller one, detaches from its stalk and withdraws its tentacles

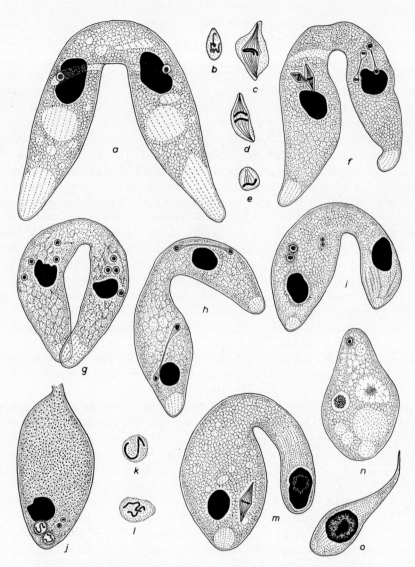

Fig. 14. Total conjugation in *Metopus sigmoides*: *a*—early stage; *b–e*—details of the first maturation division of the micronucleus showing a single chromatin thread (*b*—prophase, *c*—metaphase, *d*—anaphase, *e*—telophase); *f*—second maturation division; *g*—its four derivatives; *h*—third maturation division, beginning of cytoplasmic transfer; *i*—two functional (at left) and two degenerating pronuclei in the "recipient" conjugant; *j*—functional pronuclei before fusion (at lower left), two degenerating pronuclei at lower right (only the "recipient" conjugant shown); *k*—pronucleus; *l*—synkaryon; *m*—synkaryon division; *n*—the large exconjugant showing the micronucleus (at top), the macronuclear anlage (at left center), and remnants of the old macronucleus (at right center); *o*—the small exconjugant. Iron hematoxylin staining; *a*: ×350; *b–e*, *k*, *l*: ×700; *f–j*, *m–o*: ×300. After Noland.[184]

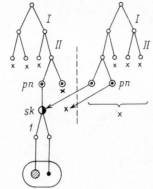

FIG. 15. Diagram of nuclear behavior during total conjugation in *Metopus sigmoides* (both partners shown). Abbreviations as in Fig. 6.

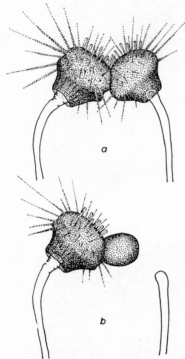

FIG. 16. Aspects of total conjugation in living *Ephelota gemmipara*: *a*—pairing of two equally large partners; *b*—the right partner detaches from its stalk and becomes microconjugant. After Grell.[96]

(Fig. 16, *b*). This partner becomes the "donor" (or microconjugant), while the partner retaining its stalk becomes the "recipient" (or macroconjugant).

Ephelota gemmipara has several micronuclei. All undergo the first maturation division (Fig. 17, *a*). A considerable number of their derivatives divide

6a*

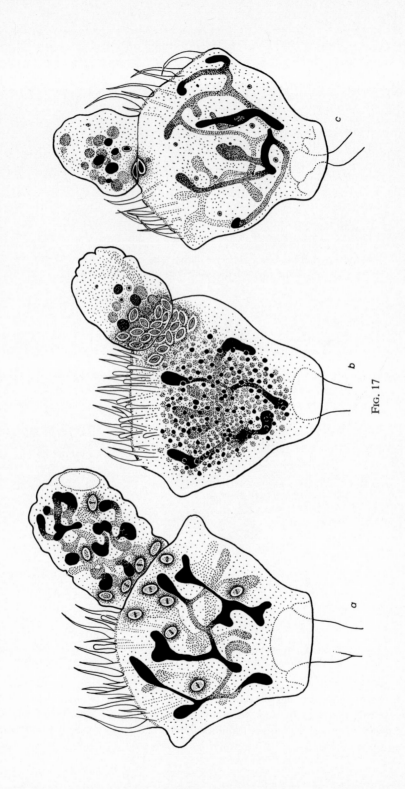

Fig. 17

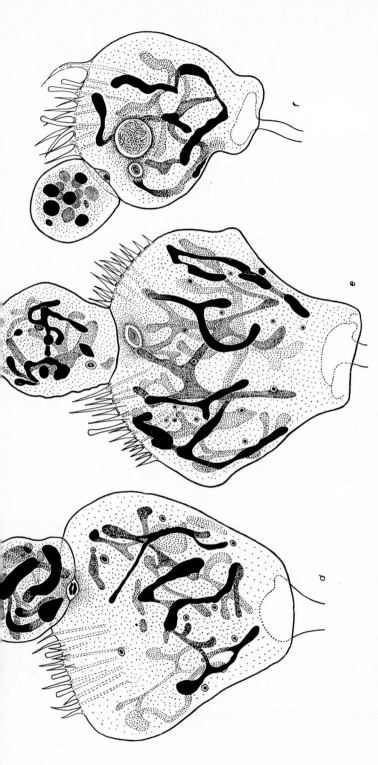

Fig. 17. Total conjugation in *Ephelota gemmipara*: *a*—first maturation division of all micronuclei; *b*—second maturation division (the granules in the macroconjugant are food inclusions); *c*—pronuclei (at boundary of the partners); *d*—synkaryon (at boundary of the partners); *e*—synkaryon division (at top of macroconjugant); *f*—a macronuclear anlage and a micronucleus in the macroconjugant (an "excess" pronucleus seen at the boundary of the conjugants). Feulgen reaction, ×530. After Grell.[96]

for the second time (Fig. 17, *b*). The third maturation division seems to be absent in this species. Only one pronucleus usually differentiates in each conjugant from the products of the second maturation division (Fig. 17, *c*); less frequently, two pronuclei differentiate. In either event, only one synkaryon forms at the boundary of the conjugants (*d*), later passes into the cytoplasm of the "recipient" and there undergoes a single division (*e*). One of the products of this division differentiates into the macronuclear anlage, the other becomes the micronucleus (*f*). If "excess" pronuclei are present, they degenerate (Fig. 17, *f*).

In the "donor" conjugant, all derivatives of the micronucleus, except the single pronucleus, degenerate; the macronucleus fragments earlier than in the "recipient" cell (Fig. 17, *f*). The main mass of the cytoplasm of the "donor" seems to flow over into the "recipient".

3. Total Anisogamontic Conjugation

This type of conjugation is characteristic of the Peritricha (see Table 3, p. 188) and exists in some Suctoria as well (Table 2). The conjugation of *Carchesium polypinum* (Figs. 18, 19), described by Maupas[165] and Popoff,[202] may serve as our example.

The swimming microconjugants arise in *C. polypinum* by two or three equal preconjugation divisions of vegetative zooids (the so-called "rosette formation", see p. 157). The macroconjugants resemble the vegetative zooids. A microconjugant attaches to the lower third of the macroconjugant's body (Fig. 18, *a*). The micronucleus of the microconjugant undergoes one extra mitotic division, the so-called preliminary division (Fig. 19, *prl*); this division is absent in the macroconjugant. Therefore two spindles take part in the first maturation division in the microconjugant, and only one in the macroconjugant (Fig. 18, *b*). All the derivatives of this division undergo the second maturation division (*c*), forming eight nuclei in the microconjugant and four in the macroconjugant (*d*). In each conjugant, only one of these nuclei undergoes the third maturation division (*e, f*); the other nuclei degenerate. The third maturation division gives rise to two pronuclei in each conjugant (*g*). At that time, the cytoplasm of the microconjugant, together with both pronuclei, begins to flow over into the macroconjugant (Fig. 18, *g, h*), so that only a small shrunken pellicular residue of the microconjugant remains (*i, l*). Of the four pronuclei present in the synconjugant, only two fuse to form the synkaryon, while the other two degenerate (Fig. 19). The single synkaryon divides three times (Figs. 18, *h–k*; 19, *1–3*) and gives rise to eight spindle-shaped derivatives (Fig. 18, *k*). Seven of them become macronuclear anlagen, and the eighth is the new micronucleus (Fig. 18, *l*).

The normal nuclear apparatus is reconstructed in *Carchesium* as a result of three synconjugant divisions, during which the micronucleus divides by mitosis, and the macronuclear anlagen segregate at first as 4 : 3 (Figs. 18, *m*;

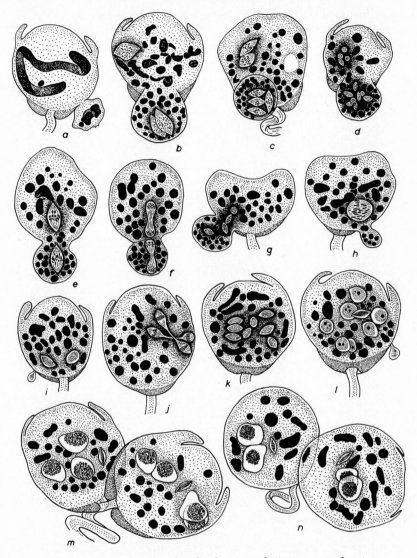

FIG. 18. Total conjugation in *Carchesium polypinum*: *a*—early stage; *b*—first maturation division, fragmentation of macronuclei; *c*—second maturation division; *d*—its derivatives; *e, f*—third maturation division; *g*—pronuclei; *h*—first synkaryon division and cytoplasmic transfer from the microconjugant to the macroconjugant; *i*—products of the first synkaryon division; *j*—second synkaryon division; *k*—eight derivatives of the third synkaryon division; *l*—seven macronuclear anlagen and a single micronucleus; *m*—descendants of the first synconjugant division (with three and four macronuclear anlagen); *n*—descendants of the second synconjugant division (with two macronuclear anlagen in each). Borax carmine staining. After Popoff.[202]

19, *a*), then as 2 : 2 : 2 : 1 (Figs. 18, *n*; 19, *b*), and finally by one in each daughter cell (Fig. 19, *c*).

The old macronucleus disintegrates very early (Fig. 18, *b*); the macronuclear fragments of the macroconjugant and of the microconjugant mix up in the synconjugant (Fig. 18, *h*) and segregate during its divisions (Fig. 18, *m, n*). The final resorption of these fragments occurs only in the descendants of the last synconjugant division which have one new macronucleus each.

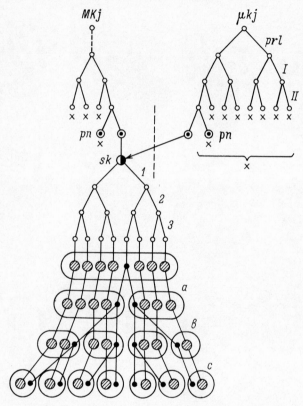

Fig. 19. Diagram of nuclear behavior during total conjugation in *Carchesium polypinum* (both partners shown). *MKj*—macroconjugant; *μkj*—microconjugant; *prl*—preliminary micronuclear division in the microconjugant; other abbreviations as in Fig. 6.

4. Patterns of Conjugation in Various Ciliate Taxa

Condensed data on nuclear phenomena during conjugation in the ciliates heretofore studied in this respect are represented in Tables 1 (Holotricha and Spirotricha), 2 (Suctoria and Chonotrichida), and 3 (Peritricha). Only sufficiently complete descriptions are included. Fragmentary and definitely erroneous data are omitted.

The following abbreviations are used in Tables 1–3:

In the column "Type of conjugation": *temp* = temporary, *tot* = total conjugation; in Table 2 also: *inn* = "inner" conjugation.

In the column "Preconjugants": "—" = no preconjugants, (+) = the conjugating individuals of smaller size, + = preconjugants morphologically different from vegetative animals, *zyg* = zygopalintomy, *μkj* = anisogamonty with microconjugants and macroconjugants.

In the column "Origin of microconjugants" (in Peritricha, Table 3 only): *uneq* = unequal division of a vegetative cell into a macroconjugant and a microconjugant; *prot* = unequal division of a vegetative cell into a macroconjugant and a protoconjugant, the latter dividing into several microconjugants; *ros* = several rapid equal divisions of a vegetative cell to form a "rosette" of microconjugants; *trf* = direct transformation of a microzooid into a microconjugant.

In the column "Number of maturation divisions": *prl* = preliminary division of the micronucleus.

In the column "Difference between pronuclei" (in Table 1 only): "—" = no difference, (+) = slight difference, + = marked difference between the stationary and the migratory pronucleus.

In the column "State of pronuclei during fusion" (in Table 1 only): *iph* = pronuclei fuse in interphase (vesicular) state; *spn* = pronuclei fuse in spindle form.

In the column "Behavior of macronuclear anlagen": *segr* = segregation of the anlagen among daughter cells during the exconjugant (or synconjugant) divisions; *fus* = fusion of two or more anlagen to form a single macronucleus; *div* = division of a single anlage into two or more definitive macronuclei; + *maf* = fusion of the anlage with fragments of the old macronucleus.†

In the column "Old macronucleus": *fr* = fragmentation without preceding skein formation; *sk* = fragmentation following skein formation; *pyc* = pycnosis without fragmentation, *exch* = exchange of macronuclear halves between the conjugants.

Data on conjugation of certain lower ciliates (Trachelocercidae, *Loxodes*) are excluded from Table 1. In these ciliates, the pattern of nuclear phenomena is so peculiar, that it needs to be considered separately (see Section VII of this chapter).

---

† As shown by Diller,[71] descriptions of such fusion are probably erroneous, at least in *Euplotes eurystomus*. Also Rao[216a] abandoned his previous view[214] that such fusion existed in *E. woodruffi*.

TABLE 1. PATTERNS OF CONJUGATION IN HOLOTRICHA AND SPIROTRICHA

Species	Type of conjugation	Preconjugants	Number of maturation divisions of the micronucleus	Number of spindles of the postmeiotic division	Difference between pronuclei	State of pronuclei during fusion	Number of synkaryon divisions	Number of macronuclear anlagen	Number of new micronuclei	Number of degenerating synkaryon derivatives	Behavior of macronuclear anlagen	Old macronucleus	References
1	2	3	4	5	6	7	8	9	10	11	12	13	14
Holotricha													
Gymnostomatida													
Enchelyidae													
Prorodon griseus	temp	?	3	1(2)	+	iph	1	1	1	—	—	pyc	Tannreuther[251]
Didiniidae													
Didinium nasutum	temp	(+)	3	1	+	iph or spn	2	2	2	—	segr or fus	fr	Prandtl[204]
Tracheliidae													
Dileptus gigas	temp	(+)	3	1	—	?	1	1	1	—	div	pyc	Visscher[255]
Amphileptidae													
Litonotus parvus	temp	?	3	1	?	?	2	2	1	1	fus	?	Prowazek[206]
Litonotus lamella	temp	?	3	1	?	?	3	1	1	4	segr	fr	Messjatzev[167]
Loxophyllum fasciola	temp	(+)	3	1	—	spn	2	2	1	1	?	fr	Maupas[165]
Bütschliidae													
Bütschlia parva	temp	+	?	?	?	spn	1	1	1	—	—	fr	Dogiel[77]
Pycnotrichidae													
Nicollella ctenodactyli	temp	(+)	3	1	—	?	2	2	1	1	fus	pyc	Chatton and Perard[23]
Collinella gundii	temp	(+)	3	1	?	?	2	2	1	1	fus	pyc	Chatton and Perard[23]

TABLE 1 (cont.)

Chlamydodontidae														
Chilodonella uncinata	temp	(+)	3	—	1	spn	1	1	1	1	—	—	pyc	Maupas,[165] Enriques,[86] MacDougall[160]
Chilodonella chattoni	temp	?	3	?	1	?	1	1	1	1	—	—	pyc	MacDougall[161]
Chilodonella caudata	temp	?	3	—	1	spn	1	1	1	1	—	—	pyc	MacDougall[161]
Chilodonella labiata	temp	?	3	—	1	spn	1	1	1	1	—	—	exch	MacDougall[161]
Chilodonella faurei	temp	?	3	—	1	?	1	1	1	1	—	—	exch	MacDougall[161]
Chilodonella cucullus	temp	?	3	—	1	spn	1	1	1	1	—	—	pyc	Ivanić,[118] Radzikowski[207a]
Trichostomatida														
Isotrichidae														
Isotricha ruminantium	temp	+	?	?	?	?	2	2	1	1	1	fus	?	Dogiel[76]
Paraisotrichidae														
Paraisotricha colpoidea	temp	+	?	—	?	spn	2	2	1	1	—	fus	?	Dogiel[78]
Balantidiidae														
Balantidium coli	temp	+	3	—	1	iph	2	2	1	1	—	fus	fr	Jameson,[119] Nelson[179]
Balantidium caviae (?)	temp	(+)	3	—	1	iph	1	1	1	1	—	—	pyc	Scott[227]
Entorhipidiidae														
Entorhipidium echini	temp	—	3	+	1	iph	3	1	1	1	6	—	pyc	Yagiu[278]
Hymenostomatida														
Tetrahymenidae														
Tetrahymena pyriformis	temp	—	3	—	1	iph	2	2	2	2(1)	0(1)	segr	pyc	Elliott and Hayes,[82] Nanney,[177] Ray[219]
Tetrahymena (= *Leucophrys*) *patula*	temp	+	3	(+)	1	iph	2	2	2	2(1)	0(1)	segr	pyc	Maupas[165]
Colpidium colpoda	temp	—	3	—	1	spn	2	2	2	2	—	segr	pyc	Maupas,[165] Dangeard,[44] Prowazek[207]
Colpidium truncatum	temp	—	3	—	1	spn	2	2	2	2(1)	0(1)	segr	pyc	Maupas,[165] Dehorne[54]
Colpidium campylum	temp	?	3	(+)	1	iph	2	2	2	1	1	?	pyc	Devidé[56]

TABLE 1 (cont.)

1 Species	2 Type of conjugation	3 Preconjugants	4 Number of maturation divisions of the micronucleus	5 Number of spindles of the postmeiotic division	6 Difference between pronuclei	7 State of pronuclei during fusion	8 Number of synkaryon divisions	9 Number of macronuclear anlagen	10 Number of new micronuclei	11 Number of degenerating synkaryon derivatives	12 Behavior of macronuclear anlagen	13 Old macronucleus	14 References
Glaucoma scintillans	temp	—	3	1	?	?	2	2	2	—	?	pyc	Maupas[165]
Glaucoma (= Dallasia) frontata	temp	—	3	1	?	?	2	2	2	—	segr	fr	Calkins and Bowling[18]
Loxocephalus sp.	temp	?	3	1	?	iph	3	6	1	—	fus or segr	pyc	Behrend[7]
Cryptochilum echini Parameciidae	temp	+	3	1	(+)	spn	2	3	1	—	segr	pyc	Dain[43]
Paramecium caudatum	temp	—	3	1	(+)	spn	3	4	1	3	segr	sk	Maupas,[165] Klitzke,[148-9] Dehorne,[54] Penn,[191] Diller,[59] Wichterman,[267] Peshkovskaja[194]
The same	temp	(+)	3	1	(+)	spn	3	4	4	—	segr	sk	Calkins and Cull[19]
Paramecium aurelia	temp	(+)	3	1	—	spn	2	2	2	—	segr	sk	Hertwig,[106] Maupas,[165] Kösciuszko[156]
Paramecium multimicronuclea um	temp	(+)	3	1	(+)	spn	5[a]	4	4	7[a]	segr	sk	Landis,[159] Müller[174]
The same	temp	—	3	1	—	?	3	4	4	—	segr	sk	Barnett[6]

TABLE 1 (cont.)

													Reference
Paramecium bursaria	temp	—	3	1	—	spn	2(b)	2	2	—	segr	pyc	Maupas,[165] Hamburger[99]
The same	temp	—	3	1	(+)	spn	3	2	2	1	segr	pyc	Chen,[25, 32, 33, 35, 36] Wichterman,[268-9, 272] Egelhaaf[81]
Paramecium putrinum (= P. trichium)	temp	—	3	1	(+)	spn	3	4	1	3	segr	sk	Wichterman,[266] Diller,[61] Doflein and Reichenow,[75] Jankowski[120]
Paramecium woodruffi	temp	—	3	1-3	?	?	4	4	4	1	segr	sk	Jankowski[121]
Paramecium polycaryum	temp	—	3	1	—	spn	3	4	4	—	segr	sk	Diller,[67] Hayashi and Takayanagi[104]
Paramecium calkinsi	temp	—	3	1	—	spn	3	2	2	1	segr	sk	Nakata[176]
Frontoniidae													
Frontonia leucas	temp	(+)	3	1	(+)	spn	?	?	?	?	?	fr	Peshkovskaja[194]
Frontonia acuminata	temp	?	3	1	?	?	1	1	1	—	—	fr	Pérez-Silva[192]
Thigmotrichida													
Hemispeiridae													
Ancistruma isseli	temp	—	3	1	?	?	3	7	1	—	segr	fr	Kidder[141]
Hysterocinetidae													
Ptychostomum (= Lada) tanishi	temp	(+)	3	1	+	spn	2	2	1	1	fus	pyc	Miyashita[170]
Conchophthiridae													
Conchophthirius mytili	temp	?	3	2	?	?	4	12-15	1-4	—	segr	?	Kidder[140]
Ancistrocomidae													
Parachaenia myae	temp	—	3	1	—	?	3	4	1	3	segr	sk	Kofoid and Bush[150]
Astomatida													
Anoplophryidae													
Dogielella sphaerii	temp	(+)	3	1	+	iph	2	2	1	1	segr	pyc	Poljansky[198]
Apostomatida													
Collinia (= Anoplophrya) branchiarum	temp	(+)	3	1	—	iph	2	1	1	2	—	exch	Collin[38]

TABLE 1 (cont.)

Species	Type of conjugation	Preconjugants	Number of maturation divisions of the micronucleus	Number of spindles of the postmeiotic division	Difference between pronuclei	State of pronuclei during fusion	Number of synkaryon divisions	Number of macronuclear anlagen	Number of new micronuclei	Number of degenerating synkaryon derivatives	Behavior of macronuclear anlagen	Old macronucleus	References
1	2	3	4	5	6	7	8	9	10	11	12	13	14
Collinia (= *Anoplophrya*) *circulans*	temp	(+)	3	1	—	iph	2	1	1	2	—	exch	Brumpt[13]
Collinia orchestii	temp	(+)	3	1	—	iph	2	1	1	2	—	exch	Summers and Kidder[249]
Foettingeriidae													
Polyspira delagei	temp	zyg	3	1	+	iph	3–4	1	1	7–9	—	?	Minkiewicz,[168] Chatton and Lwoff[22]
Spirotricha													
Heterotrichida													
Spirostomatidae													
Spirostomum teres	temp	(+)	3	1	—	iph	2	2	2	—	?	pyc	Maupas[165]
Spirostomum ambiguum	temp	?	3	1	?	?	2	2	2	—	fus	fr	Seshachar,[228] Seshachar and Bai,[228a] Rao[216]
Blepharisma americanum	temp	?	3	1	—	?	3	3–7	?	?	fus	pyc	Bhandary[9a]
Blepharisma seshachari	temp	(+)	3	1	—	?	2	3	1	—	segr	fr	Bhandary[9]
Blepharisma japonicum	temp	?	3	1	?	?	3	6	2	1	?	pyc	Inaba[115]
Blepharisma tropicum	temp	?	3	2	?	?	{2, 3}	2, 3–4	1, ?	1, ?	segr	pyc	Seshachar and Bhandary[229]

TABLE 1 (cont.)

Plagiotomidae													
Nyctotherus cordiformis	temp	+	3	1	—	iph	1	1	1	1	—	—	sk	Wichterman,[265] Golikowa[93]
Gyrocorythidae													
Metopus sigmoides	tot	—	3	1	—	iph[c]	1	1	1	1	—	—	pyc	Noland[184]
Bursariidae													
Bursaria truncatella	temp	(+)	3	1	—	spn	4[a]	2–5	1	?	6–8[a]	fus[a]	fr	Prowazek[205]
The same	temp	—	3	1	—	spn	3	4	1	4	—	segr	fr	Poljansky[199]
Stentoridae													
Climacostomum virens	temp	—	3	1	(+)	spn	3	4	1	4	—	fus	fr	Peshkovskaja[193]
Fabrea salina	temp	(+)	3	3–6	+	iph	3	4	1	4	—	fus	fr	Ellis[84]
Stentor coeruleus	temp	(+)	3	1	+	iph	2	2	1	2	—	fus	fr	Mulsow[175]
Stentor polymorphus	temp	(+)	3	1	+	iph	3	6	1	2	—	fus	fr	Mulsow[175]
Entodiniomorphida													
Diplodinium ecaudatum	temp	+	2	1	+	spn	1	1	1	1	—	—	fr	Dogiel[76]
Diplodinium triloricatum	temp	+	2	1	+	iph	1	1	1	1	—	—	fr	Dogiel[76]
Entodinium caudatum	temp	+	2	1	+	iph	1	1	1	1	—	—	fr	Poljansky and Strelkov[201]
Cycloposthium bipalmatum	temp	+	2	1	+	iph	1	1	1	1	—	—	fr	Dogiel[76]
Opisthotrichum janus	temp	µkj	2	1	+	iph	1	1	1	1	—	—	fr	Dogiel[76]
Hypotrichida													
Oxytrichidae													
Uroleptus mobilis	temp	(+)	3	2–4	(+)	spn	2	1	2	2	1	div	pyc	Calkins[15]
Kahlia sp.	temp	—	3	1	?	?	2	1	2	2	1	div	fr	Rao[215]
Urostyla hologama	tot	—	?	?	?	?[d]	2	1	3	?	—	div	?	Heckmann[105]
Urostyla polymicronucleata	tot	—	2	—	+	spn[e]	2	1	2	2	1	div	fr	Moldenhauer[172]
Pleurotricha lanceolata	temp	?	3	1	—	iph	2	1	2	2	1	div	?	Manwell[162]
Oxytricha fallax	temp	?	3	2–3	—	iph	2	1	3(2)	2	0(1)	div	fr	Gregory[95]
Oxytricha bifaria	temp	(+)	3	2	—	spn	2	1	2	2	1	div	fr	Kay[139]
Onychodromus grandis	temp	(+)	3	2	—	spn	2	1	2	2	1	div	pyc	Maupas[165]

TABLE 1 (cont.)

Species	Type of conjugation	Preconjugants	Number of maturation divisions of the micronucleus	Number of spindles of the postmeiotic division	Difference between pronuclei	State of pronuclei during fusion	Number of synkaryon divisions	Number of macronuclear anlagen	Number of new micronuclei	Number of degenerating synkaryon derivatives	Behavior of macronuclear anlagen	Old macronucleus	References
1	2	3	4	5	6	7	8	9	10	11	12	13	14
Stylonychia mytilus	temp	—	4?	1	—	?	2	1	2	1	div	fr	Peshkovskaja[195]
The same	temp	—	3	2–3	(+)	?	2	1	2(3)	1(0)	div	fr	Ammermann[2]
The same	temp	—	2	?	+	spn	2	1	2	1	div	fr	Moldenhauer[172]
Stylonychia pustulata	temp	—	3	1	—	spn	2	1	2	1	div	fr	Maupas,[165] Pieri[197]
The same	temp	(+)	4?	1	—	spn	2	1	2	1	div	fr	Prowazek,[205] Peshkovskaja[195]
Stylonychia muscorum	temp	—	3	1–2	—	?	2	1	3(2)	0(1)	div	fr	Alonso and Pérez-Silva[1]
Euplotidae													
Euplotes eurystomus	temp	—	prl+3	2	(+)	spn	2	1	1	2	+maf (a)	fr	Maupas,[165] Turner[253]
The same	temp	?	prl+3	2	?	?	2	1	1	2	—	fr	Katashima,[135–6] Diller[71]
Euplotes patella	temp	—	prl+3	2	—	?	2	1	1	2	—	fr	Katashima[138]
Euplotes harpa	temp	?	prl+3	2	?	?	2	1	1	2	—	fr	Katashima[134–5]
Euplotes minuta	temp	—	prl+3	2	?	?	2	1	1	2	—	fr	Siegel and Heckmann[232]
Euplotes charon	temp	?	prl+3	2	—	iph	2	1	1	2	—	fr	Maupas,[165] Devidé[56]

TABLE 1 (cont.)

Euplotes cristatus	temp	—	prl+3	2	—	iph	{1 2}	1	1	{2]	—	fr	Wichterman[273]
Euplotes woodruffi	temp	—	prl+3	2	—	spn	2	1	1	2	—	fr	Rao[214, 216a]
Diophrys scutum	temp	?	3	1	?	?	2	1	2	1	div	?	Ito[117]
Uronychia transfuga	temp	?	3	1	?	?	{1 2}	1	1	{1}	div	pyc	Reiff[220a]
Aspidiscidae *Aspidisca costata*	temp	?	3	2	?	?(f)	1(g)	1	2	1	—	?	Diller[72]

(a) Probably erroneous data.
(b) Clearly an error: the first synkaryon division, one derivative of which degenerates, was overlooked.
(c) Only one synkaryon forms in the synconjugant; two residual pronuclei degenerate.
(d) Two synkarya are formed in the synconjugant, but only one remains functional.
(e) Only one synkaryon forms in the synconjugant.
(f) Each conjugant forms four pronuclei and two synkarya (one by nuclear exchange, the other by autogamy).
(g) Both synkarya of an exconjugant are functional.

TABLE 2. PATTERNS OF CONJUGATION IN SUCTORIA AND CHONOTRICHIDA

Species	Type of conjugation	Preconjugants	Number of maturation divisions of the micronucleus	Presence of the postmeiotic micronuclear division	Number of pronuclei in one conjugant	For total conjugation only		Number of synkaryon divisions	Number of macronuclear anlagen	Number of new micronuclei	Number of degenerating synkaryon derivatives	Old macronucleus	References
						Number of synkarya in the synconjugant	Number of synkarya remaining functional						
1	2	3	4	5	6	7	8	9	10	11	12	13	14
Suctoria													
Podophryidae													
Podophrya parasitica	temp	—	3	+	2	...	...	1	1	1		?	Kormos and Kormos[154]
Podophrya fixa	temp	—	3	+	2	...	...	1	1	1		fr	Maupas[165]
Dendrocometidae													
Dendrocometes paradoxus	temp	—	2[a]	+	2	...	...	2	1	3		sk	Hickson and Wadsworth[109]
Dendrosomatidae													
Platophrya rotunda	temp	—	3	+	2	...	...	2	1	1	1	fr	Gönnert[94]
Platophrya sp.	temp	—	3	+	2	...	...	1	1	1		?	King[145]
Lernaeophrya capitata	tot	—	3	+	2	2	one	2	1	3		fr	Gönnert[94]
Dendrosoma radians	tot	(+)	?	?	?	?	one	2	1	3		fr	Gönnert[94]
Unassigned genus													
Phalacrocleptes verruciformis	temp	—	3	+	2	...	...	1	1	1		pyc	Kozloff[158a]

TABLE 2 (cont.)

Acinetidae													
Acineta papillifera	temp	—	3	+	2	⋯	⋯	2	1	1	2	sk	Martin,[163] Collin[39]
Tokophrya lemnarum	temp	—	3	+	2	⋯	⋯	4	1	1	7	fr	Noble[192]
Tokophrya cyclopum	tot	—	3	+	?	?	one	2	1	1	2	fr	Maupas,[165] Collin[39]
Discophryidae													
Choanophrya infundibulifera	tot	—	?	?	2	2	one	?	1	1	?	fr	Collin[39]
Cyclophrya katharinae	temp	—	3	+	2	⋯	⋯	1	1	1		fr	Kormos and Kormos[155]
Discophrya collini	tot, inn	μkj	3	+	2	2	both	1	1	3		fr	Kormos,[151] Kormos and Kormos,[153] Canella[19a]
Discophrya endogama	tot, inn	μkj	3	+	2	2	both	1	1	3		fr	Kormos and Kormos[153]
Discophrya buckei	tot, inn	μkj	3	+	2	2	both	1	1	3		fr	Kormos and Kormos[154]
Ephelotidae													
Ephelota gemmipara	tot	(+)	2?	−?	?	1	one	1	1	1	{4, 12}	fr	Collin,[39] Grell[96]
Chonotrichida													
Spirochona gemmipara	tot	—	?	?	2	2	both	{2, 3}	1	3		fr	Tuffrau[252]

(a) Probably an error (see p. 166).

TABLE 3. PATTERNS OF TOTAL CONJUGATION IN PERITRICHA

Species	Origin of microconjugants	Number of maturation divisions of the micronucleus (at left, in macroconjugants, at right, in microconjugants)	Presence of the postmeiotic micronuclear division	Number of pronuclei in one conjugant	Number of synkarya in the synconjugant	Number of synkaryon divisions	Number of macronuclear anlagen	Number of new micronuclei	Behavior of the macronuclear anlagen	Old macronucleus	References
1	2	3	4	5	6	7	8	9	10	11	12
Vorticellidae											
Vorticella microstoma	uneq	2 / prl+2	—	1	1	3	7	1	segr	sk	Maupas,[165] Finley[87]
Vorticella convallaria	uneq	?	?	?	?	3	6	2	segr	fr	Seshachar and Dass[230]
Vorticella campanula	uneq	prl+3 / 2prl+3	+	2	1[a]	3	7	1	segr	fr	Mügge[173]
Vorticella cucullus	ros	3 / prl+3	+	2	1[a]	3	7	1	segr	fr	Maupas[165]
Vorticella monilata	ros	3 / prl+3	+	2	1[a]	3	7	1	segr	fr	Maupas[165]
Vorticella nebulifera	ros	3 / prl+3	+	2	1[a]	3	7	1	segr	fr	Maupas[165]
Carchesium polypinum	ros	3 / prl+3	+	2	1[a]	3	7	1	segr	fr	Maupas,[165] Popoff[202]
Carchesium spectabile	ros	2 / prl+2	—	1	1	3	7	1	segr	sk	Dass[47]
Zoothamnium arbuscula	ros	3 / prl+3	+	2	2[b]	3	7	1	segr	fr	Furssenko[91]
Zoothamnium alternans	trf	?	?	?	?	3	7	1	segr	?	Summers[248]
Epistylidae											
Epistylis articulata	ros	2 / prl+2	—	1	1	3	7	1	segr	sk	Dass[46]
Epistylis sp.	ros	2 / prl+2	—	1	1	3	7	1	segr	sk	Dass[48]

TABLE 3 (cont.)

											Reference
Opercularia coarctata	uneq	3 / prl+3	+	2	1(a)	2	3	1	segr	sk	Enriques[85]
Rhabdostyla vernalis	prot	2 / prl+2	−	1	1	3	7	1	segr	fr	Finley and Nicholas,[90] Finley[89]
Telotrochidium (= Opisthonecta) henneguyi Ophrydiidae	uneq	2 / prl+2		1	1	3	7	1	segr	sk	Rosenberg[223]
Ophrydium versatile Lagenophryidae	ros	3 / prl+3	+	2	1(a)	3	7	1	segr	fr	Kaltenbach[132]
Lagenophrys tattersalli Urceolariidae	prot	?	?	2	1(a)	3	7	1	segr	sk	Willis[274]
Urceolaria synaptae (= "Trichodina sp.")	—	3 / 3	+	2	1(a)	3	4	1(c)	segr	sk	Hunter,[112] Colwin[40]
Trichodina patellae	?	3 / 3	?	1	1	3	7	1	segr	fr	Caullery and Mesnil,[21] Brouardel[12]
Trichodina spheroidesi	ros	2 / 2	−	1	1	3	7	1	segr	fr	Padnos,[186] Padnos and Nigrelli[187]

(a) Two "excess" pronuclei degenerate.
(b) Of the two synkarya, one degenerates.
(c) Three synkaryon derivatives degenerate.

Tables 1 to 3 show that our knowledge of conjugation in the various ciliate taxa is very uneven, thorough in Tetrahymenidae, Parameciidae, Spirostomatidae, Oxytrichidae, Euplotidae, and Vorticellidae but scanty in some other important ciliate groups: the lower Holotricha (Enchelyidae, Amphileptidae, Tracheliidae, etc.), the orders Thigmotrichida, Astomatida, Apostomatida, Chonotrichida, and the family Urceolariidae. Apparently conjugation has never been observed in such families as Colpodidae and Ophryoglenidae, which in other respects are well studied.

In some systematic groups, the cytological pattern of conjugation is rather uniform. Almost all the Tetrahymenidae have two synkaryon divisions and two macronuclear anlagen; pycnosis of the old macronucleus (without fragmentation) predominates. All Entodiniomorphida have only two maturation divisions and only one synkaryon division; there is sharp difference between their stationary and migratory pronuclei. Almost all the Hypotrichida have two synkaryon divisions and only one macronuclear anlage. Almost all Peritricha show three synkaryon divisions and seven macronuclear anlagen which segregate during three synconjugant divisions.

However, in many other taxa, the pattern of conjugation is highly varied, and the evolutionary relationships of the various patterns are by no means clear. For example, from one to three synkaryon divisions occur among the Gymnostomatida; in the single genus *Paramecium*, from two to four synkaryon divisions and either two or four macronuclear anlagen occur. The Heterotrichida may have from one to three synkaryon divisions and from one to six or seven macronuclear anlagen which in some cases fuse, in other cases segregate. Among the Suctoria, three types of conjugation exist: the temporary, the total isogamontic, and the total anisogamontic; even within the genus *Tokophrya*, both of the first two types occur, and within the family Discophryidae, all three types occur. The number of synkaryon derivatives in Suctoria varies from one to four, but the macronuclear anlage is always single.

Temporary conjugation is characteristic of all the Holotricha, of the vast majority of Heterotrichida (except *Metopus*) and Hypotrichida (except the genus *Urostyla*), of all Entodiniomorphida, and of some Suctoria. It does not exist in Peritricha. Total isogamontic conjugation occurs mainly in Suctoria and Chonotrichida, and, as isolated cases, in Spirotricha (*Metopus*, *Urostyla*). Total anisogamontic conjugation exists in some Suctoria and in all Peritricha (except, possibly, *Urceolaria synaptae*, where conjugation seems to be total isogamontic, occurring between two vegetative individuals). Thus, total conjugation (both the isogamontic and the anisogamontic) is characteristic of specialized ciliate taxa and must be considered secondary. Temporary conjugation is beyond any doubt the primitive type of conjugation among the ciliates.

IV. NORMAL PATTERN OF NUCLEAR BEHAVIOR DURING AUTOGAMY

This section deals only with autogamy in singles. Autogamy in pairs ("cytogamy") will be considered below (p. 229).

Autogamy is known most completely in *Paramecium aurelia* (Figs. 20, 21). A reorganization process ("parthenogenesis") was observed in this species by Hertwig[106, 108] but not investigated in detail. Woodruff and Erdmann[277] extensively studied this process and called it "endomixis". They maintained that no third maturation division, no formation of pronuclei, and no karyogamy occurred during "endomixis". However, Diller[58] demonstrated that the stages of pronuclear formation and fusion were overlooked by Woodruff and Erdmann, and that in reality the reorganization process was autogamy. Later on, Sonneborn[234-5, 241] produced genetic evidence of autogamy in *P. aurelia*: transition of all genes from the heterozygous to the homozygous condition during the nuclear reorganization.

During autogamy, both micronuclei of *P. aurelia* undergo the "crescent stage" of the meiotic prophase (Fig. 20, *a*). This single fact shows that the reorganization process is not diploid parthenogenesis ("endomixis"). The first (Fig. 20, *b*) and the second (*c*) maturation divisions follow. Of the eight nuclei thus formed (*d*), some enter the third maturation division (*e*) while others degenerate by pycnosis. Only one spindle usually completes the third division and gives rise to two pronuclei (*f*). The latter fuse with each other into the synkaryon (*g*), which divides twice (*h*, *i*). The pronuclei, the synkaryon, and the spindle of the first synkaryon division lie within the paroral cone (Fig. 20, *f–h*) resembling their position during conjugation (p. 221). And the resemblance continues through later stages, the exautogamonts differing no more from the exconjugants: the four derivatives of the second synkaryon division (Fig. 20, *j*) become two micronuclei and two macronuclear anlagen, the latter segregating during the only division of the exautogamont (Fig. 21). The old macronucleus disintegrates following skein formation (Fig. 20).

Thus, autogamy in *P. aurelia* is cytologically almost identical with conjugation; only lacking are the pairing of individuals and the exchange of migratory pronuclei.

Autogamy in singles of a similar type was later described in *Paramecium polycaryum*[66] and *P. jenningsi*.[169] Spontaneous autogamy is unknown in *P. bursaria*, *P. calkinsi*, and *P. putrinum*, but autogamy can be induced in these species by mixing two or several complementary mating types (*P. calkinsi*,[62] *P. putrinum*[126]), or by treatment of the ciliates with animal-free culture fluid from a complementary mating type (*P. bursaria*[31]).

Autogamy in *Tetrahymena rostrata* (Fig. 22), described by Corliss,[41, 42] occurs in cysts. The micronucleus undergoes the first and the second matura-

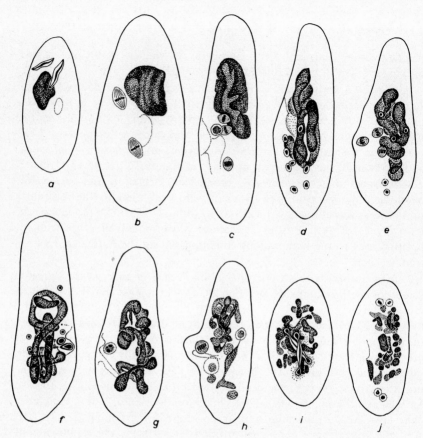

FIG. 20. Autogamy in *Paramecium aurelia*: *a*—meiotic prophase ("crescent stage") in both micronuclei; *b*—metaphase of the first maturation division; *c*—metaphase of the second maturation division; *d*—eight derivatives of the second maturation division, macronucleus transforming into skein; *e*—three spindles of the third maturation division; *f*—pronuclei in the paroral cone; *g*—synkaryon; *h*—first synkaryon division, fragmentation of the macronuclear skein; *i*—telophase of the second synkaryon division; *j*—four derivatives of the second synkaryon division. × 530. After Diller.[58]

tion divisions, the meiotic nature of which is indicated by the "crescent stage" in the prophase of the first division. Only one nucleus undergoes the third (equational) maturation division, and the pronuclei thus formed fuse with each other to form the synkaryon which divides twice. Of its four derivatives, two become macronuclear anlagen, one becomes the new micronucleus, and one degenerates. The macronuclear anlagen segregate during the division of the excysted exautogamont (Fig. 22). The old macronucleus degenerates by pycnosis.

Autogamy is known also in *Frontonia leucas*.[55] The maturation divisions proceed here as in *Tetrahymena rostrata*; the synkaryon divides four times, and five macronuclear anlagen are formed. Finally, in *Euplotes minuta*[232, 181a] autogamy in singles occurs: the micronucleus undergoes the preliminary and two meiotic divisions, and two nuclei undergo the postmeiotic

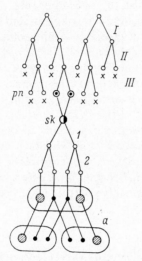

FIG. 21. Diagram of nuclear behavior during autogamy in *Paramecium aurelia* (maturation divisions of both micronuclei shown); *a*—division of the exautogamont; other abbreviations as in Fig. 6.

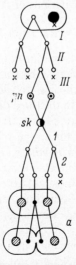

FIG. 22. Diagram of nuclear behavior during autogamy in *Tetrahymena rostrata*. Solid black circle is the old macronucleus; *a*—division of the exautogamont; other abbreviations as in Fig. 6.

maturation division; the pronuclei therefore may not be sister nuclei (see also p. 220). There are two synkaryon divisions followed by formation of one macronuclear anlage and one micronucleus; two synkaryon derivatives degenerate.

The close similarity of the nuclear phenomena in autogamy and in conjugation is stressed also by the autogamy undergone by artificially separated conjugants. It has been long known that nuclei continue to divide and macronuclear anlagen appear in conjugants separated after the start of the first maturation division (*Bursaria*,[205] *Uroleptus*,[16] *Stylonychia*[114]). Poljansky[200] first showed in *Bursaria truncatella* that pronuclei are formed in such separated conjugants and that the pronuclei fuse with each other as in autogamy. Similar observations were later made on *Paramecium aurelia*,[152] *P. caudatum*,[185] *Stylonychia mytilus*, *S. muscorum*,[128a] and *Euplotes eurystomus*.[137] Autogamy occurs in the normal (micronucleate) partners of *Paramecium caudatum* following precocious spontaneous separation of conjugants in crosses of a normal clone with an amicronucleate clone.[232a] Also the "killer" partner of *P. polycaryum* undergoes autogamy when conjugating with a sensitive partner which perishes.[250a]

V. COMPARATIVE CYTOLOGY OF SOME STAGES OF CONJUGATION AND AUTOGAMY

1. Maturation Divisions of the Micronucleus

A. NUMBER OF MATURATION DIVISIONS

Tables 1 to 3 show that the vast majority of ciliates have three maturation divisions. The first two are meiotic, the third equational (postmeiotic). In most cases, both derivatives of the first division undergo the second division, but only one nucleus takes part in the third division (Fig. 23, *a*). This pattern is the most frequent among the ciliates. It is characteristic of almost all Holotricha, Heterotrichida, Suctoria, and of the macroconjugants of some of the Peritricha, e.g. *Vorticella monilata*, *V. nebulifera*, *Carchesium polypinum*, *Zoothamnium arbuscula*, *Opercularia coarctata*, and *Ophrydium versatile* (see Table 3). All (or almost all) other patterns seem to be secondary modifications of this pattern. Sometimes only one derivative of the first division undergoes the second division, e.g. in *Paramecium bursaria* (Fig. 23, *b*).

In some cases more than one nucleus undergoes the third maturation division (Fig. 23, *c*). The pronuclei are not always sister nuclei in this case and, consequently, may not be genetically identical. This pattern is rare among Holotricha (*Conchophthirius mytili*) and among Heterotrichida

(*Blepharisma tropicum, Fabrea salina*), but fairly frequent among Hypotrichida, in the family Oxytrichidae (*Uroleptus mobilis, Oxytricha fallax, O. bifaria, Onychodromus grandis,* etc. See Table 1.)

The first maturation division may be preceded by the so-called preliminary micronuclear division, which is equational. This preliminary division is characteristic of the family Euplotidae and of the microconjugants of Peritricha. The Euplotidae pattern (Fig. 23, *d*) can be easily derived from the Oxytrichidae pattern (Fig. 23, *c*) by the addition of the preliminary division. As in the Oxytrichidae, two haploid nuclei usually undergo the postmeiotic (third maturation) division. The total number of micronuclear divisions is four. The pattern characteristic of the microconjugants of many Peritricha (Fig. 23, *e*) can be derived by addition of the preliminary micronuclear division to the pattern peculiar to the macroconjugants of the same species (Fig. 23, *a*). The total number of micronuclear division is also four in this case.

Possibly even two preliminary divisions may sometimes exist in the microconjugants of Peritricha, the total number of micronuclear divisions

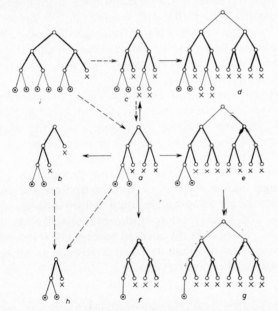

FIG. 23. Types of micronuclear maturation in ciliates and their possible evolutionary relationships. Bold lines show meiotic divisions, circles with a dot are pronuclei, crosses mark degenerating nuclei. *a—Paramecium caudatum, b—Paramecium bursaria, c—Onychodromus grandis, d—Euplotes eurystomus, e—*microconjugants of *Carchesium polypinum, f—*macroconjugants of *Vorticella microstoma, g—*microconjugants of *Vorticella microstoma, h—Cycloposthium bipalmatum, i—Tracheloraphis phoenicopterus.* Original.

reaching five—e.g. in *Vorticella campanula*.[173] The macroconjugant of *V. campanula* also shows one preliminary micronuclear division (a unique case among the peritrichs).

The postmeiotic (third maturation) division shows a tendency to disappear in many ciliates with total conjugation. The pattern of maturation of the macroconjugants of *Vorticella microstoma*, *Carchesium spectabile*, *Epistylis articulata*, *Rhabdostyla vernalis*, etc., which includes only two maturation divisions (Fig. 23, *f*), may be derived in this way from the basic pattern with three divisions (Fig. 23, *a*). Parallel with this, the maturation pattern of the microconjugants of the same species, including only three micronuclear divisions (the preliminary and two meiotic, Fig. 23, *g*), may be derived from the pattern of the peritrich microconjugants which have four micronuclear divisions (Fig. 23, *e*). When postmeiotic division is lost, each conjugant produces only one pronucleus (Fig. 23, *f*, *g*).

When vegetative individuals have several micronuclei, all of them usually take part in the first and second maturation divisions, e.g. in *Paramecium aurelia*,[106, 165] *P. polycaryum*,[67] *P. woodruffi*,[121] or *P. multimicronucleatum*.[6] In other cases not all but still many micronuclei undergo the first two divisions, e.g. in *Bursaria truncatella*,[199] *Spirostomum ambiguum*,[216, 228a] or *Uroleptus mobilis*.[45] But regardless of the initial number of micronuclei, usually only one derivative of the second maturation division divides for the third time; less frequently two to four derivatives divide, mainly among the Hypotrichida.

Quite apart is the pattern of maturation in Entodiniomorphida, which includes only two micronuclear divisions (Fig. 23, *h*). Their homology is not clear, the chromosome situation remaining unknown. There are some grounds for considering the last (second) division as postmeiotic, since it differentiates the pronuclei, as does the third division in other ciliates. Then, we have to admit either that meiosis in Entodiniomorphida is completed in a single nuclear division, or that the first meiotic division occurs in these ciliates during the preconjugation cell division giving rise to two preconjugants. The latter view, advanced by Poljansky and Strelkov,[201] seems to be more likely since the micronuclear spindle is much larger during preconjugation cell division than during vegetative division (see p. 153). But the matter needs further investigation.

B. MEIOSIS IN CILIATES

All ciliates are diplonts with gametic chromosome reduction. In 1889 Maupas[165] and Hertwig[106] drew a homology between maturation divisions of the ciliate micronucleus and maturation divisions of the metazoan ovocyte. It is firmly established now that chromosome reduction in ciliates takes place during the maturation divisions. But the details of meiosis began to clear up only recently.

The ciliate meiosis was considered atypical for a long time; it was supposed that, in one of the maturation divisions, some whole chromosomes moved to one pole, other whole chromosomes to the other pole, and that this was not preceded by bivalent formation (i.e. by pairing of homologous chromosomes). Usually the second maturation division has been considered "reductional" (*Didinium*,[204] *Collinia*,[38] *Chilodonella*,[86, 160-1] *Uroleptus*,[15, 17] *Pleurotricha*,[162] *Euplotes*,[253] *Conchophthirius*,[140] *Ancistruma*,[141] *Climacostomum*,[193] *Fabrea*,[84] etc.). Less frequently, "reduction" in the first maturation division was admitted (*Carchesium*,[202] *Oxytricha*,[95] etc.).

It is currently known, however, that meiosis in ciliates begins with pairing of the homologous chromosomes. They form bivalents which transform later into tetrads by longitudinal splitting of both chromosomes forming the bivalent. Meiosis includes not one but two nuclear divisions: at the first one, the tetrads separate into dyads, and at the second one, the dyads separate into single chromosomes, the number of which is haploid. In other words, meiosis in ciliates corresponds to the classical scheme of the metazoan meiosis.

The meiotic prophase of the first maturation division† in ciliates usually comprises one of the two alternative characteristic stages—either the "crescent stage", or the "parachute stage".

Meiosis with the "crescent stage" was first extensively studied by Calkins and Cull[19] in *Paramecium caudatum* (Fig. 24). The resting micronucleus of this ciliate consists of a chromatin mass and of an achromatic "cap". In early meiotic prophase, thread-like chromosomes appear in the micronucleus, which begins to elongate (Fig. 24, *a*). The micronucleus curves during elongation and enters the "crescent stage" (*b*). The chromosomes are oriented along the "crescent". The achromatic "cap", called "division center" by Calkins and Cull, gradually migrates from one of the extremities of the crescent to the middle of its convex side. After this, the nucleus shortens, and the extremities of the "crescent" become gradually withdrawn (*c–e*). The chromosomes are definitely paired at these stages (*d*). The chromosome pairs are probably bivalents resulting from parallel union of homologous chromosomes. If so, the "crescent stage" must be considered homologous to the zygotene stage of the classical meiosis (the stage when chromosome pairing occurs). At later stages, the bivalents seem even to show chiasmata (diplotene stage, Fig. 24, *e*).

The axis of the spindle of the first maturation division develops in perpendicular position to the long axis of the "crescent" [19, 54] (Fig. 24, *f*). The homologues seem to separate during the first maturation division: in anaphase, the chromosomes are already single (*f*, *g*). However, the chromo-

---

† The preliminary micronuclear division, when present, precedes the meiotic prophase. The preliminary micronuclear division is not called a "maturation division" in this chapter.

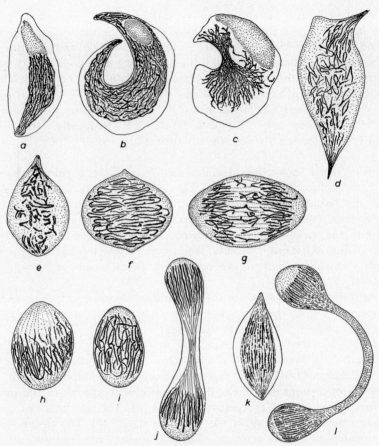

FIG. 24. Meiosis and maturation divisions in *Paramecium caudatum*:
a—micronuclear elongation; *b*—"crescent stage"; *c*—shortening of the
"crescent"; *d*—appearance of bivalents; *e*—diplotene stage; *f*, *g*—anaphase
of the first maturation division; *h*, *i*, *j*—metaphase, anaphase, and telophase
of the second maturation division; *k*, *l*—anaphase and telophase of the
third maturation division (in *l*, the lower spindle pole is the future migratory
pronucleus). Iron hematoxylin staining; *a*, *j*, *l*: ×1000; *b*–*g*: ×1100;
h: ×900; *i*: ×700; *k*: ×1200. After Calkins and Cull.[19]

somes become paired again in the metaphase of the second maturation
division (Fig. 24, *h*). This time it is probably the result of their longitudinal
splitting. The chromatids separate in the second division anaphase (*i, j*). The
chromosomes and their duplication are not distinct in the third maturation
division (Fig. 24, *k, l*).

Calkins and Cull admit, on grounds of the data just presented, that
chromosome reduction is completed in the first division, and that the second
division is purely mitotic. But it seems more likely that both homologues
are split into chromatids already at the diplotene stage (Fig. 24, *e*), i.e. that

bivalents transform into tetrads before the first division metaphase. This view is supported by the presence of chiasma-like pictures during the late meiotic prophase (Fig. 24). And chiasmata are known to arise only at the four-stranded stage, following crossing-over between two of the four chromatids of the tetrad. This interpretation implies that the second maturation division is also involved in the meiotic process.

More or less typical tetrads were first described in ciliates by Devidé and Geitler.[56, 57] They showed that no true chromosomes but only elongate aggregates of many chromosomes are visible in vegetative mitoses of many ciliates. Such aggregates frequently divide transversely, and their number is inconstant. According to Devidé and Geitler, the difficulties in studying the ciliate meiosis are related to the fact that chromosome aggregation may affect also the meiosis, more frequently the second maturation division, but sometimes also the first (*Chilodonella*). Devidé and Geitler could observe true meiotic tetrads only in some ciliates, e.g. in *Colpidium campylum* (Fig. 25.)

The meiotic prophase of *C. campylum* also includes a "crescent stage". The "crescent" is thin, almost thread-like, with dilated ends (Fig. 25, *a*). Longitudinal chromosomes appear in the "crescent" when it shortens (*b*). The chromosomes become highly condensed during diakinesis (*c*). Tetrads having one, two, or three chiasmata and respectively the form of crosses, rhombs, or eights are clearly visible at this stage. The number of the tetrads is twenty-one. In metaphase of the first maturation division, both homologous chromosomes forming a tetrad are distinctly split into two chromatids (Fig. 25, *d*, *e*). In anaphase, dyads go to the spindle poles. The dyads have either two arms (in the case of telokinetic chromosomes) or four arms (when the respective chromosomes are metakinetic or submetakinetic, Fig. 25, *f*, *g*). In the anaphase of the second maturation division, there are no four-armed figures (*h*); single elongate (telokinetic) or V-shaped (metakinetic and submetakinetic) chromosomes move to the poles.

No single chromosomes are visible in the third maturation division; they apparently unite into long chromosome aggregates (Fig. 25, *i*).

A very distinct meiosis was observed by Ray[219] in *Tetrahymena pyriformis*. The zygotene stage is also crescent-like in this species. The diploid chromosome number is ten, so that only five bivalents are formed, which transform later (in diplotene) into tetrads. Up to three chiasmata appear on some tetrads; respectively, during metaphase of the first maturation division the tetrads are cross-shaped (one chiasma), annular (two chiasmata), or in form of eights (three chiasmata). Typical dyads appear in the anaphase of the first division. They have four arms with a common kinetochore (the chromosomes are metakinetic or submetakinetic in this species). Five single chromosomes move to each pole in the second division anaphase.

The "crescent stage" was first described by Bütschli[14] in three species of *Paramecium* and extensively studied afterwards, mainly in *P. caudatum*.[19,

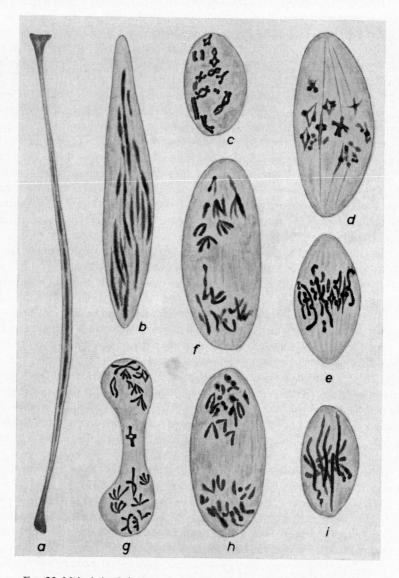

FIG. 25. Meiosis in *Colpidium campylum*: *a*—"crescent stage"; *b*—shortening of the "crescent"; *c*—diakinesis showing tetrads; *d, e*—metaphases of the first maturation division showing tetrads; *f, g*—anaphases of the first maturation division showing dyads; *h*—anaphase of the second maturation division showing single chromosomes; *i*—metaphase of the third maturation division showing chromosome aggregates. Aceto-carmine staining; *a, b, d, f, h, i,* × 2700 (after Devidé and Geitler[57]); *c, e, g,* × 2100 (after Devidé[56]).

[54, 165, 191, 194] This stage is known to exist in many other ciliates as well, especially among the Holotricha Hymenostomatida. It occurs in the meiotic prophase of *Tetrahymena* (= *Leucophrys*) *patula*,[165] *Glaucoma scintillans*,[165] *Colpidium colpoda*,[165, 207] *Loxocephalus* sp.,[7] *Frontonia leucas*,[55] *Entorhipidium echini*,[278] *Cryptochilum echini*[43] (Fig. 7, *a*), *Paramecium aurelia*[58, 106, 156, 165] (Fig. 20, *a*), *P. bursaria*,[14, 25, 32, 33, 35, 36, 81, 99, 165] *P. woodruffi*,[121] *P. polycaryum*,[67] *P. multimicronucleatum*.[159] Outside the Hymenostomatida, the "crescent stage" exists in some Suctoria, e.g. *Platophrya*,[94] and in some Peritricha, e.g. *Vorticella campanula*.[173]

Meiosis in *Nassula ornata*[213] may serve as an example of meiosis including the "parachute stage". The chromosomes show radial arrangement in the early meiotic prophase (Fig. 26, *a*) but later become polarized, their ends converging on a small chromatic body beneath the nuclear envelope (*b*). The other side of the micronucleus contains a tangle of thin thread-like chromosomes which have rather distinct chromomere structure. These chromosomes begin to pair. This stage, called the "parachute stage", apparently corresponds to the zygotene or bouquet stage of the classical meiosis. Consequently, it must also be homologous to the "crescent stage" in the meiosis of other ciliates.

Later on, the chromomeres condense into a large chromocenter (Fig. 26, *c*). Thread-like chromosomes, now distinctly paired, are obvious only in the "stalk" connecting the chromocenter with the nuclear envelope (*c*). The chromocenter soon disintegrates into separate bivalents (diakinesis stage, Fig. 26, *d*, *e*, *f*). The bivalents scatter throughout the spindle in prometaphase of the first maturation division (Fig. 26, *g*). In metaphase, they show maximum condensation and form an equatorial plate (*h*). The bivalents are now transformed into distinct tetrads having one, less frequently two chiasmata (*h*). The spindle of the first maturation division is large and has long pointed ends; it differs sharply from spindles of vegetative mitoses of *N. ornata*.[211] Dyads are indistinct in the first division telophase (Fig. 26, *i*) and in the second division prophase (*j*), but can be seen in the metaphase of the second maturation division (*k*). The spindle of the second division is much smaller than that of the first.

Meiosis comprising the "parachute stage" has a much wider distribution than meiosis with the "crescent stage". Among Holotricha, it has been described in Gymnostomatida (*Didinium*,[204] *Chilodonella*[160-1]), Astomatida (*Dogielella*[198]), Apostomatida (*Collinia*,[38, 249] Fig. 11, *a*). The "parachute stage" is characteristic of the meiosis in Heterotrichida (*Bursaria*,[199] *Fabrea*[84]) and especially in Hypotrichida, where it was first noticed by Bütschli[14] in *Stylonychia mytilus*. Calkins[15] thoroughly studied the "parachute stage" in *Uroleptus mobilis*. This stage is known in several species of *Stylonychia*[1, 2, 195, 197] and *Oxytricha*,[95, 139] in *Pleurotricha lanceolata*,[162] *Onychodromus grandis*,[165] *Kahlia* sp.,[215] and *Euplotes eurystomus*.[253]

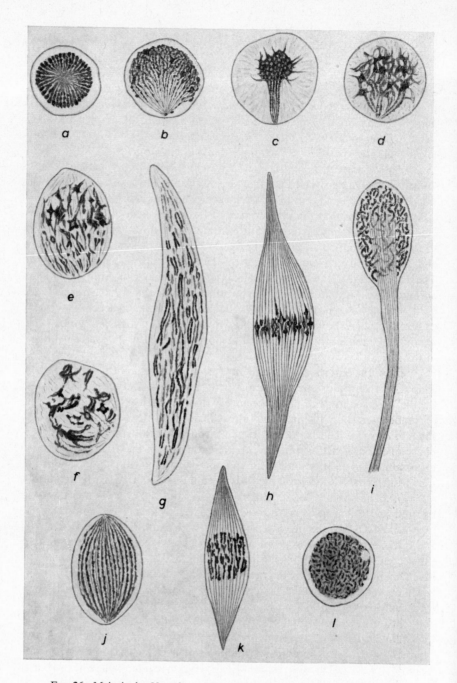

FIG. 26. Meiosis in *Nassula ornata*: *a*—early prophase; *b*—"parachute stage"; *c*—condensation of the zygotene chromosome tangle; *d–f*—diakinesis; *g*—prometaphase; *h*—metaphase; *i*—telophase of the first maturation division; *j*—prophase; *k*—metaphase of the second maturation division; *l*—haploid meiotic derivative. Feulgen reaction, ×2000. After Raikov.[213]

Besides *Colpidium, Tetrahymena,* and *Nassula,* more or less typical tetrads have been observed in *Euplotes*[56, 214] *Vorticella,*[57, 173] *Stylonychia,*[2] *Kahlia,*[215] and *Spirostomum.*[216] In these ciliates, however, further stages of meiosis remain obscure due to the small size of the chromosomes and their early aggregation, often beginning in the second maturation division.

Indirect evidence for existence in ciliates of a typical meiosis with two nuclear divisions is provided by the data of Pieri[196] and Pieri *et al.,*[197b] who measured the DNA content of the micronucleus during the maturation divisions in *Stylonychia pustulata* and *Paramecium trichium.* The DNA contents of the first division prophase, of the derivatives of the first division, and of the derivatives of the second division were related as $4:2:1$, exactly as in classical meiosis. No DNA replication occurred before the second maturation division. The third maturation division proved to be equational and preceded by DNA replication. The latter fact was demonstrated also in *Paramecium aurelia* by autoradiography.[8]

The equational nature of the third maturation division is also confirmed in many ciliates by the identity in genotype of the two exconjugants.[241] This is possible only if both pronuclei of a conjugant are derivatives of a single third division spindle, and if the third division is equational.

Especially interesting is the meiosis in *Metopus sigmoides.*[184] In prophase of the first maturation division, a single twisted thread appears in the micronucleus (Fig. 14, *b*). It shortens to form a single "chromosome" which goes to the equator of the spindle and divides longitudinally in anaphase (Fig. 14, *c, d, e*). This "chromosome" fragments into small chromatic "granules" during the second maturation division (Fig. 14, *f*). But before fusion of the pronuclei, the "granules" unite again into a single thread (Fig. 14, *j, k*); the synkaryon contains two such threads (*l*). Apparently we have to interpret the small "granules" of the second maturation division as true chromosomes, and the single thread in the pronucleus as a composite chromosome (a product of linear connection of all chromosomes of a haploid genome). Then, the two threads of the synkaryon are two haploid genomes. And the single thread appearing in the diploid micronucleus before the first maturation division may be the product of the pairing of two homologous composite chromosomes (i.e. a "composite bivalent").

A composite chromosome was found also in the micronucleus of *Nyctotherus* by Golikowa,[93] but in the third, not in the first, maturation division. These observations are important because in the same ciliates (*Nyctotherus, Metopus*) the single composite chromosome appears again in the macronuclear anlage, where it becomes polytene (see Vol. 3, p. 29).

C. CHROMOSOME NUMBERS IN CILIATES

Data on chromosome numbers in Ciliates are highly confusing. Very small chromosome numbers we re formerly indicated for some ciliates, e.g. $2n = 4$ for *Chilodonella.*[86, 160–1] But it is precisely in *Chilodonella* where Devidé and Geitler[56, 57] demonstrated that structures formerly held to be true chromosomes are in reality chromosome aggregates. They showed that such aggregates appear during mitosis (and frequently even during meiosis) in many ciliates, and thus called in question almost all the previous chromosome counts in ciliates. Devidé and Geitler argued that reliable data on chromosome numbers can be obtained in ciliates only by counting the tetrads in the prophase of the first meiotic division of the micronucleus, since the tetrads may begin to stick together even in metaphase of the first division (e.g. in *Stylonychia*). And in such forms as *Chilodonella*, where chromosome aggregation affects also the meiotic prophase and no tetrads can be seen at all,[56, 57] morphological determination of the true chromosome number seems to be impossible.

Table 4 summarizes the data on chromosome numbers in ciliates based on bivalent (or tetrad) counts during the meiotic prophase. In meiosis, the number of bivalents is known to equal the haploid number (n).

TABLE 4. HAPLOID CHROMOSOME NUMBERS IN CILIATES

Species	Chromosome number (n)	References
Gymnostomatida		
Nassula ornata	~20	Raikov[213]
Hymenostomatida		
Tetrahymena pyriformis	5	Ray[219]
Colpidium campylum	21	Devidé and Geitler[57]
Paramecium caudatum	~165	Calkins and Cull[19]
The same	~18	Penn[191]
Paramecium aurelia, syngen 4	~35–~50†	Dippell[73]
The same, syngen 1	~43–~63†	Kościuszko[156]
Paramecium bursaria, race Fd	~40	Chen[24, 26, 28]
Heterotrichida		
Spirostomum ambiguum	~18–24	Rao[216]
Hypotrichida		
Kahlia sp.	~25	Rao[215]
Stylonychia mytilus	~125	Ammermann[2]
Euplotes woodruffi	~16	Rao[214]
Peritricha		
Vorticella campanula	~75	Mügge[173]

† Different figures for several races.

Analysis of the data of Table 4 shows that rather high chromosome numbers are usual among the ciliates (except *Tetrahymena pyriformis*). In older papers, a considerable number of "small granules" or "chromomeres" was frequently claimed to appear in the meiotic prophase; later on, these "chromomeres" fused into a smaller number of elongate "chromosomes". Inspection of the corresponding illustrations leaves a strong suspicion that "chromomeres" were in reality meiotic tetrads, and that larger "chromosomes" were aggregates of many chromosomes. If this is true, it may be deduced that the haploid number is approximately 32 in *Didinium nasutum*,[204] 24 in *Uroleptus mobilis*,[15] 48 in *Oxytricha fallax* and *O. bifaria*,[95, 139] and 32 in *Euplotes eurystomus*.[253]

Of considerable interest is the racial variability of chromosome number in *Paramecium aurelia*.[73, 156] Within a single syngen the racial differences are comparatively small and so cannot be related to polyploidy. According to Dippel, several races of syngen 4 differ from each other mainly by the presence or absence of small dot-like chromosomes. This reveals the existence of an aneuploid series in *P. aurelia*. According to Kościuszko, in six races of syngen 1 of *P. aurelia* the haploid chromosome numbers are: ~ 43, ~ 45, ~ 49, ~ 51, ~ 58, ~ 63. The first two races have six large chromosomes in the haploid set, the following three races have seven, and the last race eight to ten. This also points toward a considerable role of aneuploidy in the intraspecies differentiation of *P. aurelia*. It is of interest that the progeny of crosses between races having different chromosome numbers shows a high mortality of exconjugants (or exautogamonts) after the next conjugation (or autogamy). This F_2 inviability may be related to disturbances of meiosis in the hybrids.

A sharp difference in chromosome morphology was found by Chen[36] in two clones of *Paramecium bursaria* belonging to syngen 6 and capable of conjugating with each other. Almost all the chromosomes of clone Ck-1 are dot-like during the late meiotic prophase, while the chromosomes of clone En-1 are much longer and thicker (Fig. 27). Conjugation between the two clones is nevertheless not lethal in the F_1, though accompanied by a higher proportion of abnormalities of nuclear behavior in exconjugants and by a considerable percentage of inviability of the progeny.

Finally, polyploidy (heteroploidy) of the micronucleus also exists in ciliates. The strong difference between the chromosome numbers recorded in *Paramecium caudatum* by Calkins and Cull[19] and by Penn[191] seems to be related to polyploidy. Chen[24, 26, 28] thoroughly studied the micronuclear polyploidy in *Paramecium bursaria*. He found approximately eighty chromosomes ($2n$?) in the race Fd and discovered several polyploid races with hundreds of chromosomes in each. The origin of polyploidy in *P. bursaria* is usually connected with irregularities of pronuclear behavior (see p. 228). Micronuclear heteroploidy has been described also in conjugating *Paramecium putrinum* ($=$ *trichium*).[61]

D. Mechanisms of Nuclear Differentiation during the Maturation Divisions

Long discussed has been the question as to what mechanism(s) might determine the differing fates of the originally identical derivatives of the second maturation division; under what conditions does one of them (less

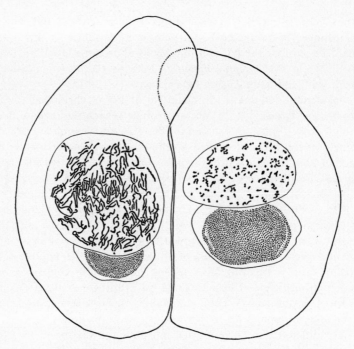

Fig. 27. Conjugation between clones En-1 (at left) and Ck-1 of *Paramecium bursaria*. Late prophase of the first maturation division showing differences in chromosome morphology between the two conjugants. ×910. After Chen.[36]

frequently, two or three) enter the third division while the others degenerate. Maupas[165] came to the conclusion that the different fates of the derivatives of the second maturation division are determined by their positions in the cytoplasm: usually the third division is undergone only by the nucleus which lies nearest to the region of the future pronuclear interchange (mouth region in *Paramecium* species, preoral region in *Colpidium*, etc.). This conclusion of Maupas has been fully confirmed.[243-4, 177] In species of *Paramecium*, the nucleus inside the cytoplasmic paroral cone undergoes the third division.[58, 59, 61, 67, 267, 244, 6] The cytoplasm of the paroral cone seems to protect from degeneration the nucleus lying inside it, while the main mass of the cytoplasm has no such protective effect. In an abnormal clone (d-59) of *P. aurelia*, none of the derivatives of the second maturation division get into the paroral cone. All the nuclei degenerate, no pronuclei being formed by the conjugant or autogamont.[243]

The cytoplasm near the region of contact between the partners is sometimes more dense, homogeneous, and rich in RNA. For example, in *Bursaria truncatella*[199] or *Nassula ornata* (Fig. 34, *a*) several nuclei usually enter the prophase of the third division, and small homogeneous zones form around each of them. However, only one spindle, the one nearest to the contact region, gets inside a large zone of homogeneous cytoplasm and completes mitosis (Fig. 34, *b*) while the others degenerate. The pronuclei and the synkaryon are also surrounded by zones of homogeneous cytoplasm (Fig. 34, *c, d*). In the suctorian *Cyclophrya katharinae*[155] and in some other ciliates an "isolating cytoplasm" around the spindle of the third maturation division has been described. In *Spirostomum ambiguum*, this "isolating cytoplasm" has been shown to be very dense, homogeneous, and strongly basophilic.[228a] In these cases, regional differences in cytoplasmic properties seem to be morphologically expressed.

More complicated is the pattern of nuclear differentiation in *Euplotes patella*, where two nuclei take part in the postmeiotic (third maturation) division.[138] The telophase spindles of the maturation divisions are here of constant length (about 26 µ) and are always oriented parallel to the boundary of the partners. After the preliminary micronuclear division (Fig. 28, *Prl*), the posterior daughter micronucleus migrates forward and comes to lie next to the anterior daughter micronucleus. The first meiotic division gives rise to two groups of nuclei, the anterior and the posterior, each with two nuclei (Fig. 28, *I*). At the second meiotic division, each of the two anterior

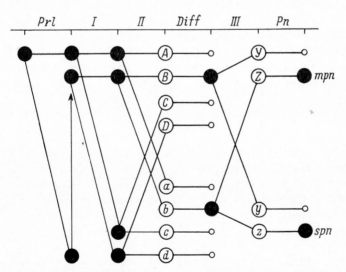

FIG. 28. Diagram of nuclear differentiation during maturation divisions and at the pronuclear stage in *Euplotes patella*. *Prl*—preliminary micronuclear division; *I, II*—first and second meiotic divisions; *A* and *a*, *B* and *b*, *C* and *c*, *D* and *d*—four pairs of sister derivatives of the second meiotic division; *Diff*—nuclear differentiation preceding the postmeiotic maturation division; *III*—postmeiotic division; *Y* and *y*, *Z* and *z*—pairs of sister derivatives of the postmeiotic division; *Pn*—differentiation of the migratory (*mpn*) and the stationary (*spn*) pronuclei. Small circles are degenerating nuclei. After Katashima.[138]

nuclei produces an anterior and a posterior nucleus, and each of the two posterior nuclei also produces one posterior nucleus and one anterior nucleus. This results in four nuclei in the anterior group and four nuclei in the posterior group (Fig. 28, *II*). Then comes nuclear differentiation: in each group, only one nucleus undergoes the postmeiotic division (Fig. 28, *Diff*). Let us designate the nuclei of the anterior group by *A*, *B*, *C*, *D*, and the nuclei of the posterior group by *a*, *b*, *c*, *d*. Usually (in 74–78 per cent of cases) it is the sister nuclei *B* and *b* which take part in the postmeiotic division. Less frequently, the nuclei *A* and *a*, or *C* and *c* divide, and almost never *D* and *d*. Sister nuclei divide in 89–96 per cent of cases, while in the remaining cases non-sister nuclei divide (e.g. *A* and *b*, or *A* and *c*, or *B* and *a*, or *B* and *c*, etc.). In 58 per cent of cases, the same pair of sister nuclei survives in both partners.

Katashima relates this pattern to the fact that the conjugants of *Euplotes* hold together at two points of union, the anterior and the posterior. The initial localization of the micronucleus and the constancy of the length of the telophase spindles assure that the nucleus *B* is usually nearest to the anterior point of union, while its sister nucleus *b* is nearest to the posterior point of union. These two nuclei usually undergo the postmeiotic division.

Centrifugation of the conjugants of *Euplotes* may displace the nuclei, and then all displaced nuclei degenerate. The union at the posterior point sometimes breaks during centrifugation. In this case all the nuclei of the corresponding (posterior) group also degenerate. Thus, initiation of the postmeiotic division in *Euplotes* seems to be induced by the cytoplasm adjacent to the two points of union of the conjugants.

E. Individual and Strain Variability of the Pattern of Maturation Divisions

Two categories of deviations from the normal, species-specific pattern of conjugation may be distinguished. In the first category, the deviations are peculiar to certain individuals among a homogeneous conjugating population (individual variability). In the second category, the deviations characterize all the individuals of certain strains or races and are inherited in these strains (strain variability).

Individual variations of the pattern of the maturation divisions are relatively rare, much rarer than variations in the exconjugation period. The following types of spontaneous individual deviations from the normal course of maturation divisions are now known:

1. Loss of synchrony of the maturation divisions in the two mates—e.g. in Entodiniomorphida [76] or *Bursaria*.[199]

2. Loss of division synchrony of the two spindles of the second maturation division in a single conjugant (or autogamont)—e.g. during induced autogamy in *Paramecium putrinum*.[126]

3. Degeneration of one of the products of the first maturation division in species where it normally survives—e.g. in *Paramecium caudatum*,[59] *P. putrinum*[61] (Fig. 44, *a*, *c*, *e*, right conjugant).

4. Non-degeneration of one of the products of the first maturation division in species where it normally degenerates—e.g. in *Paramecium bursaria*.[99, 36]

5. Degeneration of all the derivatives of the second maturation division; no pronuclei are formed—e.g. in *Bursaria truncatella*,[199] *Paramecium bursaria*,[99] *P. putrinum*,[126] or *P. caudatum*.[232a]

6. Non-degeneration of some derivatives of the second maturation division leading to an increased number of third maturation division spindles—e.g. during induced autogamy in *Paramecium putrinum*[126] or *P. caudatum*.[232a]

7. Total disappearance of the third maturation division; the pronuclei are products of the second divisions—e.g. in *Paramecium putrinum*[61] (Fig. 44, *c*, *d*).

Much more interesting is the problem of strain variability of the pattern of maturation divisions. This type of variability has been most extensively studied by Jankowski[120, 122, 125] in *Paramecium putrinum*.

In many strains of *P. putrinum*, mixing of complementary mating types is followed by normal conjugation, including three maturation divisions, exchange of migratory pronuclei, karyogamy, and three synkaryon divisions (Figs. 29; 32, *a*). Jankowski[120] calls this type of nuclear behavior "amphimictic" conjugation. But in other strains, conjugation *always* takes another course. For example, in the strain T-2 (Figs. 30; 32, *b*), after the normal first and second maturation divisions (Fig. 30, *a–c*), two nuclei get into the mouth region, and the other two nuclei into the posterior body end (Fig. 30, *d*). The posterior nuclei degenerate and the anterior ones fuse with each other to form a synkaryon (Fig. 30, *e*). There is no third maturation division and no exchange of pronuclei (Fig. 32, *b*). The synkaryon undergoes the usual three divisions (Fig. 30, *f–i*). Thus the strain T-2 shows a special inherited pattern of autogamy in pairs, with pronuclei differentiating directly from derivatives of the second maturation division.†[120]

A third pattern of conjugation is characteristic of the strains CR and QS-3.[122] The first micronuclear division (Fig. 31, *a*) is devoid of meiotic prophase and seems to be equational and homologous not to the first but to the second maturation division during amphimixis (Fig. 32, *c*). The entire process of meiosis seems to be lost together with the first maturation division. No chromosome reduction could be observed. One of the products of the first micronuclear division gets into the posterior body end and degenerates (Fig. 31, *b*), and the other product divides three times in succession

---

† It is interesting that the third maturation division is retained in singles during autogamy, which may be induced in amphimictic strains.[126]

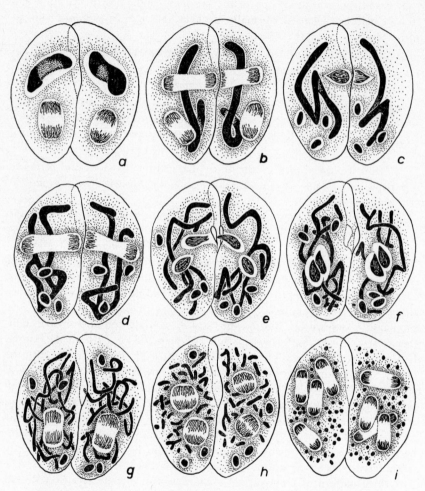

FIG. 29. Amphimictic conjugation in *Paramecium putrinum*: *a*—first, *b*—second maturation division; *c*—degeneration of three haploid nuclei; *d*—third maturation division; *e*—exchange of migratory pronuclei; *f*—pronuclear fusion; *g*—first, *h*—second, *i*—third synkaryon division. Feulgen reaction, ×600. After Jankowski.[120]

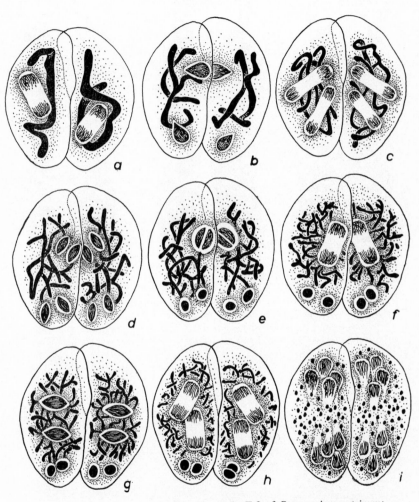

FIG. 30. Autogamy at conjugation in strain T-2 of *Paramecium putrinum*:
a—first maturation division; *b*—its derivatives; *c*—second maturation
division; *d*—its four derivatives; *e*—fusion of two anterior derivatives of
the second maturation division into a synkaryon and degeneration of the
other two haploid nuclei; *f*—first synkaryon division; *g*—its derivatives;
h—second, *i*—third synkaryon division. Feulgen reaction, ×600. After
Jankowski.[120]

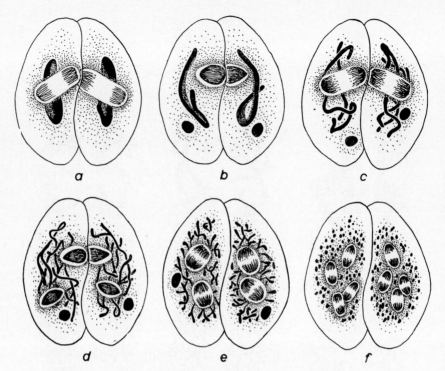

FIG. 31. Apomictic conjugation in strain CR of *Paramecium putrinum*: *a*—first micronuclear division; *b*—pycnosis of its posterior derivative; *c*—second micronuclear division; *d*—its products; *e*—third, *f*—fourth micronuclear division. Feulgen reaction, ×600. After Jankowski.[122]

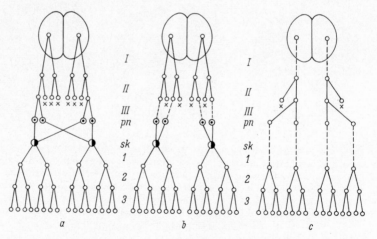

FIG. 32. Comparison of the patterns of amphimixis (*a*), autogamy (*b*), and apomixis (*c*) in *Paramecium putrinum*. *I, II, III*—the respective maturation divisions; *pn*—pronuclei; *sk*—synkarya; *1, 2, 3*—the synkaryon divisions. After Jankowski.[120] modified.

(Fig. 31, *c–f*). There are no homologues of pronuclei and no stages of nuclear exchange between the mates. Of the eight nuclei formed after the last (fourth) micronuclear division, four become macronuclear anlagen, one a micronucleus, and three degenerate, as in normal amphimictic exconjugants. Jankowski believes that one of the last three micronuclear divisions is homologous to the third maturation division, and the other two to the second and third synkaryon divisions (Fig. 32, *c*). Thus meiosis and the entire haploid phase, the first maturation division, the stages of pronuclei and of the synkaryon, and the first synkaryon division seem to be lost in this type of conjugation; conjugation has become transformed into a secondarily agamic process of the type of diploid (ameiotic) parthenogenesis. Jankowski prefers to call this process *apomixis*.† It might also be called "double endomixis" or "endomixis in pairs", if confusion were not associated with the term "endomixis" (see p. 191).

Jankowski proposed the term *mixotypes* for the various patterns of nuclear behavior during conjugation, inherited in certain strains or groups of clones. The system of mixotypes thus seems to be another form of intraspecies differentiation in *P. putrinum*, independent of the system of syngens and mating types.[122] The mixotype remains constant not only during agamic reproduction, but also after conjugation.

A peculiar "abbreviated" type of conjugation, inherited in a group of clones, has been described in *P. putrinum* (= *trichium*) by Diller.[63, 68] The micronucleus undergoes a single "maturation" division. It is not completely clear whether or not it is accompanied by chromosome reduction. Unlike the apomixis of Jankowski, both derivatives of this division remain functional and behave like pronuclei. Exchange of these nuclei may occur in some pairs while not occurring in other pairs. In both cases, the nuclei can fuse into a "synkaryon" or develop "parthenogenetically". After two or three further nuclear divisions (which are probably homologous to the synkaryon divisions), differentiation of macronuclear anlagen occurs.

The transition from one mixotype to another mixotype probably occurs in *P. putrinum* by mutation. Jankowski[125] found that isolated amphimictic pairs may suddenly appear among a mass of apomictic conjugants. All the progeny of such pairs underwent only amphimixis, i.e. showed an irreversible change of the pattern of nuclear reorganization. In comparison with amphimixis, apomixis is beyond doubt a secondarily modified method of nuclear reorganization, and Jankowski considers this spontaneous return to amphimixis a reverse mutation.

Apomictic conjugation without meiosis and without third maturation division has been described also by Suzuki[250] in a race of *Blepharisma japonicum*, under the name of "parthenogenetic" conjugation.

Meiosis and maturation divisions may not only disappear from conjuga-

† Another special case of diploid apomixis is, for example, apogamy of higher plants.

tion but may also disappear from the autogamy in singles. This would
result in an apomictic reorganization process in single animals, which would
well correspond to the initial sense of the term "endomixis" (see p. 191).
Such a process has been recently described by Jankowski[124] in *Cyclidium
glaucoma* (Fig. 33). The micronucleus divides three times in succession
(Fig. 33, *a, b, c*) and gives rise to eight identical micronuclei (*d*). No signs
of meiosis and karyogamy have been discovered. These three micronuclear
divisions seem to correspond to the synkaryon divisions, the maturation
divisions being totally lost. The old macronucleus and one of the eight micronuclei undergo pycnosis (Fig. 33, *e*); of the seven remaining nuclei, one is
the new micronucleus, and six become macronuclear anlagen (*f*). The latter
segregate during three divisions of the "apogamont" (*g, h*).

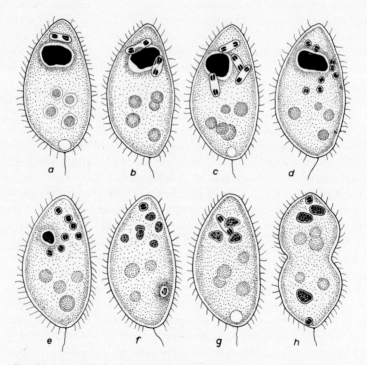

FIG. 33. Apomixis in singles of *Cyclidium glaucoma*: *a*—first, *b*—second,
c—third micronuclear division; *d*—eight derivatives of the third division;
e—pycnosis of the old macronucleus, one of the micronuclei has been
resorbed; *f*—differentiation of six macronuclear anlagen and of a single
micronucleus; *g*—offspring of the first division of the apomictic animal
(three macronuclear anlagen present); *h*—second division of the apomictic
animal showing segregation of the macronuclear anlagen. Alum hematoxylin, × 1600. After Jankowski.[124]

2. Pronuclei and Synkaryon

A. MORPHOLOGY OF THE PRONUCLEI

In many ciliates, the stationary and the migratory pronuclei are morphologically alike (see Table 1, p. 178). The pronuclei of *Chilodonella uncinata* (Fig. 9, *d*), *Collinia branchiarum* (Fig. 11, *g*), or *Dendrocometes paradoxus* (Fig. 13, *d*) may be cited as examples. Such pronuclei are usually spherical and remain in interphase during migration. In *Nassula ornata* (Fig. 34) and *Bursaria truncatella*,[199] both pronuclei of one conjugant and both pronuclei of the other conjugant lie inside a common zone of dense cytoplasm which connects the partners (Fig. 34, *c*); it is even difficult to decide which of the pronuclei are migratory, and which stationary. The four pronuclei seem to approach each other in a pairwise manner and fuse to form two synkarya inside the common intermediate zone of dense cytoplasm (Fig. 34, *d*). In most ciliates, however, the migratory pronucleus traverses a considerable path while the stationary pronucleus remains in place.

In many cases, the migratory pronucleus morphologically differs from the stationary pronucleus, but the difference is not sharp. The difference may lie in the shape of the pronuclei, their size, and the presence of specialized cytoplasmic zones around them.

In many species of *Paramecium*, the migratory pronucleus is somewhat thinner and longer than the stationary pronucleus, although both are spindle-shaped and contain longitudinal threadlike chromosomes (Figs. 5, *f*, *g*; 29, *e*; 40). This is true of *Paramecium* species having large micronuclei— *P. caudatum*,[19] *P. bursaria*,[25] *P. putrinum*.[120] It is also true of *Cryptochilum echini* (Fig. 7, *h*) and *Ptychostomum* (= *Lada*) *tanishi*.[170] In *Dogielella sphaerii*, the stationary pronucleus is spherical and the migratory lens-shaped, pressed to the membrane separating the conjugants.[198]

When pronuclei differ by their size, it is always the stationary pronucleus which is larger (*Prorodon griseus*,[251] *Paramecium multimicronucleatum*,[159] *Entorhipidium echini*,[278] *Fabrea salina* [84]).

A specialized cytoplasmic zone usually develops around the migratory pronucleus, but in some cases such zones may develop around both pronuclei. Prandtl[204] found that cytoplasmic "rays" appear around both poles of the telophase spindle of the third maturation division in *Didinium nasutum* (Fig. 35, *a*). Later on, the cytoplasmic zones of the two pronuclei become different: the "rays" become longer and sparser around the stationary pronucleus, and shorter and denser around the migratory pronucleus (Fig. 35, *b*, *c*). The migratory pronucleus is also smaller than the stationary, and has a protrusion at the place where the telophase spindle was attached (Fig. 35, *c*). The migratory pronucleus loses its "rays" at the moment of passage into the partner (*d*), so that only the stationary pronucleus has them during pronuclear fusion (*e*, *f*).

A zone of dense, finely granular cytoplasm (the so-called "attraction

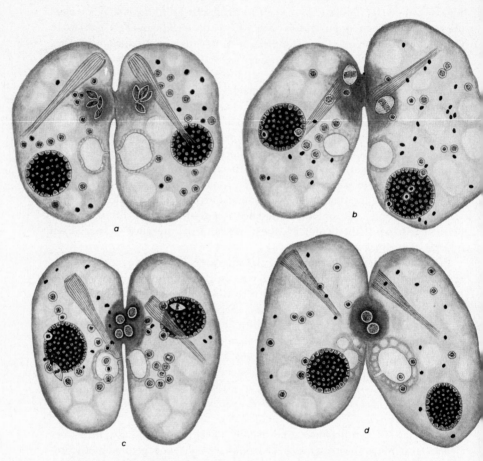

FIG. 34. Conjugation in *Nassula ornata*: *a*—four derivatives of the second maturation division inside a zone of dense cytoplasm; *b*—a single spindle of the third maturation division in each partner; *c*—two pronuclei of each conjugant in a common zone of homogeneous cytoplasm; *d*—two synkarya in the common cytoplasmic zone. All figures show also the old macronucleus, the degenerating micronuclei (derivatives of the first and the second maturation divisions), and the pharyngeal basket. Reconstructions after sections. Feulgen reaction. ×315, Original.

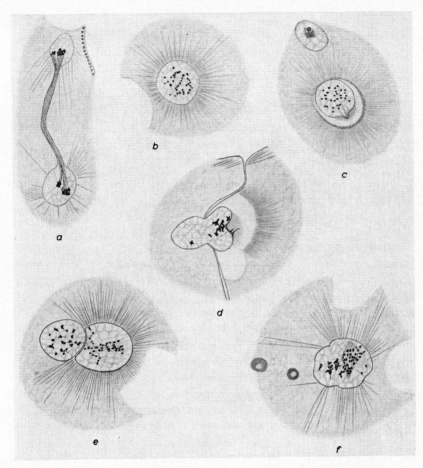

FIG. 35. Pronuclei of *Didinium nasutum*: *a*—telophase of the third maturation division (the future migratory pronucleus at top); *b*—stationary pronucleus; *c*—migratory pronucleus (a degenerating micronucleus at top);
d—passage of the migratory pronucleus into the other partner (from right
to left); *e*—pronuclei approaching each other (the migratory at left);
f—fusion of pronuclei. After Prandtl.[204]

sphere") forms near the migratory pronucleus in *Uroleptus mobilis*.[15]
The "attraction sphere" moves like a pseudopodium in front of the migratory pronucleus (Fig. 36). The stationary pronucleus, otherwise identical
with the migratory, has no such zone. The "attraction sphere" may possibly
play some role in the movement of the migratory pronucleus, but this needs
further investigation. Similar "pseudopodia" moving in front of the migratory pronuclei are known also in *Stylonychia*, *Urostyla*,[172] and *Euplotes*.[165] A zone of dense cytoplasm surrounds the migratory pronucleus in

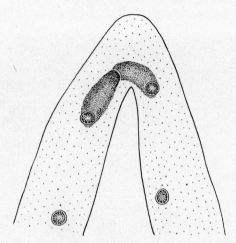

FIG. 36. Pronuclei of *Uroleptus mobilis.* "Attraction spheres" move in
front of the migratory pronuclei, ×1000. After Calkins.[15]

Stentor coeruleus, but it looks like a uniform halo, not like a pseudopodium.[175]

Sharp differences between the stationary and the migratory pronuclei
exist in Entodiniomorphida,[76] e.g. in *Cycloposthium bipalmatum* (Fig. 37).
The spindle of the last maturation division is markedly heteropolar in telophase: the future migratory pronucleus has a pointed "perforatorium"
directed against the bottom of the peristomal cavity (Fig. 37, *a*). A zone of
cytoplasmic "rays" develops later around the stationary pronucleus, and
the migratory pronucleus penetrates the peristomal pellicle and comes out
of the ciliate body into the peristomal cavity (Fig. 37, *b*). A long cytoplasmic
"tail" forms at the posterior end of the migratory pronucleus. This "tail"
seems to be a residue of the telophase spindle of the last maturation division
enclosed by the nuclear envelope (Fig. 37, *b, c*). As a result, the migratory
pronucleus surprisingly resembles a spermatozoon, both by its shape and
by the fact that it leaves the conjugant's cytoplasm and makes its way
through a section of the external medium confined inside the joined peristomal cavities of the partners (Fig. 37, *c*). The migratory pronucleus enters
the gullet of the partner (*c*) and passes through it into the partner's cytoplasm (Fig. 37, *d*), where it fuses with the stationary pronucleus.

Spermatozoon-like migratory pronuclei exist not only in advanced
entodiniomorphids like *Cycloposthium, Diplodinium,* and *Opisthotrichum,*[76]
but also in the lower representatives of this taxon, e.g. in *Entodinium caudatum.*[201]

The resemblance of the migratory pronucleus of the Entodiniomorphida
to a spermatozoon is of course pure convergence. The pronuclear "tail" is
not of flagellar origin and thus not homologous to a spermatozoon tail;

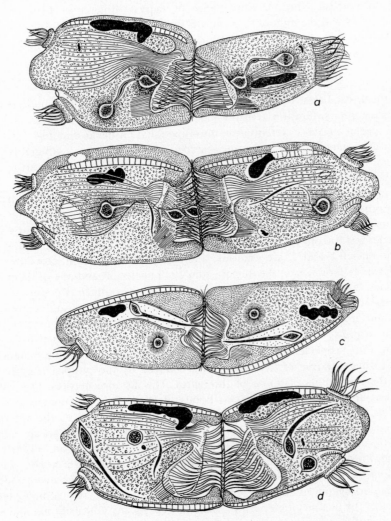

FIG. 37. Conjugation of *Cycloposthium bipalmatum*: *a*—telophases of the last maturation division in both partners; *b*—migratory pronuclei are in the free space between the conjugants, stationary pronuclei are in the endoplasm; *c*—migratory pronuclei penetrate into the partner's gullet, stationary pronuclei show cytoplasmic "rays"; *d*—migratory pronuclei have entered the partner's endoplasm. In *a*, *b* (right mate), and *d*, degenerating derivatives of the first maturation division are seen. In *b* and *d*, remnants of the telophase spindle of the last maturation division are visible. Iron hematoxylin. After Dogiel.[76]

also, it is unknown whether the "tail" plays any role in the movement of the pronucleus. But the appearance of marked differences between the stationary and the migratory pronuclei must nevertheless be considered as a manifestation of sexual differentiation at the level of pronuclei, which are homologous to the gamete nuclei in other Protozoa (see p. 274).

B. MECHANISM OF DIFFERENTIATION OF THE PRONUCLEI

In most ciliates, both pronuclei are derivatives of a single spindle of the postmeiotic maturation division and are thus sister nuclei. In these ciliates, no special nuclear differentiation into functional pronuclei and degenerating nuclei occurs after the postmeiotic division. But such a differentiation exists in species where more than one derivative of meiosis takes part in the postmeiotic maturation division.

In *Euplotes patella*,[138] two nuclei undergo the postmeiotic (third maturation) division, one nucleus from the anterior group of nuclei and one from the posterior group (see p. 207 and Fig. 28, *Diff*). Two pairs of sister nuclei result from the postmeiotic division (Fig. 28, *III*, *Y* and *y*, *Z* and *z*). The nuclei *Y* and *Z* are anterior, and one of them differentiates into the migratory pronucleus; the nuclei *y* and *z* are posterior, and one of them becomes the stationary pronucleus (Fig. 28, *Pn*). The other two nuclei, one anterior and one posterior, degenerate. The sister nuclei *Y* and *y* become pronuclei in 41 per cent of conjugants, the sister nuclei *Z* and *z*, in 50 per cent (Fig. 28, *Pn*). Pairs of non-sister nuclei (*Y* and *z*, or *Z* and *y*) differentiate into pronuclei in 9 per cent of conjugants only. Katashima maintains that differentiation of the pronuclei is induced by the cytoplasm adjacent to the two points of union of the mates. The distance between two sister nuclei and that between the two points of union is approximately equal. Thus, if the nucleus *Y*, of the anterior group, gets nearest to the anterior point of union, its sister nucleus, *y*, belonging to the posterior group, automatically becomes the nearest to the posterior point of union. Therefore sister nuclei more frequently become pronuclei than non-sister nuclei.

Exactly the reverse situation exists, however, in *Euplotes minuta* undergoing autogamy in singles.[181a] As usually in *Euplotes*, two spindles take part in the postmeiotic division. Since heterozygosity is usually maintained after autogamy in this species (in contrast to *Paramecium aurelia*, see p. 191), the synkaryon formation seems to be brought about by preferential fusion of genetically dissimilar (i.e. non-sister) derivatives of the postmeiotic division.

Another aspect of pronuclear differentiation may be expressed by the question: what conditions determine which of the two pronuclei will become migratory (and in some cases will acquire the corresponding morphological peculiarities), and which will become stationary? Regional cytoplasmic differences seem to play the main role also in this case. It was noted by Maupas[165] that the spindle of the third (postmeiotic) maturation division is usually at an angle to the plane of contact of the partners. As a result, one of the poles of this spindle gets into immediate proximity to the surface of contact of the mates, while the other pole remains deeper in the cytoplasm. The former pole always forms the migratory pronucleus, the latter pole, the stationary pronucleus (*Paramecium*—Figs. 5, *e*; 29, *d*, *e*; *Cryptochilum*—Fig. 7, *g*; *Cycloposthium*—Fig. 37, *a*). The cytoplasm of the contact zone possibly induces differentiation into the migratory pronucleus. However, the nature of this induction is obscure.

It is difficult to understand the mechanism of differentiation of pronuclei into migratory and stationary in cases where the spindle of the postmeiotic division is parallel to the plane of contact of the conjugants, both derivatives of this spindle being in apparently identical situations. This is the case in the suctorians *Cyclophrya katharinae*[155] and *Dendrocometes paradoxus* (Fig. 13, *c*, *d*).

C. EXCHANGE OF THE MIGRATORY PRONUCLEI

The mechanism of passage of the migratory pronuclei into the partner is one of the most difficult questions in the cytology of conjugation. Light microscopical data relative to this question are vague and contradictory, and electron microscopical data are still very scanty. Nevertheless, three methods of passage of the migratory pronuclei into the partner can be distinguished.

1. In most cases, there is no direct continuity of the cytoplasm (no "cytoplasmic bridge") between the mates, at least at the light microscopical level. The pronuclei passing into the other mate "break through" (or "locally dissolve") the apposed pellicles of the joined conjugants.

The simplest modification of this method occurs when, at the moment of pronuclear interchange, the migratory pronuclei remain connected by telophase spindles with their respective sister nuclei (the stationary pronuclei). This is the case in *Cryptochilum echini*[43] (Fig. 7, *g*), in several species of *Chilodonella*[86, 160-1] (Fig. 9, *c*), and in *Prorodon griseus*.[251] A "pushing" of the migratory pronuclei into the partners by the elongating spindles of the third maturation division may be supposed to occur in these species.

Much more often, however, the migratory pronuclei cross the boundary of the conjugants after the disappearance of the spindles of the third maturation division. This looks like an "active penetration" into the partner; of course, this expression only screens our complete ignorance of the forces which might ensure the movement of the pronuclei. In many *Paramecium* species, the migratory pronuclei lie inside cytoplasmic protrusions, the paroral cones;† from there they penetrate the apposed pellicles into the partner's cytoplasm (Fig. 5, *f*). A "cytoplasmic bridge" was formerly believed

† However, Vivier and André[259] consider the paroral cones an optical illusion.

to appear between the conjugants before the pronuclear exchange,[106, 19] but more recent authors deny its existence. [58, 67, 191, 25, 176, 171].

Vivier and André[3, 259, 260] confirmed electron microscopically that no broad "cytoplasmic bridge" is formed in *Paramecium caudatum*. However, many small cytoplasmic channels, about 0.25 μ in diameter, were found to form between the partners as a result of the dissolution of the pellicular membranes at the tops of the interciliary knobs (Fig. 38, *a*). These channels are scattered on a surface measuring about 80 μ in length. Later on, the line of contact of the conjugants becomes very sinuous at the place where the pronuclear interchange is to occur (Fig. 38, *b*). On electron micrographs, the migratory pronuclei show an extremely irregular amoeboid form, their envelope forming numerous protrusions and invaginations (Fig. 38, *b*). According to Vivier and André, the migratory pronucleus passes into the other partner through one of the small cytoplasmic channels. Since the channels are far narrower than the pronuclei, the latter have to undergo strong deformation at the moment of passage—so to say, squeeze through a narrow opening.

The exchange of the migratory pronuclei in *P. multimicronucleatum* proceeds in a similar way.[116] The migratory pronuclei, which are about 8 μ in diameter, squeeze through channels not wider than 1 μ, becoming deformed, though not so strongly as in *P. caudatum*. Vivier and André as well as Inaba *et al.* believe that pronuclear migration is carried out in *Paramecium* by amoeboid movement of the pronuclei involving formation of "pseudopodia".

Electron microscopy of conjugating *Paramecium aurelia*[224] also reveals many narrow cytoplasmic channels which connect the two conjugants and are scattered throughout their contact zone (Fig. 39, *a*). But in *P. aurelia* a large "cytoplasmic bridge", about 10 μ in width, also forms in the paroral region of the mates before pronuclear exchange (Fig. 39, *b*). The migratory pronuclei, which pass through this "bridge", retain their regular outlines (Fig. 39, *b*).

Small cytoplasmic channels connecting the two partners were observed electron microscopically also in *Tetrahymena pyriformis* by Elliott and Tre-

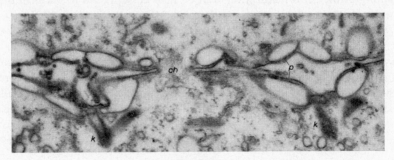

FIG. 38a

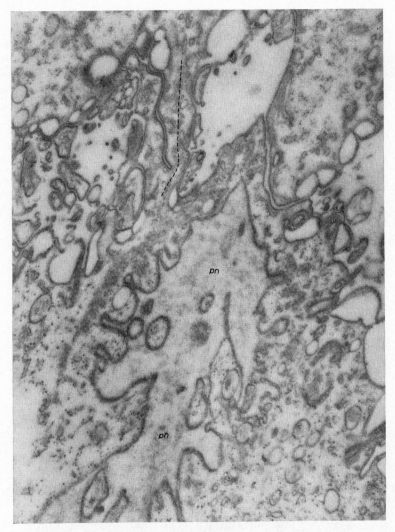

FIG. 38b

FIG. 38. Ultrastructure of the contact zone of conjugating *Paramecium caudatum*: a—surface structures in the contact zone (*k*—kinetosomes, *p*—apposed pellicular membranes of the partners, *ch*—cytoplasmic channels), ×20,000; b—migratory pronucleus (*pn*) before passage into the other mate (convoluted line of contact of the conjugants seen at top; broken line indicates a cytoplasmic channel between the conjugants), ×24,000. After André and Vivier.[3]

224   I. B. RAIKOV

mor.[83] The migratory pronuclei are much larger than the channels and have a regular shape. Their passage through the channels is unlikely. It has been shown that migratory pronuclei protrude the apposed pellicles of the mates, but the moment of their penetration into the other partner was not observed.

Pronuclei "breaking through" the apposed pellicles of the mates or passing through a narrow opening in the pellicles have been observed light microscopically also in *Stentor coeruleus*,[175] *Loxocephalus* sp., [7] *Dogielella sphaerii*,[198] *Euplotes eurystomus*,[253] *Cyclophrya katharinae*,[155] and some other ciliates. The migratory pronucleus of *Didinium nasutum* loses its halo of cytoplasmic "rays" when it squeezes through an opening in the pellicles of the partners[204] (Fig. 35, *d*).

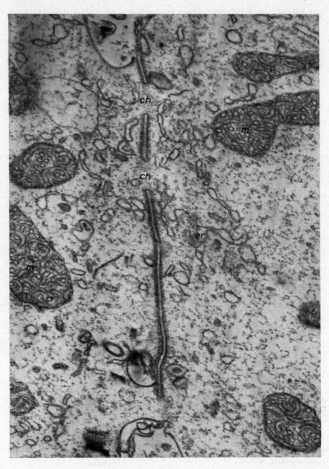

FIG. 39a

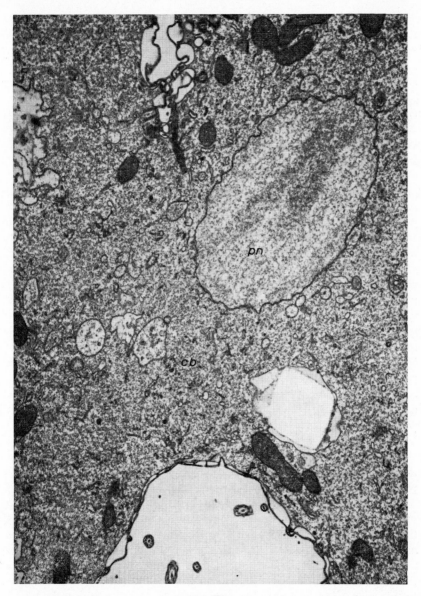

FIG. 39b

FIG. 39. Ultrastructure of the contact zone of conjugating *Paramecium aurelia*: *a*—small cytoplasmic channels (*ch*) connecting the partners, ×39,700; *b*—broad cytoplasmic bridge (*cb*) at the moment of passage of a migratory pronucleus (*pn*), ×10,000. *er*—endoplasmic reticulum, *m*—mitochondria. After Schneider.[224]

2. Another method of exchange of the migratory pronuclei is their passage through a broad cytoplasmic bridge connecting the partners. The pronuclei encounter no visible obstacles in this case. This situation exists in *Uroleptus mobilis* (Fig. 36), *Stylonychia mytilus*,[172] *Kahlia* sp.,[215] *Euplotes woodruffi*.[214] The formation of a broad cytoplasmic bridge between the conjugants has been recently confirmed electron microscopically in *Euplotes vannus* by Nobili.[181] However, in this case the bridge remains barred by two sheets of tubular microfibrils, which dissolve only just before passage of the pronuclei.

The same method of pronuclear migration is characteristic of all cases of total conjugation, e.g. in *Urostylaho logama*,[105] *U. polymicronucleata*,[172] and *Metopus sigmoides*[184] (Fig. 14, *i*). The formation of a broad cytoplasmic bridge seems to be an important prerequisite of transformation of temporary conjugation into total conjugation. A significant mass of cytoplasm of the "donor" conjugant, carrying both its pronuclei, can flow over into the "recipient" conjugant only when such a bridge is present (Fig. 14, *i*).

3. The third method of transfer of the migratory pronucleus into the partner is its passage through a preformed opening in the conjugant's body, usually through the cytostome. Here belongs the above-described penetration of spermatozoon-like migratory pronuclei through the peristomal cavity into the partner's gullet in Entodiniomorphida (Fig. 37). An exchange of migratory pronuclei through the cytostomes of the mates has been reported by Jankowski[120] in *Paramecium putrinum* (Fig. 29, *e*). According to Maupas,[165] the migratory pronucleus of *Euplotes eurystomus*, supplied with a cytoplasmic "pseudopodium", comes out of the conjugant's body through a "rupture" of the pellicle, "creeps" a little along the outer surface of the conjugant's body, and enters the newly formed cytostome of the partner. However, Turner[253] failed to confirm these observations; according to him, the migratory pronucleus of *E. eurystomus* penetrates into the partner through the wall of the old peristomal field and moves towards the stationary pronucleus in the partner's cytoplasm.

Until 1936 it was generally accepted that exchange of the migratory pronuclei *always* occurs during conjugation. However, Diller[58] showed that exchange of pronuclei may be lacking in conjugating *Paramecium aurelia*, the synkaryon being formed by autogamy in both partners. Wichterman[267] maintained that such a double autogamy ("cytogamy") was the only method of nuclear behavior in conjugating *Paramecium caudatum*. In this connection, it is useful to examine the evidence confirming the regular existence of pronuclear exchange in conjugating ciliates. This evidence was collected in the most complete form by Chen[24, 25, 29, 30, 32, 33, 35, 36] for *Paramecium bursaria*.

The first evidence is the direct observation of the migratory pronuclei at the moment of their passage into the partner. This was done by Chen on stained preparations of *P. bursaria* (Fig. 40), and by Diller[65] on those of

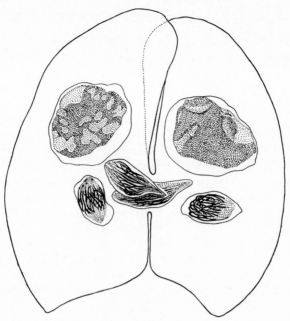

FIG. 40. Exchange of migratory pronuclei in *Paramecium bursaria*, ×1100. After Chen.[33]

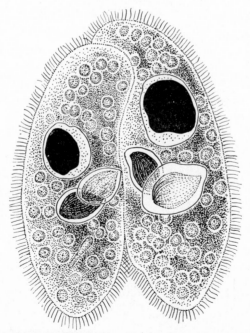

FIG. 41. Conjugation between races Gr 14 and S of *Paramecium bursaria* (stage preceding pronuclear fusion). Heteroploidy of pronuclei indicates that pronuclear exchange already took place. ×820. After Chen.[25]

P. caudatum. The exchange of migratory pronuclei has also been observed in living conjugants of *P. bursaria* by Wichterman[268-9,272], and in living pairs of *Tetrahymena pyriformis* by Ray[217].

The second evidence is that various clones of *P. bursaria* have micronuclei of unequal degree of ploidy.[24-26, 28-30, 32, 33, 35, 36] The pronuclei formed by a polyploid clone are larger and more heavily stained than pronuclei formed by a diploid (or less polyploid?) clone. Therefore, when two clones with different micronuclear ploidy degrees are crossed, the pronuclei produced by each one of the two conjugants may be easily identified. In most cases, the synkarya form in both partners by fusion of one heavily stained and one pale pronucleus (Fig. 41). This situation can obviously be brought about only by reciprocal exchange of the migratory pronuclei. Similar observations were made by Diller[61] on *Paramecium putrinum* (= *P. trichium*).

The third evidence comes from observations of unilateral pronuclear migration in *P. bursaria*.[24, 32, 33, 35, 36, 270] The transfer of a migratory pronucleus in one of the two directions is sometimes delayed, resulting in three pronuclei in one mate, and only one pronucleus in the other mate

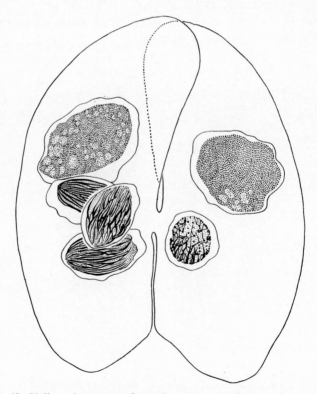

FIG. 42. Unilateral passage of a migratory pronucleus in *Paramecium bursaria*, ×1100. After Chen.[33]

(Fig. 42). The same phenomenon occurs also in *P. caudatum*.[191] The migration of one pronucleus may be not only delayed but totally omitted in some pairs. In any case, the presence of three pronuclei in one of the mates can be brought about only by migration of a pronucleus from the other mate.

The fourth evidence is given by conjugation of normal animals with amicronucleate animals in *P. bursaria*.[25, 29, 30] The migratory pronucleus of the normal partner passes into the amicronucleate partner (Fig. 59), resulting in each conjugant containing a single pronucleus. Similar observations were made by Diller[69] in *P. multimicronucleatum*.

Finally, the fifth evidence is of genetic nature. If no exchange of migratory pronuclei occurred during conjugation of two genetically different homozygous animals, both exconjugants would retain their old (parent) genotypes. Sonneborn[236, 241] showed that such a behavior is rare in *P. aurelia*. In most cases, both exconjugants become heterozygous and genetically similar to each other; this could follow only the exchange of the migratory pronuclei.

Double autogamy ("cytogamy") occurs but is usually facultative and infrequent. In *Paramecium aurelia* it occurs rather as an exception, at least at room temperature.[236, 238, 241-2]. At higher (27°C) and lower (10°C) temperatures, the proportion of pairs undergoing double autogamy is considerably higher (up to 47–60 per cent).

Double autogamy is rare also in *Paramecium bursaria*. It may be detected in crosses between heteroploid clones by the fusion of two heavily stained

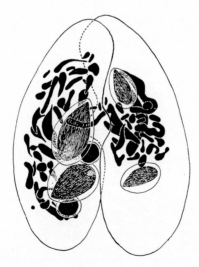

FIG. 43. Conjugation in *Paramecium putrinum*. Two products of the first synkaryon division in each partner. Heteroploidy of nuclei indicates autogamous origin of both sets of nuclei. Carmine–indulin staining, ×1080. After Diller.[61]

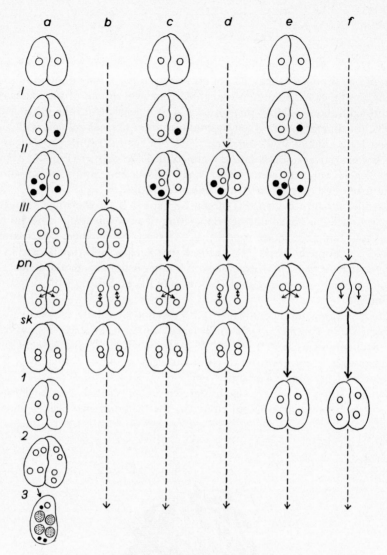

Fig. 44. Diagram illustrating variability of the nuclear behavior during conjugation in *Paramecium putrinum*. Light circles represent functional nuclei; dark circles, degenerating nuclei. Dotted circles are macronuclear anlagen. Broken lines indicate a nuclear behavior similar to that of the column to the left of it; solid lines mean that a respective stage has been omitted.

Column *a*—reciprocal conjugation with three maturation divisions; *b*—autogamy with three maturation divisions; *c*—reciprocal conjugation with two maturation divisions; *d*—autogamy with two maturation divisions; *e, f*—differentiation of only one pronucleus per conjugant following two maturation divisions, and parthenogenetic development of single pronuclei (*e*—with pronuclear exchange, *f*—without exchange).

I, II, III—the respective maturation divisions; *pn*—stage of pronuclei; *sk*—synkaryon formation; *1, 2, 3*—the respective synkaryon divisions. After Diller.[61]

pronuclei in one mate, and of two pale pronuclei in the other mate.[25, 30, 32, 33, 35, 36] The results of Wichterman,[267] who found autogamy in 100 per cent of the living investigated pairs of *Paramecium caudatum*, probably can be attributed to abnormal conditions in the compression chamber into which the conjugants were placed for observation. Diller[59] observed double autogamy only in a small proportion of conjugating pairs of *P. caudatum*.

Autogamy in pairs appears to be somewhat more frequent in *Paramecium putrinum*.[61] Here, as in *P. bursaria*, it can be detected due to the heteroploidy of the micronuclei (Fig. 43). According to Diller, the process of conjugation is highly variable in *P. putrinum* (Fig. 44). Differentiation of the pronuclei may occur after the third (Fig. 44, *a*, *b*) or after the second maturation division (*c*, *d*). In both cases, the synkaryon may form either after reciprocal exchange of migratory pronuclei (*a*, *c*), or without nuclear exchange, i.e. by autogamy (*b*, *d*). Finally, only one pronucleus may differentiate after the second maturation division (*e*, *f*); in this case also, the partners may sometimes exchange their single pronuclei (*e*), or single pronuclei develop "parthenogenetically" in the mates where they originated (*f*).

It has been stated above (p. 209) that double autogamy (with differentiation of pronuclei after the second maturation division) may be a hereditary trait of some strains of *P. putrinum*[120] (Fig. 30). Double autogamy occurs also in *P. polycaryum*.[67]

Autogamy in pairs seems to be rare outside the genus *Paramecium*. According to genetic data of Kimball,[143] it sometimes occurs in *Euplotes*

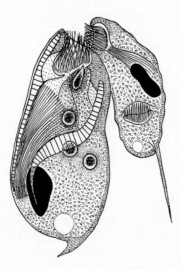

Fig. 45. Abnormal passage of both pronuclei of the microconjugant into the macroconjugant in *Opisthotrichum janus*. Iron hematoxylin. After Dogiel.[76]

patella. Katashima[134] found facultative autogamy in conjugants of *E. harpa*. Wichterman[273] admits that it may sometimes occur in *E. cristatus*. An abnormality of pronuclear behavior, consisting of passage of both the stationary and the migratory pronuclei of one partner into the other partner, may sometimes occur in species which normally show temporary conjugation. This has been observed by Dogiel[76] in *Opisthotrichum janus* (Fig. 45), and by Chen[25] in *Paramecium bursaria*. The fate of the conjugant deprived of pronuclei is unknown; probably it degenerates. Evolutionary fixation of this abnormality will probably lead to the establishment of total conjugation. Such may be the origin of total conjugation in *Metopus* and *Urostyla* (see pp. 169, 226).

D. FORMATION OF THE SYNKARYON

The synkaryon is formed by fusion of the stationary pronucleus with the migratory pronucleus which came from the other mate (reciprocal conjugation) or originated in the same individual (autogamy in singles or in pairs). In autogamy, the synkaryon usually arises by fusion of two genetically identical haploid sister nuclei and thus becomes homozygous at all loci (Sonneborn[234-5, 241]), but exceptions to this rule also exist[181a] (see also p. 220).

Two morphological types of pronuclear fusion may be distinguished. In the first case, the pronuclei fuse in interphase; they have vesicular structure without visible chromosomes. This type of synkaryon formation occurs, for example, in *Collinia* (Fig. 11, *h, i*) and *Nassula* (Fig. 34, *c, d*). In the second case, the pronuclei either have no interphase stage at all (e.g. species of *Paramecium*) or acquire the form of prophase or metaphase spindles before fusion (*Bursaria*[199]). The spindles of the two pronuclei become apposed parallel or at a small angle to each other and fuse into the common spindle of the synkaryon, which contains a diploid number of chromosomes. This type occurs, for example, in *Paramecium caudatum* (Fig. 5, *g, h*), *P. putrinum* (Fig. 29, *f*), *P. bursaria* (Figs. 41; 57, *b*), *Cryptochilum echini* (Fig. 7, *i*), and *Diplodinium ecaudatum* (Fig. 49, *a, b*).

The distribution of both types of synkaryon formation is indicated in Table 1 (p. 178). However, transitions between the two types also occur. For example, the pronuclei of *Didinium* may fuse either in interphase, or at various phases of spindle development (Fig. 35, *e, f*). In *Chilodonella*, the fusing pronuclei are vesicular but chromosomes (or their aggregates—see p. 203) nevertheless remain condensed (Fig. 9, *d*). In *Metopus*, the pronuclei are also vesicular during fusion, but a single composite chromosome (see p. 203) is clearly seen in each pronucleus (Fig. 14, *j–l*).

During autogamy in singles, the synkaryon arises in the same way as during conjugation. It is interesting that even the paroral cytoplasmic cone containing the pronuclei and, later, the synkaryon, is formed during autogamy in singles—e.g. in *Paramecium aurelia* (Fig. 20, *f*, *g*), *P. polycaryum*,[66] *P. putrinum*,[126] and *P. caudatum*.[232a]

More or less frequent abnormalities in the pronuclear behavior may occur during the period of synkaryon formation. One of them is non-fusion of the pronuclei, the so-called *gonomery*. Gonomery has been found, for example, in 8 per cent of conjugating pairs of *Cycloposthium bipalmatum* by Dogiel.[76] After normal pronuclear exchange, the migratory pronucleus approaches the stationary one; however, they do not fuse into the synkaryon but both, independently, undergo a single mitotic division (corresponding to the single synkaryon division in normal exconjugants). In the gonomerous exconjugant, each of the two spindles gives rise to a macronuclear anlage and to a micronucleus (Fig. 46).

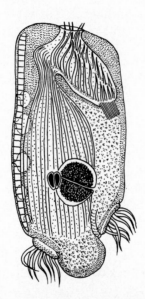

Fɪɢ. 46. Gonomerous exconjugant of *Cycloposthium bipalmatum*. Iron hematoxylin. After Dogiel.[76]

Cases of non-fusion of pronuclei were recorded also by Chen[36] in conjugation between young and old clones of *Paramecium bursaria*; each pronucleus divided independently. Such pronuclei, capable of developing further without fusion into a synkaryon, are called *hemikarya*.[24, 25, 28, 30] During conjugation between old and young clones, non-fusion of pronuclei was sometimes preceded by unilateral passage of one migratory pronucleus (cf. Fig. 42). Then, the nuclear apparatus of one exconjugant was reconstructed by independent development of three hemikarya, and that of the other exconjugant, by development of a single hemikaryon.[36]

Reconstruction of the nuclear apparatus from a single hemikaryon, following unilateral pronuclear migration, has been observed in one of the two exconjugants of *Paramecium bursaria* by Hamburger.[99] This process was extensively studied by Chen,[24, 28] who showed that the hemikaryon behaves in *P. bursaria* exactly like the synkaryon. However, the ploidy degree of the micronucleus in the progeny of such exconjugants is obviously halved, since no compensation of the meiotic reduction of the chromosome number takes place. Development of a single hemikaryon ("parthenogenesis of a pronucleus") has been found also by Diller[59] in *Paramecium caudatum*. In *Paramecium putrinum* (= *trichium*), the same process may occur in both conjugants in cases when the third maturation division is absent and only one pronucleus differentiates in each partner[61] (Fig. 41, *e*, *f*). Finally, single hemikarya may develop in both partners after conjugation of a normal cell with an amicronucleate cell (see below, p. 258, Fig. 59).†

Another abnormality consists of the fusion of more than two pronuclei into a single synkaryon which obviously becomes polyploid or increases its polyploidy. Fusion of three pronuclei may occur in one of the two partners after unilateral passage of one migratory pronucleus (Fig. 42). This occurs, for example, in *Paramecium bursaria*[24, 28] and *P. putrinum*.[61] In *P. bursaria*, fusion of several (three, four, and even six) pronuclei into a single synkaryon may also follow differentiation of "excess" pronuclei in cases of non-degeneration of some derivatives of the maturation divisions, especially in individuals which initially had two micronuclei.[24, 28]

Up to eight pronuclei may differentiate during autogamy in *P. caudatum*, induced by contact with an amicronucleate individual of the complementary mating type. These pronuclei sometimes fuse pairwise, forming two or three synkarya in a single autogamont. However, all the synkarya, except one, later degenerate. Also more than two pronuclei may fuse to form a single synkaryon, and excess pronuclei may develop as hemikarya.[232a]

E. PRONUCLEI AND SYNKARYON DURING TOTAL CONJUGATION

A single pair of pronuclei and a single synkaryon would be sufficient to reconstruct the nuclear apparatus of the synconjugant after total conjugation. However, the actual diversity of the nuclear phenomena during karyogamy in totally conjugating ciliates is much greater. Some peculiarities of the nuclear behavior seem to be recapitulations of the ancestral temporary conjugation.

Four types of pronuclear behavior during total conjugation, diagrammatically shown in Fig. 47, may be distinguished among the ciliates.[173, 154]

---

† Hoyer[111] and Dehorne[52, 53] maintained that nuclear reconstruction in both conjugants of *Colpidium colpoda* and *Paramecium caudatum* always occurred by development of solely the migratory pronuclei; the stationary pronuclei were said to degenerate. This idea is clearly erroneous; it has been disproved by other authors[44, 45, 207] and abandoned by Dehorne himself in his later paper[54] dealing with both *Colpidium* and *Paramecium*.

In the first type (Fig. 47, a_1 and a_2), each conjugant produces two pronuclei, as in temporary conjugation. After fusion of the cytoplasms of the partners, the four pronuclei unite pairwise forming two synkarya. Both synkarya remain functional, i.e. undergo the synkaryon divisions. Consequently, the synkaryon derivatives differentiating into macronuclei and into micronuclei need not originate from a single synkaryon. This type is rare and still insufficiently studied. It exists in two modifications. The first modification, as yet described only by Tuffrau[252] in *Spirochona gemmipara*, is characterized by external, initially isogamontic conjugation (Fig. 47, a_1). In the second modification, conjugation is anisogamontic, the small microconjugant penetrating entirely inside the macroconjugant (Fig. 47, a_2). This so-called "inner conjugation" has been studied by Kormos and Kormos[153-4] in three suctorian species, *Discophrya collini*, *D. endogama*, and *D. buckei*, and by Canella[19a] in *D. collini* (see Table 2, p. 186). Persistence of two functional synkarya appears to be a clear recapitulation of the ancestral temporary conjugation, when two exconjugants developed.

In the second type (Fig. 47, b), this recapitulation is not so strong. Here, although each conjugant produces two pronuclei and two synkarya are formed in the synconjugant, one of them soon degenerates, only the other

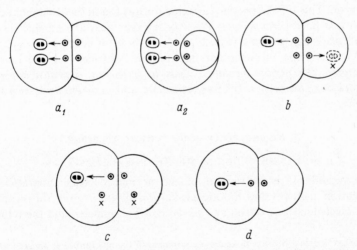

$$a_1 \qquad a_2 \qquad b$$

$$c \qquad d$$

Fig. 47. Types of pronuclear behavior during total conjugation. Circles with a dot are pronuclei, ovals with two dots are synkarya, crosses indicate degenerating nuclei. a—each conjugant produces two pronuclei, and two synkarya form in the synconjugant, both synkarya remaining functional (a_1—usual conjugation, a_2—"inner" conjugation); b—each conjugant forms two pronuclei, and two synkarya arise, but one of them later degenerates; c—each conjugant produces two pronuclei, but only one synkaryon forms, and two "excess" pronuclei degenerate; d—each conjugant produces a single pronucleus, and only one synkaryon forms in the synconjugant. After Kormos and Kormos,[154] modified.

remaining functional, dividing, and giving rise to new macronuclei and micronuclei of the synconjugant. This type exists in *Urostyla hologama*,[105] in the suctorians *Choanophrya infundibulifera* and *Lernaeophrya capitata*.[39,94] Among the peritrichs it is the rule in *Zoothamnium arbuscula*[91] and occurs in exceptional cases in *Vorticella campanula*,[173] but in the main the peritrichs are characterized by the third and fourth types of pronuclear behavior.

In the third type (Fig. 47, *c*), each conjugant still produces two pronuclei (the postmeiotic maturation division still exists). But only one pair of pronuclei fuses to form the single synkaryon of the synconjugant, the other two pronuclei degenerating without undergoing karyogamy. Such is the pronuclear behavior in *Metopus sigmoides* (Fig. 14, *i, j*)[184] and in approximately half the peritrichs, e.g. in *Carchesium polypinum* (Fig. 18, *g, h*), *Vorticella campanula*, *V. cucullus*, *V. monilata*, *V. nebulifera*, *Opercularia coarctata*, *Ophrydium versatile*, *Lagenophrys tattersalli*, *Urceolaria synaptae* (for references see Table 3, p. 188).

Finally, in the fourth type (Fig. 47, *d*), no recapitulation of temporary conjugation occurs. The postmeiotic maturation division is lost; in each conjugant, only one meiotic derivative directly differentiates into a pronucleus. In the synconjugant, the two pronuclei fuse to form the single synkaryon. This type of pronuclear behavior was found by Moldenhauer[172] in *Urostyla polymicronucleata*. Possibly it exists also among the Suctoria, e.g. in *Ephelota gemmipara* (Fig. 17, *c, d*),[96] but this needs to be confirmed. It is characteristic of the other half of the Peritricha, i.e. of *Vorticella microstoma*, *Carchesium spectabile*, *Epistylis articulata*, *Rhabdostyla vernalis*, *Telotrochidium henneguyi*, *Trichodina patellae*, and *T. spheroidesi* (see Table 3, p. 188).

3. Reconstruction of the Nuclear Apparatus†

A. CLASSIFICATION OF THE RECONSTRUCTION TYPES

The period of reconstruction of the normal nuclear apparatus after conjugation includes the mitotic synkaryon divisions, the differentiation of the macronuclear anlagen and of the new micronuclei, and the return to

---

† The cytological phenomena of the development of the macronuclear anlagen will not be considered in this chapter, since a special paragraph was concerned with this question in the chapter "Macronucleus of ciliates" (Vol. 3, p. 26). This applies also to the cytology of degeneration of the old macronucleus (see Vol. 3, p. 70). However, it seems useful to name here some important papers published after conclusion of the chapter "Macronucleus of ciliates" and dealing with development of the macronuclear anlagen. These contributions include: observations on morphology of polytene chromosomes in the anlagen;[1a, 192a,b, 207a, 215] tracer studies of DNA synthesis [2a, 93a] and DNA destruction[2a, b] in polytene chromosomes; studies on DNA and RNA synthesis at later stages of anlagen development[2a, 197a, 216], and electron microscopical studies of macronuclear anlagen.[131a, 204a]

the species-specific number of nuclei either by segregation of the macronuclear anlagen during the exconjugant divisions, or by fusion of the anlagen with each other. Sometimes division of a single anlage into several macronuclei also occurs. In temporary conjugation, these processes usually take place after separation of the mates.

The nuclear phenomena show a much greater diversity during the exconjugant period than during the period of maturation divisions or during the period of karyogamy. Classifications of the types of post-conjugation nuclear reconstruction were given by Calkins,[15] Dogiel,[76] Kidder,[141] and Turner.[254] The classification given below is a modified and supplemented form of Dogiel's classification. The exconjugation periods are divided into types according to the number of synkaryon divisions, and into sub-types according to the number and fate of the macronuclear anlagen.

The number of synkaryon divisions (before differentiation of the macronuclear anlagen) varies in ciliates from one to four. According to the chromosome behavior, the synkaryon divisions are ordinary mitoses. However, the spindles of the first synkaryon division are often larger than those of vegetative mitoses, and sometimes show dual structure resulting from incomplete fusion of the spindles of the pronuclei (Fig. 5, *h*; 56, *b*).

FIG. 48. Diagram of nuclear reconstruction with a single synkaryon division. The semi-black circle is the synkaryon, light circles are its non-differentiated derivatives, the small dark circle is the new micronucleus, the hatched large circle is the macronuclear anlage.

Type I. One synkaryon division (Fig. 48). In this simplest (but not necessarily primitive!) type of the exconjugation period, one of the two derivatives of the single synkaryon division becomes the macronuclear anlage, the other derivative, the micronucleus. This is the type of exconjugation period in *Prorodon griseus, Bütschlia parva*, all the *Chilodonella* species (Fig. 9, *e–g*), *Balantidium caviae, Frontonia acuminata, Metopus sigmoides* (Fig. 14, *m, n*), *Nyctotherus cordiformis* (for references see Table 1, p. 178). A single synkaryon division exists in all Entodiniomorphida heretofore studied.[76, 183, 201] The differentiation of the macronuclear anlage begins in entodiniomorphids very early, in the telophase of the synkaryon division, so that the spindle of this division becomes heteropolar (Fig. 49, *d–f*).

Finally, this type is common among the Suctoria (see Table 2, p. 186), e.g. in *Podophrya parasitica, P. fixa, Platophrya* sp. *Cyclophrya katharinae, Ephelota gemmipara* (Fig. 17, *e, f*), and *Phalacrocleptes verruciformis*. A

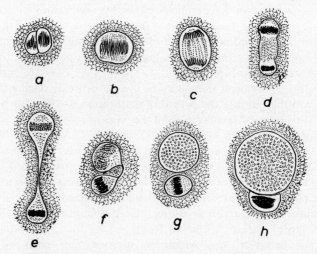

FIG. 49. Nuclear reconstruction in *Diplodinium caudatum*: *a*—fusion of pronuclei; *b*—metaphase, *c, d*—anaphase, *e–f*—heteropolar telophase of the single synkaryon division; *g–h*—macronuclear anlage (at top) and new micronucleus. Iron hematoxylin. After Dogiel.[76]

single synkaryon division also characterizes three species of *Discophrya*,[153-4] but since in this case two synkarya divide independently (see p. 235), the synconjugant receives not two but four derivatives of this division. One of them becomes the macronuclear anlage, and three become micronuclei.

The exconjugation period of *Dileptus gigas*[255] seems to be a modification of type I. In this species a single macronuclear anlage and a single micronucleus differentiate but the anlage later fragments into many vegetative macronuclei, and the micronucleus continues to multiply by mitoses.

Type II. Two synkaryon divisions (Fig. 50). This type comprises twelve sub-types.

Sub-type *Cryptochilum* (Fig. 50, *a*). Of the four derivatives of the second synkaryon division, three become macronuclear anlagen, and one a micronucleus. The macronuclear anlagen segregate at two exconjugant divisions, during which the micronucleus mitotically divides. This sub-type is peculiar to *Cryptochilum echini*[43] (Fig. 7, *j–n*), *Blepharisma seshachari*,[9] and possibly also *Opercularia coarctata*.[85]

Sub-type *Colpidium* (Fig. 50, *b*). Two macronuclear anlagen and two micronuclei are formed; both types of nuclei segregate at the single exconjugant division. Representatives: *Colpidium colpoda, Glaucoma* (= *Dallasia*) *frontata*. This sub-type occurs facultatively (as one of several possible patterns of nuclear reconstruction) also in *Didinium nasutum, Tetrahymena pyriformis, T. patula,* and *Colpidium truncatum* (for references see Table 1, p. 178).

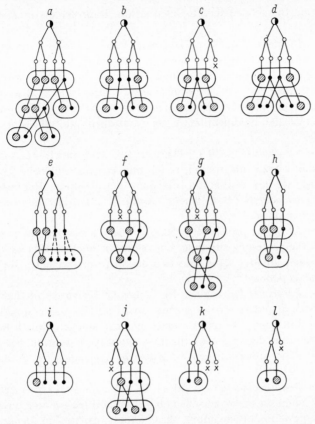

FIG. 50. Diagrams of reconstruction of the nuclear apparatus including two synkaryon divisions: *a*—sub-type *Cryptochilum*; *b*—sub-type *Colpidium*; *c*—sub-type *Dogielella*; *d*—sub-type *Paramecium aurelia*; *e*—sub-type *Didinium*; *f*—sub-type *Nicollella*; *g*—sub-type *Litonotus parvus*; *h*—sub-type *Paraisotricha*; *i*—sub-type *Dendrocometes*; *j*—sub-type *Stylonychia*; *k*—sub-type *Euplotes*; *l*—sub-type *Platophrya rotunda*. Symbols as in Fig. 48 (crosses indicate degenerating synkaryon derivatives).

Sub-type *Dogielella* (Fig. 50, *c*). Here also, two macronuclear anlagen are formed, but only one micronucleus, one derivative of the second synkaryon division degenerating. The macronuclear anlagen segregate and the micronucleus divides during the single exconjugant division. This sub-type is characteristic of conjugation in *Dogielella sphaerii*[198] and of autogamy in *Tetrahymena rostrata*.[42] It also occurs facultatively, together with the preceding sub-type, in *Colpidium truncatum*, *Tetrahymena pyriformis*, and *T. patula*, and together with type III, in *Blepharisma tropicum* (see Table 1, p. 178).

Sub-type *Paramecium aurelia* (Fig. 50, *d*). There are two macronuclear anlagen and two micronuclei; the anlagen segregate and both micronuclei divide during the single exconjugant division. Representative: *Paramecium aurelia.*[106, 156, 165]

Sub-type *Didinium* (Fig. 50, *e*). Two macronuclear anlagen and two micronuclei are formed. There are no special exconjugant divisions. Both macronuclear anlagen fuse into a single macronucleus. This is the most frequent pattern of nuclear reconstruction in *Didinium nasutum.*[204] A modification of this sub-type exists in *Spirostomum ambiguum*[216, 228] and *Stentor coeruleus;*[175] the modification is the continued multiplication of both micronuclei.

Sub-type *Nicollella* (Fig. 50, *f*). There are two macronuclear anlagen which later fuse into a single macronucleus, but only one micronucleus, one derivative of the second synkaryon division degenerating. Representatives: *Nicollella ctenodactyli,*[23] *Collinella gundii,*[23] *Isotricha ruminantium,*[76] *Ptychostomum tanishi.*[170]

Sub-type *Litonotus parvus* (Fig. 50, *g*). The nuclear reconstruction proceeds as in the preceding sub-type, but the macronucleus, formed by fusion of two anlagen, divides again into two definitive macronuclei. Representative: *Litonotus parvus.*[206]

Sub-type *Paraisotricha* (Fig. 50, *h*). Only one derivative of the first synkaryon division undergoes the second division. The single spindle of the second division gives rise to two macronuclear anlagen, which later fuse. The non-dividing derivative of the first synkaryon division becomes the micronucleus. Representatives: *Paraisotricha colpoidea,*[78] *P. magna,*[78] *Balantidium coli.*[179]

Sub-type *Dendrocometes* (Fig. 50, *i*). Of the four synkaryon derivatives, only one becomes a macronuclear anlage, while three become micronuclei. This sub-type is common among the Suctoria, existing in *Dendrocometes paradoxus* (Fig. 13, *f–h*), *Lernaeophrya capitata,*[94] and *Dendrosoma radians.*[94]

Sub-type *Stylonychia* (Fig. 50, *j*). Of the four synkaryon derivatives, one becomes a macronuclear anlage, its sister nucleus degenerates, and two derivatives become micronuclei. The macronuclear anlage divides into two or more definitive macronuclei. This sub-type is almost ubiquitous in the family Oxytrichidae (genera *Uroleptus, Urostyla, Pleurotricha, Oxytricha, Onychodromus, Stylonychia*—see Table 1, p. 178). It occurs also in *Diophrys scutum.*[117]

Sub-type *Euplotes* (Fig. 50, *k*). Of the four synkaryon derivatives, one becomes a macronuclear anlage, and one a micronucleus. These are sister nuclei. The other pair of sister nuclei degenerate. This sub-type is characteristic of all species of *Euplotes* (see Table 1, p. 178) and of *Collinia*[13, 38, 249] (Fig. 11, *j–q*), of the suctorians *Acineta papillifera*[39, 163] and *Tokophrya cyclopum.*[39, 165]

Sub-type *Platophrya rotunda* (Fig. 50, *l*). Only one derivative of the first synkaryon division undergoes the second division; the other derivative degenerates. Of the two products of the second division, one becomes the macronucleus, the other, the micronucleus. Representative: *Platophrya rotunda*.[94]

Type III. Three synkaryon divisions (Fig. 51). This type comprises eight sub-types, arranged in the order of the diminishing number of macronuclear anlagen.

Sub-type *Carchesium* (Fig. 51, *a*). Of the eight derivatives of the third synkaryon division, seven become macronuclear anlagen, and one a micronucleus. The anlagen segregate and the micronucleus divides at three successive synconjugant divisions. This sub-type is characteristic of the Peritricha, existing in the genera *Vorticella*, *Carchesium* (Fig. 18, *h–n*), *Zoothamnium*, *Epistylis*, *Rhabdostyla*, *Telotrochidium*, *Ophrydium*, *Lagenophrys*, and

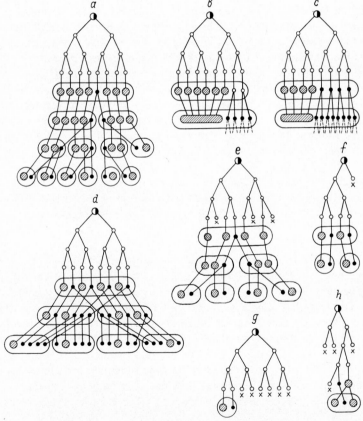

FIG. 51. Diagrams of reconstruction of the nuclear apparatus including three synkaryon divisions: *a*—sub-type *Carchesium*; *b*—sub-type *Stentor polymorphus*; *c*—sub-type *Climacostomum*; *d*—sub-type *Bursaria*; *e*—sub-type *Paramecium caudatum*; *f*—sub-type *Paramecium bursaria*; *g*—sub-type *Entorhipidium*; *h*—sub-type *Litonotus lamella*. Symbols as in Figs. 48 and 50.

Trichodina (see Table 3, p. 188). However, only six macronuclear anlagen and two micronuclei are said to form in *Vorticella convallaria*.[230] Also *Opercularia coarctata* and *Urceolaria synaptae* fall out from this sub-type (see type II, sub-type *Cryptochilum*, and type III, sub-type *Paramecium caudatum*, respectively).

Outside the Peritricha, only *Ancistruma isseli*,[141] *Cryptochilum nigricans*,[165] and *Cyclidium glaucoma*[165] belong to this sub-type.

Sub-type *Stentor polymorphus* (Fig. 51, *b*). There are six macronuclear anlagen, which fuse into a single macronucleus, and two micronuclei, which continue to multiply. The only representative is *Stentor polymorphus*.[175]

Sub-type *Climacostomum* (Fig. 51, *c*). There are four macronuclear anlagen which fuse, and four micronuclei, which continue to multiply. Representatives: *Climacostomum virens*[193] and *Fabrea salina*.[84]

Sub-type *Bursaria* (Fig. 51, *d*). Here also four macronuclear anlagen and four micronuclei differentiate. The micronuclei continue to multiply. But unlike the preceding sub-type, the macronuclear anlagen segregate at two exconjugant divisions. Representatives: *Bursaria truncatella*,[199] *Paramecium multimicronucleatum*,[6] and *Paramecium polycaryum*.[67]

Sub-type *Paramecium caudatum* (Fig. 51, *e*). Four macronuclear anlagen and one micronucleus differentiate; three synkaryon derivatives degenerate. The anlagen segregate and the micronucleus divides at two exconjugant divisions. Representatives: *Paramecium caudatum* (Fig. 5, *h–n*), *P. putrinum* (Fig. 29, *g–i*), *Parachaenia myae*,[150] and *Urceolaria synaptae*.[40]

Sub-type *Paramecium bursaria* (Fig. 51, *f*). One of the derivatives of the first synkaryon division degenerates; this results in only four derivatives of the third synkaryon division. Two of them become macronuclear anlagen, and two become micronuclei. Both types of nuclei segregate at the single exconjugant division. Representatives: *Paramecium bursaria, P. calkinsi* (for references see Table 1, p. 178).

Sub-type *Entorhipidium* (Fig. 51, *g*). Of the eight synkaryon derivatives, only one becomes a macronuclear anlage, and one, a micronucleus. The other six derivatives degenerate. The only representative is *Entorhipidium echini*.[278]

Sub-type *Litonotus lamella* (Fig. 51, *h*). Only one derivative of the second synkaryon division undergoes the third division, the other three derivatives degenerating. Of the two products of the third division, one becomes a macronuclear anlage, which further divides into two definitive macronuclei. The other product divides once more into the functional micronucleus and a degenerating nucleus. The only representative is *Litonotus lamella*.[167]

Type IV. Four synkaryon divisions (Fig. 52). Three sub-types are recognizable.

Sub-type *Conchophthirius* (Fig. 52, *a*). Of the sixteen derivatives of the fourth synkaryon division, twelve to fifteen become macronuclear anlagen, and one to four become micronuclei. The anlagen segregate at four exconjugant divisions. The only representative is *Conchophthirius mytili*.[140]

Sub-type *Paramecium woodruffi* (Fig. 52, *b*). One derivative of the first synkaryon division degenerates. This results in only eight products of the fourth synkaryon division. Four of them become macronuclear anlagen and four become micronuclei. The anlagen segregate and the micronuclei divide at two exconjugant divisions. The only representative is *Paramecium woodruffi*. [121]

Sub-type *Tokophrya lemnarum* (Fig. 52, *c*). Only one derivative of the third synkaryon division undergoes the fourth division, while the remaining seven derivatives degenerate. Of the two products of the fourth synkaryon division, one becomes the macronucleus, the other the micronucleus. The only representative is *Tokophrya lemnarum*.[182]

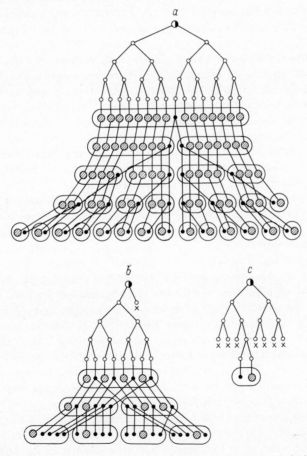

FIG. 52. Diagrams of reconstruction of the nuclear apparatus including four synkaryon divisions: *a*—sub-type *Conchophthirius*; *b*—sub-type *Paramecium woodruffi*; *c*—sub-type *Tokophrya lemnarum*. Symbols as in Figs. 48 and 50.

B. Mechanism of Differentiation of Macronuclear Anlagen

It has been repeatedly demonstrated that the different fates of the derivatives of the last synkaryon division depend on their different positions in the cytoplasm.

The spindles of the last synkaryon division are often directed parallel to each other and along the exconjugant's body, so that derivatives of these spindles lie in two groups, the anterior and the posterior. The nuclei of the anterior group usually develop into macronuclear anlagen, and nuclei of the posterior group into micronuclei (e.g. in *Paramecium caudatum*, Fig. 5, *k*). The nuclei may soon change their positions in the cytoplasm (Fig. 5, *l*), but this does not affect their fate. The synkaryon derivatives seem to be "sensitive" to the regional cytoplasmic influences during a short time only; once the differentiation has set in, it becomes irreversible and independent from the surrounding cytoplasm (although it may not yet be morphologically expressed).

Differentiation of the anterior synkaryon derivatives into macronuclear anlagen was reported by Maupas[165] in *Paramecium caudatum*, *P. aurelia*, *Colpidium colpoda*, and *Tetrahymena* (= *Leucophrys*) *patula*. The same pattern was later found in *Colpidium truncatum*,[54] *Dogielella sphaerii*,[198] *Bursaria truncatella*,[199] *Climacostomum virens*,[193] *Paramecium aurelia*,[243-4] *P. bursaria*,[81] *Tetrahymena pyriformis*,[82, 177, 219] etc. Displacement of synkaryon derivatives, brought about by centrifugation in *Tetrahymena*, *Paramecium*, or *Euplotes*, is followed by deviations from the normal pattern of differentiation of macronuclear anlagen.[177, 244, 138]

It might be supposed, however, that the differentiating factor is not in the cytoplasm but in the nuclei themselves, and that this factor (gene?) segregates during the last synkaryon division. In other words, the last division might be considered differential, i.e. giving rise to nuclei which are unequal from the very beginning of their existence. Egelhaaf[81] demonstrated that this is not the case in *Paramecium bursaria*. He studied exconjugants which were transversely cut immediately after the third synkaryon division. If factors responsible for determination of macronuclear anlagen and of micronuclei were segregated during this division, the anterior exconjugant halves could produce only macronuclear anlagen, and the posterior ones, only micronuclei. In reality, both types of halves produced both macronuclear anlagen and micronuclei. This reveals that immediately after the last synkaryon division all its derivatives are identical, and that new cytoplasmic differentiating gradients arise in both halves of the exconjugant.

The evidence presented above does not exclude, however, the existence in some ciliates of other mechanisms of differentiation of the macronuclear anlagen. For example, in *Carchesium* and other peritrichs, the single future micronucleus occupies no special place in the cytoplasm in relation to the seven macronuclear anlagen (Fig. 18, *l*). What prevents this nucleus from differentiating into an eighth macronuclear anlage is unknown.

C. Variability of the Nuclear Phenomena in Exconjugants

The period of post-conjugation reconstruction of the nuclear apparatus is so variable in some ciliates that it is difficult to hold one reconstruction pattern as "normal", and other patterns as "abnormal". Several more or less equivalent ways of nuclear reconstruction were described by Prandtl[204] in *Didinium nasutum* (Fig. 53). Within a single population, some exconjugants

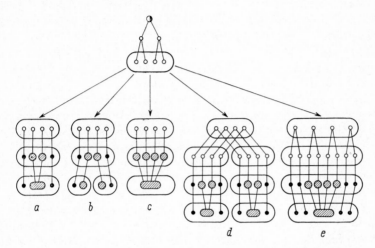

FIG. 53. Five ways of reconstruction of the nuclear apparatus following two synkaryon divisions in *Didinium nasutum*. Drawn after the data of Prandtl.[204] Symbols as in Fig. 48.

produce two macronuclear anlagen and two micronuclei; the anlagen may fuse (the most frequent way, Fig. 53, *a*), or segregate (Fig. 53, *b*). Or all four synkaryon derivatives may become macronuclear anlagen fusing with each other (*c*); this way seems to be abnormal since it gives rise to amicronucleate progeny. In some exconjugants, a third synkaryon division may occur, accompanied by division of the exconjugant itself (*d*), and further reconstruction may proceed either as in Fig. 53, *a*, or as in Fig. 53, *b*. Finally, the third synkaryon division may not be accompanied by cytokinesis; in this case, four macronuclear anlagen differentiate and later fuse into a single macronucleus (Fig. 53, *e*).

Five equivalent ways of nuclear reconstruction exist in *Loxocephalus*.[7] Here, three synkaryon divisions are followed by differentiation of a single micronucleus and of six macronuclear anlagen. Of the latter, four, three, or two may fuse into the definitive macronucleus, the remaining anlagen degenerating. Or four anlagen may fuse pairwise into two macronuclei segregating at the first division of the exconjugant. Finally, two anlagen may segregate at the exconjugant division, while four degenerate.

Euplotes cristatus has two equivalent patterns of the exconjugation period,[273] In the first case, there are two synkaryon divisions (type II, sub-type *Euplotes*; see p. 240). In the second case, the nuclear reconstruction is comprised of a single synkaryon division (type I, see p. 237). A similar situation exists in *Uronychia transfuga*,[220a] but here the two-division pattern corresponds to the sub-type *Platophrya rotunda* (see p. 240).

In most ciliates, however, one method of nuclear reconstruction may be considered normal, and the other less frequent methods regarded as individual variations, as more or less abnormal methods. Such individual variations may be classified as follows.

1. With additional synkaryon divisions, usually followed by formation of supernumerary macronuclear anlagen. In *Bütschlia*[77] and *Balantidium caviae*,[227] which normally have one synkaryon division, a second division may occur, and two or three macronuclear anlagen differentiate instead of a single anlage. An abnormal third synkaryon division has been reported in some species which normally have two synkaryon divisions, e.g. in *Colpidium colpoda*,[165] *Collinia branchiarum*, [38] *Paramecium aurelia*,[106, 165] *Pleurotricha lanceolata*,[162] and *Stylonychia pustulata*.[195] "Excess" macronuclear anlagen usually arise in such exconjugants (four instead of two in *Colpidium colpoda* and in *Paramecium aurelia*, etc.). An abnormal fourth synkaryon division may occur in ciliates usually having three divisions, e.g. in *Bursaria truncatella*,[199] *Paramecium caudatum*.[59, 165, 185a] Instead of the usual four, five to nine macronuclear anlagen may appear in the latter species following conjugation,[59] and one to ten anlagen, following induced autogamy.[185a]

Exconjugants with an increased number of nuclei are frequent in *Paramecium bursaria* (Fig. 54, *c*, *d*). They can have up to twenty synkaryon derivatives and up to six macronuclear anlagen.[99, 36, 81] However, it is possible that some of these nuclei originated from hemikarya (see p. 233); this possibility is indicated, for example, by the unequal sizes of the micronuclei of the exconjugant shown in Fig. 54, *d*.

2. With an abnormal ratio of macronuclear anlagen to the micronuclei, the total number of synkaryon derivatives being, however, normal. Deviations of this kind obviously follow disturbances in the mechanism of determination of the macronuclear anlagen. Chen[36] showed them to be frequent in exconjugants resulting from crosses between a young and an old clone of *Paramecium bursaria* (Fig. 54, *b*). The ratio of macronuclear anlagen to micronuclei may change towards an increase of the number of micronuclei at the cost of that of the anlagen (Fig. 54, *b*), as well as in the opposite direction.

Two macronuclear anlagen sometimes differentiate in *Stylonychia pustulata*,[165] *Euplotes eurystomus*,[135] *E. cristatus*,[273] *Dendrocometes paradoxus*,[109] and *Lernaeophrya capitata*,[94] which normally have only one anlage. Six macronuclear anlagen (instead of four) may form in *Paramecium*

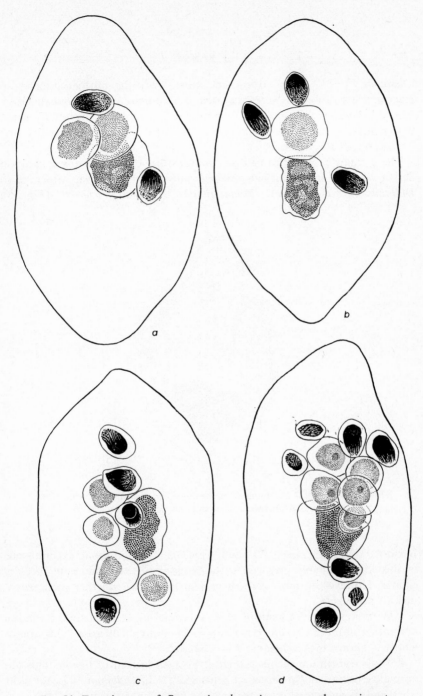

FIG. 54. Exconjugants of *Paramecium bursaria*: *a*—normal exconjugant showing two micronuclei, two macronuclear anlagen, and the old macronucleus; *b*—abnormal exconjugant (only one macronuclear anlage differentiated); *c*—abnormal exconjugant showing a double set of nuclei (four micronuclei, four macronuclear anlagen) in addition to the old macronucleus; *d*—abnormal exconjugant with seven micronuclei of unequal size, five macronuclear anlagen, and the old macronucleus. ×910. After Chen.[36]

caudatum.[165, 191] Very strong deviations from the normal number of macronuclear anlagen were reported in *Paramecium multimicronucleatum,*[174] *Bursaria truncatella,*[199] and *Fabrea salina.*[84] In the latter species, the number of macronuclear anlagen varied from one to twenty-five, the normal number being four.

The extreme case of this category is transformation of all the synkaryon derivatives into macronuclear anlagen. This abnormality was observed in *Didinium*[204] (Fig. 53, *c*), *Balantidium,*[227] *Paraisotricha*[78] (Fig. 55),

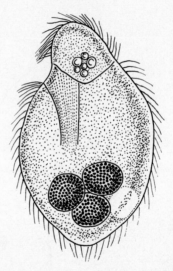

FIG. 55. Abnormal exconjugant of *Paraisotricha colpoidea* showing three macronuclear anlagen and no micronuclei. After Dogiel.[78]

Bursaria,[199] *Entodinium,*[201] and *Ephelota.*[96] Abnormal exconjugants of this kind probably give rise in some cases to amicronucleate races of ciliates. However, in most cases it remains unknown whether such exconjugants are viable.

3. With non-fusion of macronuclear anlagen (in species where the anlagen normally fuse). As a result, bimacronucleate individuals arise. This abnormality is known in *Nicollella* and in *Collinella.*[23]

4. With segregation of macronuclear anlagen instead of fusion. This abnormality occurs in *Paraisotricha colpoidea.*[78] In *Didinium*, it is not clear whether this modification of the nuclear behavior is normal or abnormal (Fig. 53, *b*).

5. With fusion of macronuclear anlagen instead of segregation. Maupas[165] believed this abnormality to be induced in *Paramecium caudatum* by starvation of the exconjugants. However, this suggestion was never

thoroughly verified. Abnormal fusion of macronuclear anlagen has also been observed in *Paramecium bursaria*[81] and *Tetrahymena* (= *Leucophrys*) *patula*.[165]

6. With irregular segregation of the macronuclear anlagen during the exconjugant divisions. Segregation of four anlagen in the ratio of three to one has been described in *Paramecium caudatum*.[59, 165] However, in the same species there is a tendency towards equal segregation even when the number of macronuclear anlagen is abnormal; e.g. ten anlagen segregate usually 5:5.[185a] In *Paraisotricha colpoidea* both macronuclear anlagen of the exconjugant may get into one daughter individual.[78]

7. With division of the exconjugant before differentiation of the macronuclear anlagen. In *Bütschlia*[77] the first synkaryon division is sometimes accompanied by plasmotomy; in this case, an additional (second) synkaryon division ensues in both daughter cells, followed by differentiation of a single macronuclear anlage and of a single micronucleus (as in normal exconjugants). Cytokinesis before differentiation of the anlagen occurs also in *Didinium* (Fig. 53, *d*).

8. With degeneration of an excess number of synkaryon derivatives. In *Paramecium caudatum*, one of the two derivatives of the first synkaryon division, or one of the four derivatives of the second synkaryon division may sometimes degenerate. This results in a smaller number of synkaryon products at the time of nuclear differentiation.[59] Degeneration of all synkaryon derivatives followed by macronuclear regeneration may occur in the same species after induced autogamy.[185a]

Parallel with the individual variability of the exconjugation period, strain-specific peculiarities of the nuclear behavior, inherited in certain strains or races, also exist. However, they are still little known. Diller[64] described a race of *Paramecium caudatum* where four synkaryon divisions always occur instead of the normal three divisions. But one of the derivatives of the first synkaryon division degenerates, so that the number of synkaryon derivatives becomes normal again at the moment of differentiation of the macronuclear anlagen (eight derivatives, four anlagen). In fact, the exconjugation period of this race proceeds according to type IV, sub-type *Paramecium woodruffi* (see p. 243).

In one of the strains of *Tetrahymena pyriformis*, there is only one synkaryon division instead of two, and only one macronuclear anlage and a single micronucleus are later formed.[177]

4. Cytoplasmic Changes during Conjugation

It has been long known that many cytoplasmic organelles undergo profound reorganization during conjugation, and that inclusions of reserve nutrient materials also change. The cytoplasmic phenomena accompanying conjugation are, however, incomparably less known than the nuclear phenomena.

A. BUCCAL ORGANELLES AND SOMATIC CILIATURE

Conjugation is usually accompanied by full or at least by partial dedifferentiation of the buccal ciliary organelles, and by disappearance of the peristome, buccal cavity, cytostome, and cytopharynx. The buccal organelles form anew in exconjugants, according to the pattern of stomatogenesis peculiar to the post-division buccal reorganization of the respective species.

In *Chilodonella*[86, 160-1], the mouth disappears, and the pharyngeal basket migrates deep into the cytoplasm where it becomes gradually resorbed (Fig. 9, *b*, *c*). At the stage of pronuclear interchange, each conjugant forms an anlage of the new pharyngeal basket (Fig. 9, *c*) which develops only in exconjugants (Fig. 9, *e*–*g*). The behavior of the pharyngeal basket of *Nassula* (Fig. 34) and of other gymnostomatids is similar.

The reorganization of the buccal organelles during conjugation has been extensively studied in *Paramecium*.[106, 165, 271, 221-2, 203, 101] The buccal cavity (the so-called "gullet"), carrying ciliary organelles (quadrulus, peniculi), persists in *Paramecium* for a rather long time. It dedifferentiates only at the stage of pronuclei. During late conjugation stages or in early exconjugants, a new buccal cavity begins to develop from an anlage which arose from a small invagination of the wall of the old buccal cavity.[101, 106, 221-2] The buccal ciliary organelles develop in the same manner as during cell division, i.e. from anarchic fields of kinetosomes present in the walls of the buccal anlage. During autogamy, the behavior of the buccal organelles of *P. aurelia* is the same as during conjugation.[221-2] No stomatogenesis occurs following lethal conjugation between amicronucleate paramecia.[70a]

The reorganization of the buccal organelles during conjugation is much more conspicuous in Heterotrichida, which usually have voluminous peristomes and strong adoral zones of membranelles. In *Bursaria truncatella*, the peristomal cavity fills up with cytoplasm and disappears during the prophase of the first maturation division.[205, 199] The main part of the adoral zone resorbs at the same time; only a small anterior section of the adoral zone persists until the third maturation division. Redifferentiation of the adoral zone and of the peristome occurs only in exconjugants.

In *Nyctotherus cordiformis*, the peristome and the adoral zone gradually resorb, starting from the anterior end, and disappear during the third maturation division. An anlage of the new adoral zone appears behind the old adoral zone before the latter is resorbed. The new peristomal cavity forms in early exconjugants.[265]

The most complex reorganization of the buccal organelles occurs in conjugating hypotrichs. It is appropriate to examine these changes together with the reorganization of the somatic ciliature (of the cirral apparatus), since both processes are in close correlation.

The replacement of the ciliary organelles of the hypotrichs during conjugation was described by Stein[247] a century ago. Later, Maupas[165] showed that all the cirri and the membranelles of the adoral zone are replaced twice during conjugation in *Onychodromus grandis* and *Euplotes eurystomus* ("deux mues ciliaires"). These observations proved to be correct.[70, 71, 100]

In many Oxytrichidae the adoral zones of the two partners undergo partial reduction immediately after mating. The persisting parts of the adoral zones, different in the two conjugants, fuse to form a common adoral zone of the pair.[95, 172] This common zone later dedifferentiates. In *Euplotes* there is no fusion of the adoral zones of the partners;[253, 100, 71] the posterior sections of the adoral zones quickly resorb, the anterior sections persisting longer.

In both conjugants of *Oxytricha fallax*,[70] adoral zones and sets of cirri of a second (provisional) generation develop; however, these second-generation peristomes are incomplete, lacking undulating membranes. The second-generation ciliary organelles are functional in early exconjugants. In later exconjugants, a new set of cirri, a new adoral zone of membranelles, and an undulating membrane are again laid down. This is the third (definitive) generation of ciliary organelles; after its development, the exconjugants start feeding. In lethal conjugation between amicronucleate *O. fallax*, only the second-generation ciliary organelles but no third generation organelles develop.[70 a]

In *Euplotes*,[71] the second-generation adoral zone is laid down during the meiotic prophase. During the second maturation division, anlagen of second generation cirri appear lacking, however, one of the frontal cirri. In early exconjugants, the old cirri and the remnants of the old adoral zone are replaced by corresponding second-generation organelles. But the second generation peristome lacks the undulating membrane and has an incomplete number of membranelles, which assures no feeding. In exconjugants having a macronuclear anlage, a third-generation peristome differentiates and fuses with a section of the second-generation peristome into the definitive peristome. Somewhat later, a third set of cirri is formed, replacing the second generation cirri. Also the contractile vacuole pore and canal are replaced two times in succession in exconjugants of *Euplotes*.[72 a] A double replacement of the buccal and of the somatic ciliary organelles occurs also during conjugation in *Aspidisca costata*.[72]

Outside the Hypotricha, very little is known about reorganization of the somatic ciliature during conjugation. According to Poljansky[199] a gradual replacement of the somatic cilia occurs in exconjugants of *Bursaria*. Dedifferentiation of the cilia and the kinetosomes on the contact surfaces of the mates has been reported in *Paramecium bursaria*[271] and *P. aurelia*.[203] However, electron microscopic studies demonstrated that only trichocysts and cilia disappear but kinetosomes persist on the contact surfaces throughout conjugation in *Paramecium caudatum*,[259, 3] *P. aurelia*[224, 131 a] and *P. multimicronucleatum*.[116] Fusion of the silverline systems of the two conjugants into a common system has been described by Klein,[147] Hammond,[100] and other authors in a variety of ciliates.

B. Exchange of Cytoplasm

Described above (p. 222) are the thin cytoplasmic channels connecting the endoplasms of the partners, which form in the contact zone of conjugating *Tetrahymena* and *Paramecium*. Electron microscopy has shown that both cilia and kinetosomes disappear from the contact surfaces in *Tetrahymena*; possibly the cytoplasmic channels form at the points previously occupied by cilia.[83] However, in *Paramecium caudatum*, according to ultrastructural data of Vivier and André,[259] only cilia disappear, the kinetosomes persisting and the cytoplasmic channels forming by fusion of the pellicular membranes of the partners at the tops of the ectoplasmic ridges, i.e. between the ciliary insertions (Fig. 38, *a*). Broader cytoplasmic bridges between the conjugants were found electron microscopically in *Paramecium aurelia*[224] and *Euplotes vannus*.[181] The existence of cytoplasmic continuities between the mates makes possible an exchange of more or less important cytoplasmic masses.

Cytoplasmic exchange was first demonstrated in *Paramecium aurelia* with genetic methods.[238–41] Sometimes, especially when separation of the mates is delayed, the conjugants of *P. aurelia* belonging to syngens of the group B may exchange their cytoplasmic mating type determinants, and in some cases also their kappa-particles. As a result, one of the exconjugants may change its mating type into that of its partner, or become killer (if previously it was sensitive). In *Paramecium bursaria*, a transfer of antigens from one mate to another occurs in 95 per cent of pairs; sometimes, symbiotic zoochlorellae are also transferred. This suggests that cytoplasmic exchange is a regular phenomenon accompanying conjugation of this species.[102]

Direct autoradiographic evidence of cytoplasmic exchange has been obtained in conjugating *Tetrahymena pyriformis*. The cells of one complementary mating type were protein-labeled with C^{14}-aspartic acid.[225] At early stages of conjugation (5 hours after mixing the complementary cultures) the label remained in only one conjugant of each pair. After the pronuclear exchange, the label spread to the other conjugant in 67 per cent of pairs; however, the activity was usually lower in the originally unlabeled partner. Schooley concludes from these data that cytoplasmic exchanges become intensive in *Tetrahymena* only during pronuclear interchange.

Somewhat different are the results of McDonald[166] with the same species. Before mating, one of the partners was either RNA-labeled (with H^3-uridine) or protein-labeled (with H^3-histidine). Intensive exchange of cytoplasm, expressed in transfer of the label to the other partner, was observed during the first maturation division. During late stages of conjugation, the labeled RNA became distributed between the partners in approximately equal amounts. The labeled protein was less mobile; its concentration remained approximately two times higher in the initially labeled partner, even at late stages of conjugation. This seems to be related to localization of the RNA in cytoplasmic ribosomes which can readily pass through the channels connecting the partners, while much protein is bound to immobile components of the cytoplasm (e.g. to fibrils), or to particles too large to pass through the channels (e.g. to mitochondria).

The passage of ribosome-like granular osmiophilic material through the channels which connect the two conjugants of *Paramecium caudatum* was observed electron microscopically by Vivier.[257]

C. CHANGES IN ENDOPLASM, MITOCHONDRIA, AND GOLGI APPARATUS

The cytoplasm often becomes denser, more homogeneous, and less vacuolated during conjugation, e.g. in *Bursaria*.[199] An increase of the RNA content of the cytoplasm occurs in *Paramecium caudatum* at late stages of conjugation and especially in exconjugants, i.e. during resorption of the fragments of the old macronucleus.[231, 98] This may be related to an increased production of RNA by the macronuclear fragments, rather than to "direct conversion" of DNA into RNA, as supposed by Gromova. In *Paramecium aurelia*, there is at first a net decrease and then a strong increase of the cytoplasmic RNA content during conjugation, which continues in the exconjugants.[275]

Poljansky[199] discovered sharp changes of localization of the mitochondria during conjugation in *Bursaria*. In vegetative animals, the mitochondria are uniformly dispersed throughout the cytoplasm. In late prophase of the first maturation division, almost all mitochondria gather at the boundary between the endoplasm and the ectoplasm and this peripheral localization persists until the stage of macronuclear anlagen. Some mitochondria also gather around the homogeneous zones of cytoplasm surrounding the pronuclei, the synkaryon, its derivatives, and the young macronuclear anlagen.

The Golgi apparatus undergoes definite ultrastructural changes during conjugation of *Tetrahymena pyriformis*.[83a] Flattened Golgi saccules, single in vegetative cells, become stacked in the oral region before mating and remain stacked throughout the conjugation. During mating, the edges of the Golgi saccules separate numerous vesicles which may contain substances active in the mating reaction.

D. RESERVE NUTRIENT SUBSTANCES

Changes in the reserve substances during conjugation are still little known. Zweibaum[279] showed the glycogen content of *Paramecium caudatum* to be minimal at early stages of conjugation and to increase in the exconjugants with macronuclear anlagen.

Zweibaum[279] reported a decrease of the size of the fat droplets in conjugating *P. caudatum* and also concluded from some cytochemical tests that their chemical composition changed. In vegetative cells neutral fats predominated, and in conjugants the inclusions consisted mainly of fatty acids. Similar changes were observed by Zweibaum in vegetative paramecia exposed to asphyxia or starvation, the fatty acids arising as a by-product of degeneration of proteins and carbohydrates (the so-called "degeneration fat" or "hunger fat"). However, Zweibaum's observations need to be verified with better methods.

Clear changes of the fat inclusions during conjugation were described by Poljansky[199] in *Bursaria*. The animals entering conjugation are comparatively poor in fat. During the first maturation division, a rapid accumulation of fat droplets begins. The fat content remains high until the exconjugants reconstruct their peristomes and resume feeding. Since conjugants do not feed, their fat inclusions may also represent "degeneration fat".

VI. ABNORMAL PATTERNS OF CONJUGATION

1. Conjugation of Three or More Animals

Conjugation of three or sometimes of more individuals (multiconjugation) has been repeatedly observed among the ciliates. Multiconjugation is never obligatory and usually occurs with low frequency. Conjugation of three animals has been reported by Maupas[165] in *Paramecium caudatum, Loxophyllum fasciola*, and *Spirostomum teres*, by Prandtl[204] in *Didinium nasutum*, by Collin[38] in *Collinia branchiarum*, by Mulsow[175] in *Stentor polymorphus*. Multiconjugation is presently known to occur also in *Blepharisma americanum*,[92, 262] *B. japonicum*,[115] *Ptychostomum* (= *Lada*) *tanishi*,[170] *Paramecium putrinum* (= *trichium*),[61] *P. caudatum*,[110, 256] *Ephelota gemmipara*.[96] Two or more microconjugants may attach to a single macroconjugant in Peritricha, e.g. in *Vorticella campanula*.[173] However, these observations largely concern the external picture of multiconjugation; at best, they report that maturation divisions of the micronuclei proceed synchronously and that macronuclear anlagen are formed in all the three mates.

Chen[27, 32] first extensively studied the nuclear phenomena during triple conjugation in *Paramecium bursaria*. He crossed races of paramecia having micronuclei of different ploidy degrees (Fd $\times$ McD$_3$). Therefore it was easy to identify both conjugants and pronuclei (Fig. 56). Two types of union into "threes" occur. In the first type, all three cells unite by their anterior body regions (approximately as in Fig. 57, *a*). In the second, more frequent type, two conjugants unite in the usual way, and the third conjugant attaches to the posterior end of the partner of complementary mating type (e.g. in Fig. 56, *a*, the third conjugant, belonging to the race Fd, is attached to the McD$_3$ partner of the anterior pair). Conjugation of four animals was sometimes also observed by Chen; additional conjugants either attach to both partners of a normal pair, or the fourth conjugant attaches to the posterior end of the third one, thus forming a chain.

The maturation divisions of the micronucleus proceed quite normally and synchronously in all three mates. Each conjugant produces two pronuclei, but an exchange of pronuclei occurs only between the two anterior partners (Fig. 56, *a*). In the third conjugant, the pronuclei also differentiate into the larger stationary and the smaller migratory but the migratory pronucleus never moves to the place of contact between the third conjugant and its mate. It moves inside its paroral cytoplasmic cone, i.e. to the place where it could pass into the partner, were the partner there (Fig. 56, *a*). Obviously, the migratory pronucleus of the third conjugant never leaves its own body.

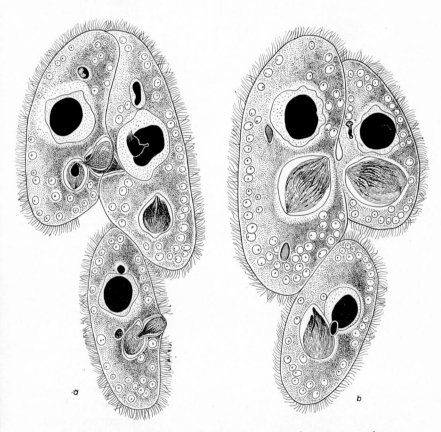

FIG. 56. Conjugation of three animals in *Paramecium bursaria*: *a*—exchange of migratory pronuclei between the two anterior conjugants, the migratory pronucleus of the third conjugant has moved to the mouth region; *b*—formation of synkarya (by reciprocal karyogamy in the two anterior conjugants and by autogamy in the third conjugant). Iron hematoxylin, ×660. After Chen.[32]

Synkarya are formed in all three conjugants: in the two anterior cells, they arise by fusion of the stationary pronucleus with the migratory pronucleus which came from the partner, and in the third cell it forms by fusion of two sister pronuclei, i.e. by autogamy (Fig. 56, *b*). All three mates undergo a normal exconjugation period including three synkaryon divisions (see Fig. 51, *f*).

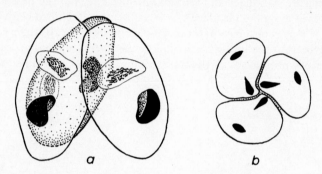

FIG. 57. Conjugation of three animals in *Paramecium putrinum*: *a*—mutual position of the partners during first maturation division prophase; *b*—schematic transverse section of the three mates showing positions of the pronuclei. After Jankowski.[127]

Another type of triple conjugation was investigated by Jankowski[127] in *Paramecium putrinum*. All three conjugants unite parallel to each other (Fig. 57, *a*); transverse sections show that all partners have equivalent positions (Fig. 57, *b*). The maturation divisions proceed simultaneously in all conjugants; each forms a stationary and a migratory pronucleus. Unlike *P. bursaria*, the migratory pronuclei of all three mates pass into the next partner in a circular manner (Fig. 57, *b*). No autogamy occurs in any conjugant.

Triple conjugation of a similar type (with symmetrical union of the mates and "circular" migration of the pronuclei) has been briefly described also in *Tetrahymena pyriformis* by Nanney.[177]

2. Conjugation between Double Monsters and Single Animals

In some ciliates, monstrous double animals are known. They usually originate by incomplete binary fission. Conjugation of double monsters with single individuals was first noticed by De Garis[51] in *Paramecium caudatum*.

Conjugation in double monsters of *Euplotes patella* was studied by Kimball.[142] Two types of double monsters exist in this species: both zooids may be micronucleate (homonucleate monsters) or only one, the other being amicronucleate (heteronucleate monsters). The double monsters can conjugate with each other or with singles. Most frequent is the mating of a single animal with one of the two zooids of a monster (sometimes with the micro-nucleate, sometimes with the amicronucleate). Conjugation of two singles with both zooids of a monster also occurs.

During conjugation of a single animal with one of the micronucleate zooids of a homonucleate double monster, the maturation divisions proceed synchronously in all three. During conjugation of a single animal with the amicronucleate zooid of a heteronucleate monster, the micronucleate zooid shows the typical nuclear changes, forming two pronuclei. It is evident that the cytoplasmic "signal", triggering the maturation divisions, acts on both zooids of a double monster.

During conjugation of a single animal with one of the micronucleate zoo-ids of a homonucleate monster, an exchange of migratory pronuclei occurs between the single and its partner, and synkarya are formed in both. In the zooid having no partner, the synkaryon forms by autogamy. In both zooids, the synkarya divide and give rise to normal macronuclear anlagen.

During conjugation of a single animal with the only micronucleate zooid of a heteronucleate double monster, maturation divisions and an exchange of migratory pronuclei proceed normally. The micronucleate zooid and the single form synkarya and, later, macronuclear anlagen. The unmated zooid remains amicronucleate and forms no macronuclear anlage.

Finally, during conjugation of a single animal with the amicronucleate zooid of a heteronucleate double monster, the single gives away its migra-tory pronucleus and receives nothing in exchange. In the single exconjugant and in the formerly amicronucleate zooid, the nuclear apparatus is recon-structed from division products of single pronuclei (hemikarya). The micro-nucleate zooid, having no partner, undergoes autogamy. This results in the formation in the double exconjugant of two macronuclear anlagen of diverse origin: one from a hemikaryon, the other from a synkaryon.

Another extensive study of conjugation of double monsters was done by Chen[34, 35] in *Paramecium bursaria*. The monsters here look like incom-pletely divided cells. Both zooids are micronucleate. Single animals may con-jugate with either zooid of the monster, more frequently with the anterior ("threes", Fig. 58, *a*), or with both zooids ("fours"). As in *Euplotes*, attach-ment of a single conjugant to the anterior zooid induces maturation divisions in both zooids. Each zooid forms two pronuclei. An exchange of migratory pro-nuclei occurs between the anterior zooid and the single conjugant (Fig. 58, *a*). In the posterior zooid, the synkaryon forms by autogamy (as in conjugation of three animals).

In double exconjugants of *P. bursaria*, both synkarya divide and give rise to eight derivatives (instead of the four in normal exconjugants). The con-striction separating the two zooids seems to smooth out at this time, and derivatives of the two synkarya mingle (Fig. 58, *b*). The number of macro-nuclear anlagen varies in double exconjugants from two to five (normal excon-jugants usually form two anlagen).

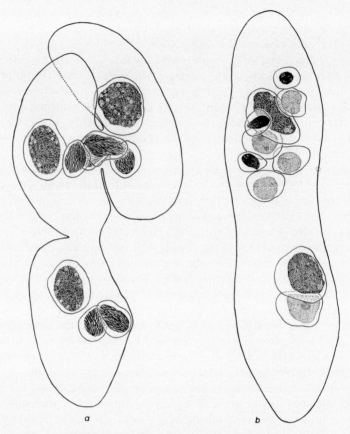

FIG. 58. Conjugation between double monsters and single animals in *Paramecium bursaria*: *a*—exchange of migratory pronuclei between the single animal and the anterior zooid of the double monster, the migratory pronucleus of the posterior zooid has moved to the mouth region; *b*—double exconjugant with three micronuclei, five macronuclear anlagen, and two old macronuclei. ×710. After Chen.[35]

3. Conjugation of Normal Animals with Amicronucleates and Haplonts

Conjugation between normal and amicronucleate *Paramecium bursaria* was investigated by Chen.[24, 25, 29, 30] The micronucleus of the normal mate undergoes the usual maturation divisions and gives rise to two pronuclei. The amicronucleate partner naturally forms no pronuclei. The migratory pronucleus of the normal mate usually passes into the amicronucleate mate (Fig. 59); thereafter, the nuclear apparatus is reconstructed from hemikarya in both partners. Each hemikaryon behaves exactly like a synkaryon, i.e. undergoes three divisions giving rise to four derivatives, of which two

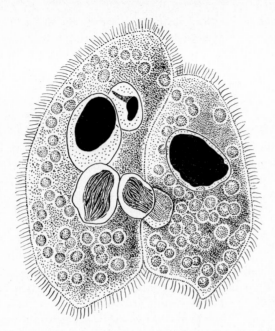

FIG. 59. Passage of a migratory pronucleus from the normal conjugant into the amicronucleate conjugant in *Paramecium bursaria*. ×820. After Chen.[25]

usually become macronuclear anlagen (see diagram, Fig. 51, *f*). In rare cases, there is no pronuclear migration, and the normal mate undergoes autogamy (its two pronuclei fuse into a synkaryon).

The latter pattern, occurring in *P. bursaria* only exceptionally, is regular during conjugation between normal and amicronucleate *Paramecium caudatum*.[232a] The conjugants unite for only a short time, separating before formation of pronuclei in the normal mate. Later, the micronucleate animal undergoes autogamy but with a high proportion of abnormalities, e.g. block of the second maturation division, formation of "excess" pronuclei and of "excess" synkarya, or variation in the number of macronuclear anlagen.

Regular passage of a single pronucleus from the micronucleate to the amicronucleate partner and development of hemikarya in both partners were reported by Diller[69] in *P. multimicronucleatum*. This resulted in diminution of micronuclear ploidy in the progeny. The descendants of such crosses survived two back-crosses with amicronucleates, but died after a third. In *Euplotes patella*,[142] when a single amicronucleate animal unites with the micronucleate zooid of a double monster, it receives a migratory pronucleus and reconstructs its nuclear apparatus from a hemikaryon.

Conjugation of two amicronucleate animals appears to be always lethal[188, 69, 70a, 165, 50, 70a, 276, 142].

Ciliates irradiated with ultraviolet or X-rays behave at conjugation like amicronucleates. A unilateral pronuclear migration from the normal cell into the irradiated one occurs during such conjugation in *Tetrahymena pyriformis*. Both exconjugants reconstruct their nuclear apparatus from single hemikarya; the micronuclei of such exconjugants and of their progeny thus become haploid.[37] In later crosses with diplonts, the haplonts produce no migratory pronuclei, and again the pronuclear migration is unilateral (from the diplont to the haplont). Crosses of two haplonts result in high inviability of the exconjugants. Similar observations were made by Reiff[220a] in *Uronychia transfuga*.

No passage of a migratory pronucleus from haplont to diplont occurs in *Paramecium aurelia* either. The pronuclei produced by the haplont seem to be aneuploid as a result of abortive meiosis.[144]

4. Lethal Conjugation

Lethal conjugation is rather frequent among the ciliates. It may follow various conditions, e.g. aging of clones (*Stylonychia pustulata*,[165] *Paramecium bursaria*,[131] *Tetrahymena pyriformis*,[263] etc.), inbreeding (*Tetrahymena pyriformis*[178]), and various types of incompatibility.

In most cases, the cytological phenomena of lethal conjugation are unknown. These phenomena were first studied in detail by Chen[33] in conjugation between the syngens II and IV of *Paramecium bursaria*. These syngens have diverse geographical origins: II, from the U.S.A., and IV, from the U.S.S.R. Conjugation is normal within each syngen and yields viable exconjugants; consequently, the lethal effect of conjugation between the two syngens is due to their incompatibility, not to abnormality of any of the races used in the crosses.

In these lethal crosses, the nuclear phenomena become abnormal in both conjugants during the meiotic prophase, and are blocked in the first maturation division. After a normal "crescent stage", the micronuclear chromosomes stick together in irregular clumps (Fig. 60, *b*), and sometimes even coalesce into a single chromatic body (Fig. 60, *c*). When a metaphase or anaphase of the first maturation division is attained, it is highly abnormal: no individual chromosomes are visible, the chromatic masses being irregularly distributed along the spindle. The first maturation division is never completed. The abnormal micronuclear spindles may strongly elongate, bend, or become tripartite (Fig. 60, *a*). Such stretched nuclei are sometimes partially transferred into the other mate, but the exchange is never complete (Fig. 60, *a*). The cytoplasmic properties seem to permit nuclear exchange at this time, but no pronuclei exist, and in their place the abnormal spindles of the first maturation division begin to pass into the other partner.

According to Chen, the lethality of conjugation is induced here by influence of one partner on the nuclei of the other partner through the cytoplasm.

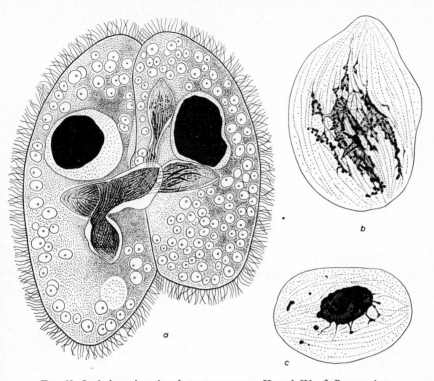

FIG. 60. Lethal conjugation between syngens II and IV of *Paramecium bursaria*: *a*—partial exchange of tripartite nuclei, which are abnormal stages of the first maturation division (×980); *b, c*—abnormal micronuclei from syngen II conjugants showing two degrees of chromosome fusion (×1980). Iron hematoxylin. After Chen.[33]

This influence becomes apparent very early, long before any nuclear exchange.

It is interesting that similar cytological disturbances occur after X-irradiation (dose 75 kr) of recently united conjugants of *Paramecium putrinum*.[157-8] Irradiation induces chromosome agglutination in the meiotic prophase and irregular segregation of chromatin clumps in anaphase of the first maturation division. In this case, second and third maturation divisions occur and pronuclei are formed. But all these stages are abnormal. The synkaryon usually does not divide and goes entirely to a single descendant of the exconjugant. This descendant is inviable. Other descendants receiving no synkaryon nor its derivatives, undergo macronuclear regeneration (see Vol. 3, p. 70) and survive. The defective synkaryon is apparently unable to give rise to a normal nuclear apparatus, but is still active in inhibiting macronuclear regeneration.

Lethal conjugation frequently occurs in *Tetrahymena pyriformis* and may be accompanied by various nuclear aberrations. In syngen 9, irregularities of meiosis are mainly apparent, including agglutination of chromosomes, formation of chromosome bridges and chromosome fragments. The general pattern of nuclear phenomena during conjugation is, however, normal.[218] On the other hand, lethal conjugation in syngen 5 is accompanied by strong disturbances of the mechanisms of nuclear differentiation.[220] Finally, during lethal conjugation between inbred[178a] or aged[263] clones of syngen 1, there are both meiotic chromosome irregularities and later disturbances of the mechanism of nuclear differentiation. Abnormal inbred clones also induce nuclear misbehavior (mainly degeneration of pronuclei) in its mate when crossed with a normal strain, leading to inviability of both exconjugants.[178a]

5. Reconjugation

Reconjugation is the mating of exconjugants which have not yet completed nuclear reconstruction. Reconjugation is usually unilateral (i.e. mating of one exconjugant with a normal animal, Fig. 61, *a*), less frequently bilateral (i.e. mating of two exconjugants, Fig. 61, *b*).

Reconjugation appears to be possible in species which have no immaturity period after conjugation, and also in species which normally have an appreciable immaturity period under conditions which bring about a strong shortening of that period.[128]

The cytology of reconjugation is insufficiently known. In most cases, authors merely state the fact of reconjugation, as in *Paramecium aurelia*,[58, 165, 233] *P. putrinum* (= *trichium*),[61] *P. caudatum*,[148] *P. multimicronucleatum*,[174] *Chilodonella uncinata*,[86] *Ch. caudata*,[161] *Dogielella sphaerii*,[198] *Euplotes longipes*,[146] and *Vorticella campanula*.[173]

Diller[60] studied reconjugation in *Paramecium caudatum* in some detail. In one of his races (P), the exconjugants entered reconjugation only at the stage of eight synkaryon derivatives, four macronuclear anlagen differentiating in the reconjugant a little later (as in normal exconjugants). There was no degeneration of synkaryon derivatives, and thus four micronuclei of the reconjugant entered the meiotic prophase (the "crescent stage"). The macronuclear anlagen of the reconjugant soon degenerated. The micronuclei underwent the first and the second maturation divisions; one of the second division derivatives completed the third maturation division.

In other races (G and U), exconjugants could enter reconjugation at any stage, e.g. at the stage of four macronuclear anlagen (Fig. 61, *a*, left partner). These anlagen later degenerated. The further behavior of the reconjugants is unknown.

Reconjugation has been studied in considerable detail in *Paramecium putrinum*.[128] Jankowski observed both unilateral and bilateral (Fig. 61, *b*) reconjugation. Exconjugants with four young macronuclear anlagen reconjugated most frequently; their first generation descendants, having only two macronuclear anlagen (Fig. 61, *b*), reconjugated less frequently. The micronucleus of the reconjugant underwent the usual maturation divisions and gave rise to pronuclei. The macronuclear anlagen of the early reconjugant stretched into thin strands (Fig. 61, *b*) and later fragmented. After this, reconjugants did not differ from normal conjugants. The ultimate fate of the reconjugants remains unknown.

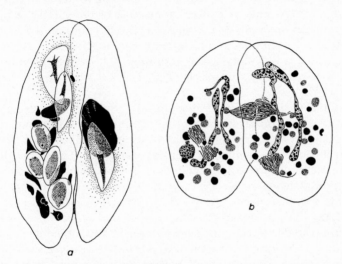

Fig. 61. Reconjugation in *Paramecium*: *a—P. caudatum*, reconjugant with four macronuclear anlagen, fragments of old macronucleus, and two micronuclei in "crescent stage" at left, normal conjugant with micronucleus in "crescent stage" at right (Carmine–indulin staining, ×450, after Diller[60]); *b—P. putrinum*, double reconjugation, each partner contains two elongated macronuclear anlagen, fragments of the old macronucleus, and micronuclear derivatives (alum hematoxylin, after Jankowski[128]).

VII. CONJUGATION IN LOWER CILIATES

Diploid macronuclei are at present known to occur in more than 100 species of lower Holotricha. These macronuclei are incapable of any kind of division, in contrast to the highly polyploid macronuclei of most ciliates. At each cell division, the number of diploid macronuclei is supplemented solely by transformation of a number of micronuclei into macronuclei (for more detail see Vol. 3, Chapter 1, p. 87).

Conjugation is known only in a few representatives of this group of lower ciliates, namely in two species of *Loxodes* and in four members of the family Trachelocercidae. Most of the data concerning conjugation are incomplete.

Nevertheless, it is clear that conjugation in lower ciliates (with diploid macronuclei) is peculiar and difficult to bring into full accord with the classical scheme of conjugation. Therefore data on conjugation of *Loxodes* and Trachelocercidae are excluded from Table 1 (p. 178).

1. Conjugation in Loxodes

Conjugation in *Loxodes striatus* was first observed by Kasanzeff,[133] and conjugation of *L. magnus* ("*L. rostrum*", according to the old terminology) by Penard.[190] A more extensive investigation of the nuclear phenomena during conjugation in *Loxodes striatus* was done by Bogdanowicz.[11]

Conjugation in *Loxodes striatus* is temporary isogamontic. The mode of union (Fig. 62) is well described by Kasanzeff.[133] Differences in body size may exist between the two partners (Fig. 62, *a, c, f*), but they appear to be fortuitous.

According to Bogdanowicz[11] a distinct generation of preconjugants exists in *Loxodes striatus*. The individuals which enter conjugation are much smaller than vegetative animals and have approximately half as many "Müller's vacuoles" containing mineral inclusions. They also differ in nuclear apparatus: vegetative animals have two macronuclei and two micronuclei, preconjugants have one macronucleus, one macronuclear anlage, and either one micronucleus (Fig. 62, *a*, right mate) or two (Fig. 62, *a*, left mate). Later studies on the division cycle of this ciliate[210] showed that both forms capable of conjugating are stages of the normal division cycle. These individuals have incompletely reconstructed their nuclear apparatus after the last (preconjugation) cell division. In both forms, transformation of the macronuclear anlage into an adult macronucleus is incomplete, and in the form with a single micronucleus, the second micronuclear mitosis, normally occurring at late stages of cell division, is absent.

The micronucleus undergoes three maturation divisions. In bimicronucleate conjugants, only one micronucleus takes part in these divisions. The meiotic prophase of the first division is the "parachute" type (Fig. 62, *b*). One of the products of the first division degenerates, the other undergoes the second maturation division (Figs. 62, *c*; 63). At this time, the functional nuclei become surrounded with a zone of dense cytoplasm (Fig. 62, *c*; cf. also p. 206).

Both derivatives of the second maturation division take part in the third maturation division (Fig. 62, *d*). The third division spindles, adjacent to the post-oral contact region of the mates, are oriented along the conjugant's body. They give rise to four nuclei (Fig. 62, *e*), of which two become pronuclei and two degenerate. The anterior derivative of each spindle degenerates, and the posterior remains functional. Thus, the two pronuclei of a conjugant of *L. striatus* are always derivatives of different third division spindles, i.e. non-sister nuclei (Fig. 63). The stationary and the migratory pronuclei are morphologically identical.

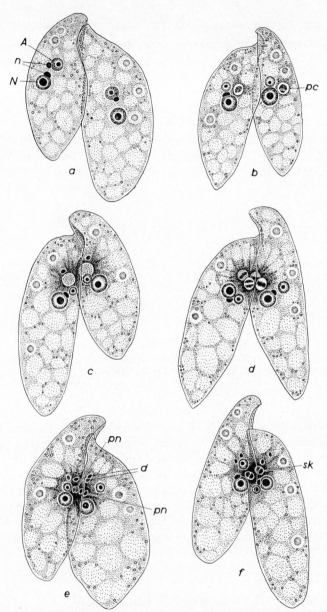

FIG. 62. Conjugation in *Loxodes striatus*: *a*—early stage (*N*—macronucleus, *n*—micronuclei, *A*—macronuclear anlage); *b*—"parachute stage" (*pc*) of the first maturation division prophase (a non-dividing micronucleus seen in the left partner); *c*—prophase of the second maturation division (both partners contain a degenerating derivative of the first division, and the left partner, also a non-dividing micronucleus); *d*—two metaphase spindles of the third maturation division in each partner; *e*—pronuclei (*pn*) and degenerating derivatives of the third maturation division (*d*); *f*—synkarya (*sk*) in both partners. After Bogdanowicz.[11]

After exchange of the migratory pronuclei, a synkaryon forms in each conjugant (Fig. 62, *f*). The pronuclei fuse in interphase. Thereafter, the partners separate.

Unfortunately, in his paper of 1930, Bogdanowicz did not describe the nuclear phenomena in exconjugants, limiting himself to a statement that no degeneration of the old macronuclei occurred even in the latest exconjugants he studied. Bogdanowicz's data on exconjugants of *Loxodes* were never published, and his preparations and original drawings were destroyed during World War II.

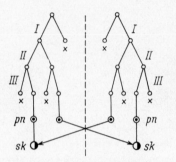

Fig. 63. Diagram of nuclear behavior during conjugation of *Loxodes striatus* (both partners shown until the synkaryon stage). *I, II, III*—maturation divisions, *pn*—pronuclei, *sk*—synkarya; crosses indicate degenerating nuclei.

Bogdanowicz[11] also briefly described conjugation in *Loxodes magnus*, a multinucleate form called "*L. rostrum*" by Bogdanowicz. Preconjugants of *L. magnus* are distinctly smaller than vegetative animals and have approximately half as many nuclei, namely six to fourteen macronuclei and three to ten micronuclei. Of the latter, usually only one, rarely two micronuclei undergo the first maturation division. What mechanism might differentiate the micronuclei into functional and degenerating is not known. All three maturation divisions are similar to those in *L. striatus*. Bogdanowicz failed to find degeneration of the old macronuclei in exconjugants of this species also.

2. Conjugation in Trachelocercidae

Among the Trachelocercidae, the full pattern of conjugation is known only in *Tracheloraphis* (= *Trachelocerca*) *phoenicopterus*.[208-9] Vegetative animals of this species have a single *complex nucleus*, a product of fusion of the entire nuclear apparatus. The complex nucleus consists of twelve macronuclei, fused into a "complex macronucleus", and six micronuclei, which pass to the *inside* of the complex nucleus during macronuclear fusion (see Vol. 3, Chapter 1, Fig. 50).

As in *Loxodes*, conjugation in *T. phoenicopterus* is temporary isogamontic. There is a distinct positive correlation of the body sizes of the two partners.

A generation of preconjugants exists in *T. phoenicopterus*. The preconjugants differ from vegetative cells by their significantly smaller body size, by smaller variability of the body length, and most sharply by absence of the complex nucleus. The six macronuclear anlagen, formed during the last (preconjugation) cell division, do not complete their development and do not fuse with each other and with the fragments of the old complex macronucleus into a new complex nucleus, as they do in usual cell division. As a

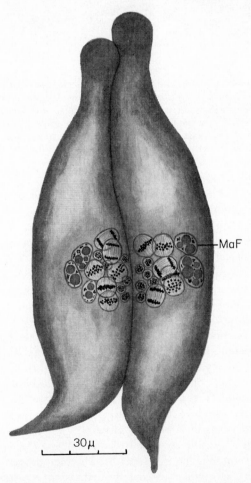

FIG. 64. Conjugation in *Tracheloraphis phoenicopterus*. Each mate contains six metaphase and anaphase spindles of the first maturation division, two or three fragments of the old complex macronucleus (*MaF*), and six macronuclear anlagen (near the boundary of the mates). Reconstruction from sections. After Raikov.[209]

9a*

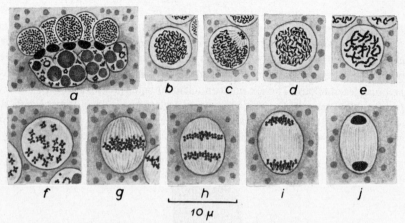

FIG. 65. First maturation division of *Tracheloraphis phoenicopterus*: *a*—nuclear apparatus of a preconjugant (fragments of the old complex macronucleus at bottom, six micronuclei at center, six macronuclear anlagen at top); *b*—leptotene stage; *c*—"parachute stage"; *d, e*—pachytene stage; *f*—diakinesis stage of the meiotic prophase; *g*—metaphase; *h, i*—anaphase; *j*—telophase. Iron hematoxylin. After Raikov.[209]

result, the nuclear apparatus of animals beginning to conjugate consists of several fragments of the old complex macronucleus, of six free micronuclei, and of six incompletely developed macronuclear anlagen (Fig. 65, *a*).

Soon after mating, all the six micronuclei undergo the first maturation division (Fig. 64). This division begins with a typical meiotic prophase comprising a leptotene stage (Fig. 65, *b*), a "parachute" (i.e. zygotene) stage (Fig. 65, *c*), a pachytene stage (Fig. 65, *d, e*), and a diakinesis stage (Fig. 65, *f*). Seventeen tetrads are distinct in each micronucleus during diakinesis; almost all are cross-shaped, i.e. have a single chiasma (Fig. 65, *f*). The haploid chromosome number of this species is thus 17.

The tetrads assemble into an equatorial plate in metaphase of the first maturation division (Figs. 64; 65, *g*), and divide into dyads in anaphase (Figs. 64; 65, *h, i*). Seventeen dyads migrate to each spindle pole.

All twelve products of the first division usually take part in the second maturation division (Fig. 66, *a*). The dyads separate into single chromosomes, haploid in number. Thus a rather typical meiosis comprising two nuclear divisions exists in *T. phoenicopterus*.

Not all products of the second division undergo the third (equational) maturation division (Fig. 66, *b*); some haploid meiotic derivatives degenerate. The number of spindles of the third division is variable but always considerable (from 5 to 13).

The most peculiar feature of conjugation of *T. phoenicopterus* is multiple formation of pronuclei. It is not clear whether all derivatives of the third maturation division become pronuclei or some degenerate. At any rate, each partner forms 7 to 22 pronuclei, which are larger than degenerating nuclei (Fig. 66, c). All the pronuclei are at first alike; later some of them differentiate into migratory pronuclei, which acquire a drop-like shape (Fig. 66, d). The number of migratory pronuclei in a single conjugant varies from three to six; this number (and also the number of stationary pronuclei) may be unequal in the two members of a pair. All the migratory pronuclei synchronously pass into the other partner; each of them approaches a stationary pronucleus and fuses with it. Thus, two to six synkarya (Fig. 67, Sk) may be formed in each conjugant. Since stationary pronuclei are usually more numerous than the migratory, some of them become "residual" and later degenerate (Fig. 67, Pn). The conjugants separate immediately after karyogamy.

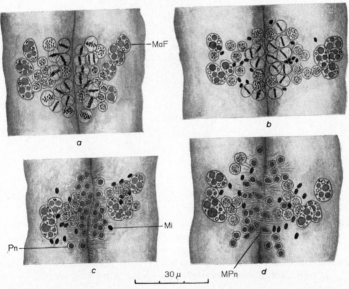

FIG. 66. Conjugation in *Tracheloraphis phoenicopterus*: a—metaphases of the second maturation division (*MaF*—macronuclear fragments); b—metaphases of the third maturation division; c—stage of pronuclei (*Pn*—pronuclei, *Mi*—degenerating derivatives of maturation divisions); d—exchange of migratory pronuclei (*MPn*). Reconstructions from sections. After Raikov.[209]

However, only one synkaryon (Fig. 68, a, Sk) remains functional in the exconjugant, the others undergoing pycnosis (D.Sk) together with the "excess" pronuclei (Pn). The functional synkaryon divides four times (Fig. 68, b, c, d, e). The synkaryon divisions resemble vegetative mitoses and show a diploid chromosome number (34). Of the sixteen synkaryon derivatives (Fig. 68, f), six become macronuclear anlagen (Fig. 68, g, A), six become micronuclei, and four degenerate.

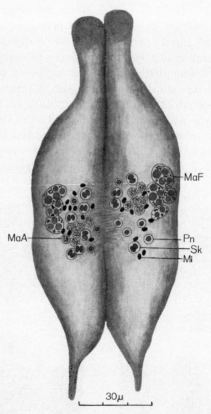

FIG. 67. Conjugation in *Tracheloraphis phoenicopterus*. The left partner contains six, the right partner, four synkarya (*Sk*). *Pn*—"excess" pronuclei, *Mi*—degenerating derivatives of maturation divisions, *MaF*—macronuclear fragments, *MaA*—macronuclear anlagen persisting from the preconjugant. Reconstruction from sections. After Raikov.[209]

The fragments of the old complex macronucleus (Figs. 64; 66, *a*; 67, *MaF*) and the six macronuclear anlagen of the preconjugant (Fig. 67, *MaA*) persist unchanged almost throughout conjugation. The fragments of the old complex macronucleus never degenerate (Fig. 68, *g*, *F*); only the macronuclear anlagen, formed during the preconjugation cell division, undergo pycnosis in late exconjugants (Fig. 68, *g*, *O*). They become replaced by the new macronuclear anlagen, originating from synkaryon derivatives (Fig. 68, *g*, *A*). The final stage of nuclear reorganization of the exconjugant is fusion of the new macronuclear anlagen with each other and with the fragments of the old complex macronucleus. A new complex nucleus is thus formed, in which the six micronuclei are enclosed (Fig. 68, *h*). This complex nucleus is consequently of dual origin: it consists of elements of the old complex

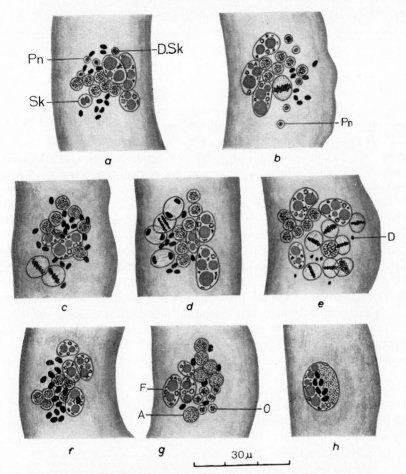

FIG. 68. Nuclear phenomena in exconjugants of *Tracheloraphis phoenicopterus*: *a*—persisting synkaryon (*Sk*) and pycnosis of the other synkarya (*D.Sk*) and of "excess" pronuclei (*Pn*); *b*—first, *c*—second, *d*—third, *e*—fourth synkaryon division (*Pn*—degenerating pronucleus, *D*—degenerating derivatives of maturation divisions); *f*—sixteen derivatives of the fourth synkaryon division; *g*—formation of six new macronuclear anlagen (*A*) and degeneration of the old macronuclear anlagen (*O*), macronuclear fragments (*F*) persisting; *h*—fusion of new macronuclear anlagen and fragments of the old macronucleus into a complex nucleus enveloping also the six micronuclei. Reconstructions from sections. After Raikov.[209]

macronucleus and of new macronuclei, which may be genetically different from the old elements.

Conjugation in *Trachelocerca coluber* was studied in less detail.[212] The nuclear apparatus of this species normally consists of four macronuclei and two micronuclei (no complex nucleus is formed). As in the preceding species, mating occurs between preconjugants differing from vegetative animals by incomplete development of two newly formed macronuclei which developed from micronuclei during the last cell division preceding conjugation.

Both micronuclei undergo the first maturation division, comprising a "parachute"-like meiotic prophase. In metaphase of the first division, twenty-six to twenty-eight tetrads are visible. All the four derivatives of the first division undergo the second maturation division, but only four of its eight products take part in the third maturation division.

Four pronuclei differentiate in each conjugant of *T. coluber* from the derivatives of the third maturation division. Later on, two synkarya appear in each conjugant. The moment of pronuclear exchange was not observed, and thus not excluded is the possibility of autogamous origin of the synkarya.

The later behavior of the synkarya remains unknown. Both the "adult" and the "underdeveloped" macronuclei persist unchanged at least until the stage of karyogamy.

Incomplete are also the data on conjugation in *Tracheloraphis caudatus*.[80] This species is multinucleate and has from 28 to 200 macronuclei and from 14 to 100 micronuclei. All the micronuclei take part in the first maturation division; twenty-two tetrads are distinct in late prophase and in metaphase of this division. Only some products of the first division undergo the second maturation division. The third maturation division was not observed with certainty. Further, *several* pronuclei differentiate in each conjugant; their number usually varies between five and ten. The stage of multiple synkarya was also observed (three to four synkarya in each conjugant). The nuclear phenomena in the exconjugants were not studied.

Finally, in *Tracheloraphis margaritatus* only the very beginning of conjugation is known, namely, the prophase of the first maturation division of the numerous micronuclei.[80] Afterwards, the meiotic micronuclei undergo a strange fragmentation. The significance of this is still unknown.

3. General Peculiarities of Conjugation in Lower Ciliates

The following features are peculiar to conjugation in all or at least several species among the small number of lower ciliates investigated.

(a) *Presence of a generation of preconjugants.* As shown above (p. 153), preconjugants with a nuclear apparatus different from that of vegetative animals are rare among the free-living ciliates, being characteristic mainly of parasitic forms. But in *Loxodes striatus*, as well as in *Tracheloraphis phoenicopterus* and *Trachelocerca coluber*, the preconjugants differ from vegetative animals by the incomplete development of the nuclear apparatus following the last (preconjugation) cell division; especially by the apparent blocking of the development of macronuclear anlagen formed during this division.

(b) *Typical meiosis.* The first two maturation divisions are meiotic in all lower ciliates heretofore studied. A "parachute stage" characterizes the meiotic prophase. In Trachelocercidae, tetrads are distinct during late prophase and metaphase of the first maturation division. The third maturation division is, as usually, equational.

(c) *Formation of multiple pronuclei and synkarya.* As shown above, only one pair of pronuclei normally forms in each conjugant in species having polyploid macronuclei (and frequently in totally conjugating species, only one pronucleus differentiates). Formation of multiple pronuclei and synkarya is thus the most striking feature of conjugation in lower ciliates, until now, however, described only among the Trachelocercidae. The number of pronuclei and of synkarya is inconstant in *Tracheloraphis phoenicopterus* and *T. caudatus*, but seems to be fixed in *Trachelocerca coluber* (four pronuclei, two synkarya in each conjugant). It is not yet clear whether in some species of lower ciliates all synkarya divide and give rise to nuclei of the exconjugant, or whether it is always only one synkaryon which remains functional (as in *T. phoenicopterus*).

Formation of multiple pronuclei in Trachelocercidae may represent a phylogenetic rudiment of gametogenesis. The conjugation of ciliates might have originated from gamontogamy involving formation of multiple gametes which then fused. If so, pronuclei of the ciliates are homologous to gamete nuclei, and in higher ciliates (with polyploid macronuclei) each conjugant produces, so to say, two gametes, a male one and a female one, without, however, separation of cytoplasm around them. Such a small number of gametes may be considered secondary, and it may be supposed that a tendency towards multiple fertilization is a primitive trait of the Trachelocercidae. A diagram of the maturation divisions and of formation of pronuclei in Trachelocercidae is shown in Fig. 23, *i*.

The tendency towards multiple formation of pronuclei is apparent also among higher ciliates. It is expressed in the production, during maturation divisions, of more nuclei than necessary for pronuclear formation.[199] The "excess" nuclei later degenerate.

However, Diller[72] recently described formation of four pronuclei in *each* conjugant in a representative of the higher ciliates, namely, in *Aspidisca costata*. Each conjugant forms only one migratory pronucleus; nevertheless, two synkarya form in each partner, one by fusion of a stationary pronucleus with the migratory pronucleus which came from the other mate, the other by autogamous fusion of the two remaining pronuclei. Most surprisingly, both synkarya remain functional in *Aspidisca*: both undergo a single division, and of the four nuclei thus formed in the exconjugant, one becomes a macronuclear anlage, two become micronuclei, and one degenerates. This behavior implies that either the macronucleus may be genetically different from both micronuclei, or the micronuclei may differ from each other, only one of them having the same genotype as the macronucleus.

It is not clear whether we have to consider the double fertilization in *Aspidisca* a secondary complication of the nuclear reorganization typical

of the other Hypotrichida (and consider the similarity with the Trachelocercidae a pure convergence), or we have to consider it an ancestral feature
persisting, strangely enough, in one of the most advanced ciliate groups.
The former supposition appears more likely, since conjugation of *Aspidisca*
resembles conjugation in Euplotidae (especially that of *Diophrys scutum*,
see Table 1, p. 178) in some other respects, e.g. in participation of two
nuclei in the postmeiotic maturation division, and in the number of macronuclear anlagen and of micronuclei differentiating in the exconjugant.

(d) *Non-degeneration of the old macronuclei.* This feature seems to be
common to both Loxodidae and Trachelocercidae. The old macronuclei
persist in the exconjugants and in their nearer descendants. They only
become gradually "diluted" by new macronuclei originating from the
synkaryon. The old macronuclei, being unable to divide, segregate between
the daughter cells at every division of the exconjugant's offspring, and at
every division new macronuclei develop from micronuclei and supplement
the set of macronuclei. The macronuclei of a single animal must consequently
be genetically diverse in the first generations following conjugation. Some
of them, the old ones, must have the parent genotype, whereas other macronuclei, the new ones, must have the new genotype which originated by
karyogamy.

In higher ciliates, conjugation results in two important events: (1) alteration of the genotype of the conjugant's progeny, and (2) replacement of the
old macronucleus. These two events are usually united into a common
process of conjugation. On the other hand, these two events of conjugation
are partially separated in *Loxodes* and in Trachelocercidae. The immediate
result of conjugation is here only alteration of the micronuclear genotype,
brought about by genetic recombination. The renewal of the macronuclear
apparatus is not directly connected with conjugation; it proceeds continuously, following every cell division. It may be supposed that such relationships are primitive and correspond to lower steps of evolution of the nuclear
dualism in ciliates.

VIII. CONCLUSION. ON SEXUAL DIFFERENTIATION IN CILIATES

The problem of sexual differentiation in ciliates has already been considered in Vol. 2 of this series, in the chapter by Grell, whose conclusions
appear perfectly justified to the author of the present review. Despite this,
it seems useful to summarize the data presented above in a somewhat diverse
aspect, relative to the problem of sexual differentiation. This is of interest
mainly because the literature shows no uniformity in answering the question
as to what ciliate phenomena can be considered manifestations of sexual
differentiation.[76, 237, 254, 226, 264, 49, 97, 103, 20, 79, 258, 180]

By a variety of authors, the significance of "sexual differentiation" has been ascribed to the following types of differentiation occurring in ciliates.

First, the different behavior and sometimes also the different morphology of the stationary and migratory pronuclei. Probably conjugation arose from the pairing of gamonts, both of which produced gametes fusing later on. If so, the difference between pronuclei is homologous to the difference between the gametes of the ancestors of the ciliates, and the migratory pronucleus can be considered male, the stationary, female.

Second, the more or less pronounced morphological difference between the two conjugants (gamonts) of some ciliates. While anisogamonty is not obligatory in such forms as *Opisthotrichum*, where two macroconjugants may pair (see p. 154), it becomes obligatory in Peritricha where macroconjugants pair only with microconjugants, and the conjugation itself becomes total. These phenomena may also be considered a manifestation of sexual differentiation operating at the level of gamonts; we may regard the macroconjugants as female gamonts (or simply females), and microconjugants, as male gamonts (males). In the most advanced cases (see p. 235), each conjugant produces a single pronucleus, which may be considered female or male, respectively (though morphological differences of the pronuclei are usually weak here).

Third, the differentiation of the mating types in isogamontic ciliates. Mating type differentiation also operates at the level of gamonts but, unlike anisogamonty, it is not accompanied by a tendency towards total conjugation; mutual fertilization of the gamonts persists. Another difference from the former two types is that differentiation of the mating types need not be bipolar; in fact, it is often multipolar (e.g. in *Paramecium bursaria*). Consequently, the notions of "male" or "female" are inapplicable to this type of differentiation.

There are two main hypotheses of sexual differentiation in ciliates.

The first hypothesis[76, 254, 49, 97, 79] recognizes that bipolar sexual differentiation in ciliates occurs at two levels, that of pronuclei (gametes) and that of conjugants (gamonts). The difference between pronuclei is considered the primary (and obligatory) sexual differentiation; this differentiation ensures karyogamy not only during reciprocal conjugation but also during autogamy. Hartmann[103] regards the difference of pronuclei as a typical example of haplophenotypical sex determination. As to the sexual differentiation at the level of conjugants, it is considered secondary and not obligatory. It is absent in most ciliates; in cases of temporary isogamontic conjugation, the two partners may be regarded, according to this hypothesis, as two similar monoecious (hermaphroditic) individuals, each producing both a male and a female pronucleus and mutually fertilizing each other. In cases of total conjugation, a suppression of the male sex in macroconjugants and of the female sex in microconjugants occurs. Thus, sexual differentiation at the level of gamonts leads, according to this hypothesis, to secondary dioecy (gonochorism). The sexual differentiation is bipolar at both levels, that of pronuclei and that of conjugants. This hypothesis is thus in agreement with Hartmann's[103] theory of sexuality, a fundamental statement of which is the law of bipolarity of the sexual differences, considered applicable to the whole organic world.

As to the mating type differences between isogamonts, no significance of sexual differentiation is ascribed to them by partisans of the first hypothesis. These differences are considered a phenomenon *sui generis*, a special mechanism excluding or limiting the inbreeding, a peculiar "superstructure over sexuality", comparable with the phenomenon of self-sterility in plants (see also Grell's chapter, Vol. 2). The partisans of the first hypothesis recognize the main principles of Hartmann's theory of sexuality and deny even the possibility of existence of multipolar sexuality.

The second hypothesis has been expressed in the clearest form by Vivier.[258] According to this hypothesis, sexual differentiation exists in ciliates only at the level of conjugants (gamonts) and is manifested in mating type differentiation (or sometimes in morphological anisogamonty). The second hypothesis refuses to ascribe any significance of sexuality to the differences between the stationary and the migratory pronuclei. The different behavior of the two pronuclei is assumed to depend not on their own properties but on the surrounding cytoplasm. A corollary of this point of view is rejection of Hartmann's bipolar theory of sexuality. According to the second hypothesis, the mating types, including the multiple ones, must be considered sexes or "sexual types".

The author of this review is inclined to follow the first hypothesis, which regards the ciliates as hermaphroditic animals and recognizes the sexual nature of pronuclear differentiation. Several objections may be raised against the second hypothesis.

1. It is known that the primary type of sex differentiation outside the ciliates is always differentiation of gametes, and that this differentiation is always bipolar. Sexual differentiation is extended to gamonts only in some protozoan groups (see Grell's chapter, Vol. 2). Consequently, the second hypothesis implies, in the evolutionary aspect of the question, a rejection of the homology between differentiation of the pronuclei into the stationary and the migratory and bisexual differentiation of gametes. But this homology is most clear exactly in the lower ciliates, e.g. in *Tracheloraphis* (see p. 269), where each conjugant produces several stationary and several migratory pronuclei.

2. If we assume the mating types to be sexes, the ciliates would become unique organisms in the whole organic world, both animal and plant: organisms with multipolar sexuality. Everywhere else, the sex differentiation is only bipolar.[103]

3. The second hypothesis does not account for pronuclear behavior during autogamy. According to this hypothesis, autogamy must be con-

sidered a sexual process without sexuality at any level. Or we should have to consider autogamy an agamic process!

From these considerations, it seems more adequate to regard as manifestations of sexual differentiation in ciliates only the former two groups of phenomena listed above, i.e. the differences between the stationary and the migratory pronuclei and the anisogamonty occurring mainly in totally conjugating ciliates.

REFERENCES

1. ALONSO, P., and PÉREZ-SILVA, J. (1965) Conjugation in *Stylonychia muscorum* Kahl. *J. Protozool.* **12**, 253–8.

1a. ALONSO, P., and PÉREZ-SILVA, J. (1966) Cromosomas politénicos en ciliados hipotricos. *Bol. Real Soc. Española Hist. Nat. (Biol.)* **64**, 171–6.

2. AMMERMANN, D. (1965) Cytologische und genetische Untersuchungen an dem Ciliaten *Stylonychia mytilus* Ehrenberg. *Arch. Protistenk.* **108**, 109–52.

2a. AMMERMANN, D. (1968) Synthese und Abbau der Nucleinsäuren während der Entwicklung des Makronucleus von *Stylonychia mytilus* (Protozoa, Ciliata). *Chromosoma* **25**, 107–20.

2b. AMMERMANN, D. (1969) Release of DNA breakdown products into the culture medium of *Stylonychia mytilus* exconjugants (Protozoa, Ciliata) during the destruction of the polytene chromosomes. *J. Cell Biol.* **40**, 576–7.

3. ANDRÉ, J., and VIVIER, E. (1962) Quelques aspects ultrastructuraux de l'échange micronucléaire lors de la conjugaison chez *Paramecium caudatum*. *J. Ultrastruct. Res.* **6**, 390–406.

4. BALBIANI, E. G. (1858) L'existence d'une génération sexuelle chez les Infusories. *J. Physiologie* **1**, 347–52.

5. BALBIANI, E. G. (1861) Recherches sur les phénomènes sexuels des Infusoires. *J. Physiologie* **4**, 102–30, 194–220, 431–48, 465–520.

6. BARNETT, A. (1964) Cytology of conjugation in *Paramecium multimicronucleatum*, syngen 2, stock 11. *J. Protozool.* **11**, 147–53.

7. BEHREND, K. (1916) Zur Konjugation von *Loxocephalus*. *Arch. Protistenk.* **37**, 1–5.

8. BERGER, J. D. (1967) Incorporation of tritiated thymidine into *Paramecium aurelia* during conjugation. *J. Protozool.* **14** (Suppl.), 32.

9. BHANDARY, A. V. (1959) Cytology of an Indian race of *Blepharisma undulans* (Stein). *J. Protozool.* **6**, 333–9.

9a. BHANDARY, A. V. (1960) Conjugation in *Blepharisma undulans americanum*. *J. Protozool.* **7**, 250–5.

10. BOJEVA-PETRUSHEVSKAJA, T. N. (1933) (Materials on the life-cycle of *Nyctotherus cordiformis*). *Trudy Leningrad. Obsz. Estestvoisp.* **42**, 27–40 (in Russian).

11. BOGDANOWICZ, A. (1930) Über die Konjugation von *Loxodes striatus* (Engelm.) Penard und *Loxodes rostrum* (O.F.M.) Ehrenb. *Zool. Anz.* **87**, 209–22.

12. BROUARDEL, J. (1936) Phénomènes nucléaires de la conjugaison chez *Trichodina patellae* Cuénot. *C.R. Acad. Sci. Paris* **203**, 818–20.

13. BRUMPT, E. (1913) Études sur les Infusoires parasites. I. La conjugaison d'*Anoplophrya circulans* Balbiani, 1885. *Arch. Parasitologie* **16**, 187–210.

14. BÜTSCHLI, O. (1876) Studien über die ersten Entwicklungsvorgänge der Eizelle, die Zellteilung und die Conjugation der Infusorien. *Abhandl. Senckenberg. Naturforsch. Ges.* **10**, 213–464.

15. CALKINS, G. N. (1919) *Uroleptus mobilis*, Engelm. I. History of the nuclei during division and conjugation. *J. Exper. Zool.* **27**, 293–357.

16. CALKINS, G. N. (1921) *Uroleptus mobilis*, Engelm. IV. Effects of cutting during conjugation. *J. Exper. Zool.* **34**, 449–70.

17. CALKINS, G. N. (1930) *Uroleptus halseyi* Calk. III. The kinetic elements and the micronucleus. *Arch. Protistenk.* **72**, 49–70.

18. CALKINS, G. N., and BOWLING, R. (1929) Studies on *Dallasia frontata*. Cytology, gametogamy, and conjugation. *Arch. Protistenk.* **66**, 11–32.

19. CALKINS, G. N., and CULL, S. W. (1907) The conjugation of *Paramecium aurelia* (*caudatum*). *Arch. Protistenk.* **10**, 375–415.

19a. CANELLA, M. F. (1957) Studi e ricerche sui tentaculiferi nel quadro della biologia generale. *Ann. Univ. Ferrara*, n. ser., Sez. III, *Biol. Anim.*, vol. 1, 259–716.

20. CANELLA, M. F. (1957–9) Biologia degli infusori e ipotetici raffronti con i Metazoi. Problematica dell'anfimissi. *Monit. Zool. Ital.* **65**, 164–83; **66**, 198–228; **67**, 143–89.

21. CAULLERY, M., and MESNIL, F. (1915) Sur *Trichodina patellae* Cuénot (Symbiose avec des Zooxanthellae; structure, division, conjugaison). *C.R. Soc. Biol. Paris* **78**, 674–7.

22. CHATTON, E., and LWOFF, A. (1935) Les Ciliés apostomes. *Arch. Zool. Expér. Génér.* **77**, 1–453.

23. CHATTON, E., and PERARD, CH. (1921) Les Nicollelidae, Infusoires intestinaux des Gondis et des Damans, et le "cycle évolutif" des Ciliés. *Bull. Biol. France Belg.* **55**, 87–153.

24. CHEN, T. T. (1940) Polyploidy and its origin in *Paramecium*. *J. Hered.* **31**, 175–84.

25. CHEN, T. T. (1940) Conjugation in *Paramecium bursaria* between animals with diverse nuclear constitutions. *J. Hered.* **31**, 185–96.

26. CHEN, T. T. (1940) A further study on polyploidy in *Paramecium*. *J. Hered.* **31**, 249–51.

27. CHEN, T. T. (1940) Conjugation of three animals in *Paramecium bursaria*. *Proc. Nat. Acad. Sci. U.S.A.* **26**, 231–8.

28. CHEN, T. T. (1940) Polyploidy in *Paramecium bursaria*. *Proc. Nat. Acad. Sci. U.S.A.* **26**, 239–40.

29. CHEN, T. T. (1940) Evidences of exchange of pronuclei during conjugation in *Paramecium bursaria*. *Proceed. Nat. Acad. Sci. U.S.A.* **26**, 241–3.

30. CHEN, T. T. (1940) Conjugation in *Paramecium bursaria* between animals with very different chromosome numbers and between animals with and without micronuclei. *Proc. Nat. Acad. Sci. U.S.A.* **26**, 243–6.

31. CHEN, T. T. (1945) Induction of conjugation in *Paramecium bursaria* among animals of one mating type by fluid from another mating type. *Proc. Nat. Acad. Sci. U.S.A.* **31**, 404–10.

32. CHEN, T. T. (1946) Conjugation in *Paramecium bursaria*. I. Conjugation of three animals. *J. Morphol.* **78**, 353–95.

33. CHEN, T. T. (1946) Conjugation in *Paramecium bursaria*. II. Nuclear phenomena in lethal conjugation between varieties. *J. Morphol.* **79**, 125–262.

34. CHEN, T. T. (1949) Conjugation between double monsters and single animals in *Paramecium bursaria*. *Proc. Nat. Acad. Sci. U.S.A.* **35**, 108–11.

35. CHEN, T. T. (1951) Conjugation in *Paramecium bursaria*. III. Nuclear changes in conjugation between double monsters and single animals. *J. Morphol.* **88**, 245–91.

36. CHEN, T. T. (1951) Conjugation in *Paramecium bursaria*. IV. Nuclear behavior in conjugation between old and young clones. *J. Morphol.* **88**, 293–360.

37. CLARK, G. M., and ELLIOTT, A. M. (1956) Nuclear behavior in haploid clones of *Tetrahymena pyriformis*. *J. Protozool.* **3** (Suppl.), 3.

38. COLLIN, B. (1909) La conjugaison d'*Anoplophrya branchiarum* (Stein) (*A. circulans* Balbiani). *Arch. Zool. Expér. Génér.*, Ser. 5, **1**, 345–88.

39. COLLIN, B. (1912) Étude monographique sur les Acinétiens. II. Morphologie, physiologie, systématique. *Arch. Zool. Expér. Génér.* **51**, 1–457.

40. COLWIN, L. H. (1944) Binary fission and conjugation in *Urceolaria synaptae* (?) type II (Protozoa, Ciliata) with special reference to the nuclear phenomena. *J. Morphol.* **75**, 203–49.

41. CORLISS, J. O. (1952) Le cycle autogamique de *Tetrahymena rostrata. C.R. Acad. Sci. Paris* **235**, 399–402.
42. CORLISS, J. O. (1965) L'autogamie et la sénescence du Cilié hyménostome *Tetrahymena rostrata* (Kahl). *Anneé Biol.* **4**, 49–69.
43. DAIN, L. (1930) Die Conjugation von *Cryptochilum echini* Maupas. *Arch. Protistenk.* **70**, 192–216.
44. DANGEARD, P.-A. (1911) Sur la conjugaison des Infusoires ciliés. *C.R. Acad. Sci. Paris* **152**, 1032–5.
45. DANGEARD, P.-A. (1911) Sur la fécondation des Infusoires ciliés. *C.R. Acad. Sci. Paris* **152**, 1703–5.
46. DASS, C. M. S. (1953) Studies on the nuclear apparatus of peritrichous ciliates. I. The nuclear apparatus of *Epistylis articulata* From. *Proc. Nat. Inst. Sci. India* **19**, 389–404.
47. DASS, C. M. S. (1954) Studies on the nuclear apparatus of peritrichous ciliates. II. The nuclear apparatus of *Carchesium spectabile* Ehrbg. *Proc. Nat. Inst. Sci. India* **20**, 174–86.
48. DASS, C. M. S. (1954) Studies on the nuclear apparatus of peritrichous ciliates. III. The nuclear apparatus of *Epistylis* sp. *Proc. Nat. Inst. Sci. India* **20**, 703–15.
49. DASS, C. M. S. (1955) Evolution of sex in ciliate Protozoa. *Bull. Nat. Inst. Sci. India* No. 7, 173–7.
50. DAWSON, J. A. (1919) An experimental study of an amicronucleate *Oxytricha*. I. Study of the normal animal with an account of cannibalism. *J. Exper. Zool.* **29**, 473–513.
51. DE GARIS, C. F. (1935) Heritable effects of conjugation between free individuals and double monsters in diverse races of *Paramecium caudatum. J. Exper. Zool.* **71**, 209–56.
52. DEHORNE, A. (1911) La non-copulation du noyau échangé et du noyau stationnaire et la disparition de ce dernier dans la conjugaison de *Paramecium caudatum. C.R. Acad. Sci. Paris* **152**, 922–5.
53. DEHORNE, A. (1911) La permutation nucléaire dans la conjugaison de *Colpidium colpoda. C.R. Acad. Sci. Paris* **152**, 1354–7.
54. DEHORNE, A. (1920) Contribution à l'étude comparée de l'appareil nucléaire des Infusoires ciliés (*Paramecium caudatum* et *Colpidium truncatum*), des Euglènes et des Cyanophycées. *Arch. Zool. Expér. Génér.* **60**, 47–176.
55. DEVI, R. V. (1961) Autogamy in *Frontonia leucas* (Ehrbg.) *J. Protozool.* **8**, 277–83.
56. DEVIDÉ, Z. (1951) Chromosomes in Ciliates (Euciliata and Opalinidae). *Bull. Internat. Acad. Yougosl. Sci. et Beaux-Arts Zagreb*, n. sér., **3**, 75–114.
57. DEVIDÉ, Z., and GEITLER, L. (1947) Die Chromosomen der Ciliaten. *Chromosoma* **3**, 110–36.
58. DILLER, W. F. (1936) Nuclear reorganization processes in *Paramecium aurelia*, with description of autogamy and "hemixis". *J. Morphol.* **59**, 11–67.
59. DILLER, W. F. (1940) Nuclear variation in *Paramecium caudatum. J. Morphol.* **66**, 605–33.
60. DILLER, W. F. (1942) Re-conjugation in *Paramecium caudatum. J. Morphol.* **70**, 229–48.
61. DILLER, W. F. (1948) Nuclear behavior of *Paramecium trichium* during conjugation. *J. Morphol.* **82**, 1–52.
62. DILLER, W. F. (1948) Induction of autogamy in single animals of *Paramecium calkinsi* following mixture of two mating types. *Biol. Bull.* **95**, 265.
63. DILLER, W. F. (1949) An abbreviated conjugation process in *Paramecium trichium. Biol. Bull.* **97**, 331–43.
64. DILLER, W. F. (1950) An extra postzygotic nuclear division in *Paramecium caudatum. Trans. Am. Microscop. Soc.* **69**, 309–16.
65. DILLER, W. F. (1950) Cytological evidence for pronuclear interchange in *Paramecium caudatum. Trans. Am. Microscop. Soc.* **69**, 317–23.

66. Diller, W. F. (1954) Autogamy in *Paramecium polycaryum*. *J. Protozool.* **1**, 60–70.
67. Diller, W. F. (1958) Studies on conjugation in *Paramecium polycaryum*. *J. Protozool.* **5**, 282–92.
68. Diller, W. F. (1959) Possible origin of hypoploidy in *Paramecium trichium*. *J. Protozool.* **6** (Suppl.), 19.
69. Diller, W. F. (1963) Nuclear activity in crosses of micronucleate and amicronucleate strains of *Paramecium multimicronucleatum*. In: *Progress in Protozoology* (*Proc. First Internat. Congr. Protozool. Prague*, 1961), 105–10.
70. Diller, W. F. (1965) Ciliary changes during the life cycle of *Oxytricha*. In: *Progress in Protozoology* (Abstr. 2nd Internat. Conf. Protozool., London), 230–1.
70a. Diller, W. F. (1965) The relation of amicronuclearity to stomatogenic activity during conjugation in several ciliates. *J. Morphol.* **116**, 51–64.
71. Diller, W. F. (1966) Correlation of ciliary and nuclear development in the life cycle of *Euplotes*. *J. Protozool.* **13**, 43–54.
72. Diller, W. F. (1967) Aspects of conjugation in *Aspidisca costata*. *J. Protozool.* **14** (Suppl.), 24.
72a. Diller, W. F. (1969) Morphogenetic studies on the contractile vacuole systems of *Euplotes* and *Paramecium*. In: *Progress in Protozoology* (Abstr. Third Internat. Congr. Protozool. Leningrad), 93–94.
73. Dippell, R. V. (1954) A preliminary report on the chromosomal constitution of certain variety 4 races of *Paramecium aurelia*. *Caryologia*, Suppl. (Atti IX Congr. Internaz. Genet.), 1109–11.
74. Doflein, F. (1907) Beobachtungen und Ideen über die Conjugation der Infusorien. *Sitz.-Ber. Ges. Morphol. Physiol. München* **23**, 107–14.
75. Doflein, F., and Reichenow, E. (1953) *Lehrbuch der Protozoenkunde*. 6-te Auflage. VEB Gustav Fischer Verlag, Jena.
76. Dogiel, V. (1925) Die Geschlechtsprozesse bei Infusorien (speziell bei den Ophryoscoleciden). Neue Tatsachen und theoretische Erwägungen. *Arch. Protistenk.* **50**, 283–442.
77. Dogiel, V. (1928) Über die Conjugation von *Bütschlia parva* (Holotricha). *Arch. Protistenk.* **62**, 1–25.
78. Dogiel, V. (1930) Die prospektive Potenz der Syncaryonderivate an der Conjugation von *Paraisotricha* erläutert. *Arch. Protistenk.* **70**, 497–516.
79. Dogiel, V. A. (1965) *General Protozoology*, 2nd ed., revised by G. Poljansky and E. Cheissin. Clarendon Press, Oxford.
80. Dragesco, J., and Raikov, I. (1966) L'appareil nucléaire, la division, et quelques stades de la conjugaison de *Tracheloraphis margaritatus* (Kahl) et *T. caudatus* sp. nov. (Ciliata, Holotricha). *Arch. Protistenk.* **109**, 99–113.
81. Egelhaaf, A. (1955) Cytologisch-entwicklungsphysiologische Untersuchungen zur Konjugation von *Paramecium bursaria* Focke. *Arch. Protistenk.* **100**, 447–514.
82. Elliott, A., and Hayes, R. (1953) Mating types in *Tetrahymena*. *Biol. Bull.* **105**, 269–84.
83. Elliott, A. M., and Tremor, J. W. (1958) The fine structure of the pellicle in the contact area of conjugating *Tetrahymena pyriformis*. *J. Biophys. Biochem. Cytol.* **4**, 839–40.
83a. Elliott, A. M., and Zieg, R. G. (1968) Golgi apparatus associated with mating in *Tetrahymena pyriformis*. *J. Cell Biol.* **36**, 391–8.
84. Ellis, I. M. (1937) The morphology, division, and conjugation of the salt-marsh ciliate *Fabrea salina* Henneguy. *Univ. Calif. Publ. Zool.* **41**, 343–88.
85. Enriques, P. (1907) La coniugazione e il differenziamento sessuale negli Infusori. *Arch. Protistenk.* **9**, 195–296.
86. Enriques, P. (1908) Die Conjugation und sexuelle Differenzierung der Infusorien. 2. Wiederconjugante und Hemisexe bei *Chilodon*. *Arch. Protistenk.* **12**, 213–76.
87. Finley, H. E. (1943) The conjugation of *Vorticella microstoma*. *Trans. Amer. Microscop. Soc.* **62**, 97–121.

88. FINLEY, H. E. (1946) Patterns of sexual reproductive cycles in ciliates. *Biodynamica* **6**, 31–79.
89. FINLEY, H. E. (1952) Sexual differentiation in peritrichous ciliates. *J. Morphol.* **91**, 569–605.
90. FINLEY, H. E., and NICHOLAS, P. A. (1950) Sexual differentiation in *Rhabdostyla vernalis. Proceed. Nat. Acad. Sci. U.S.A.* **36**, 588–90.
91. FURSSENKO, A. (1929) Lebenszyklus und Morphologie von *Zoothamnium arbuscula* Ehrenberg (Infusoria Peritricha). *Arch. Protistenk.* **67**, 376–500.
92. GIESE, A. C. (1938) Size and conjugation in *Blepharisma. Arch. Protistenk.* **91**, 125–34.
93. GOLIKOWA, M. N. (1965) Der Aufbau des Kernapparates und die Verteilung der Nukleinsäuren und Proteine bei *Nyctotherus cordiformis* Stein. *Arch. Protistenk.* **108**, 191–216.
93a. GOLIKOWA, M. N., and NILOVA, V. K. (1967) (On DNA synthesis in the macronuclear anlage of *Nyctotherus cordiformis* Stein). *Cytology (USSR)* **9**, 465–9 (in Russian with English summary).
94. GÖNNERT, R. (1935) Über Systematik, Morphologie, Entwicklungsgeschichte und Parasiten einiger Dendrosomidae nebst Beschreibung zweier neuer Suktorien. *Arch. Protistenk.* **86**, 113–54.
95. GREGORY, L. (1923) The conjugation of *Oxytricha fallax. J. Morphol.* **37**, 555–81.
96. GRELL, K. G. (1953) Die Konjugation von *Ephelota gemmipara* R. Hertwig. *Arch. Protistenk.* **98**, 287–326.
97. GRELL, K. G. (1956) *Protozoologie.* Springer Verlag, Berlin–Göttingen–Heidelberg.
98. GROMOVA, E. N. (1948) (Dynamics of nucleic acids during conjugation in *Paramecium caudatum). Doklady Acad. Nauk SSSR* **63**, 73–75 (in Russian).
99. HAMBURGER, C. (1904) Die Konjugation von *Paramecium bursaria* Focke. *Arch. Protistenk.* **4**, 199–239.
100. HAMMOND, D. M. (1937) The neuromotor system of *Euplotes patella* during binary fission and conjugation. *Quart. J. Microscop. Sci.* **79**, 507–57.
101. HANSON, E. D., and GILLIES, C. (1966) Oral structures during conjugation in *Paramecium aurelia. J. Protozool.* **13** (Suppl.), 26.
102. HARRISON, J. A., and FOWLER, E. H. (1945) Serologic evidence of cytoplasmic interchange during conjugation in *Paramecium bursaria. Science* **102**, 377–8.
103. HARTMANN, M. (1956) *Die Sexualität,* 2te Auflage. G. Fischer Verlag, Stuttgart.
104. HAYASHI, S., and TAKAYANAGI, T. (1962) Cytological and cytogenetical studies on *Paramecium polycaryum.* IV. Determination of the mating system based on some experimental and cytological observations. *Jap. J. Zool.* **13**, 357–64.
105. HECKMANN, K. (1965) Totale Konjugation bei *Urostyla hologama* n. sp. *Arch. Protistenk.* **108**, 55–62.
106. HERTWIG, R. (1889) Über die Conjugation der Infusorien. *Abhandl. Math.-Phys. Classe der Königl. Bayer. Akad. Wiss. München* **17**, 151–233.
107. HERTWIG, R. (1904) Über Konjugation von *Dileptus gigas. Sitz.-Ber. Ges. Morphol. Physiol. München* **20**, 1–3.
108. HERTWIG, R. (1914) Über Parthenogenesis der Infusorien und die Depressionszustände der Protozoen. *Biol. Zbl.* **34**, 557–81.
109. HICKSON, S. J., and WADSWORTH, J. T. (1902) *Dendrocometes paradoxus.* Part I. Conjugation. *Quart. J. Microscop. Sci.* **45**, 325–62.
110. HIWATASHI, K. (1960) An aberrant selfing strain of *Paramecium caudatum* which shows multiple unions of conjugation. *J. Protozool.* **7** (Suppl.), 20.
111. HOYER, H. (1899) Über das Verhalten der Kerne bei der Conjugation des Infusors *Colpidium colpoda* St. *Arch. Mikr. Anat.* **54**, 95–134.
112. HUNTER, L. N. (1936) Some nuclear phenomena in the *Trichodina* (Protozoa, Ciliata, Peritrichida) from *Thyone briareus* (Holothuroidea). *Biol. Bull.* **71**, 406.

113. ILOWAISKY, S. A. (1916) (Sur la copulation d'*Urostyla flavicans* Wrzésniow.). *Dnevnik Zool. Otdel. Imper. Obsz. Ljubit. Estestvozn., Antropol. i Etnogr.*, n. ser., 3, No. 4, 2–6 (in Russian with French summary).

114. ILOWAISKY, S. A. (1926) Über die Kernprozesse der getrennten Conjuganten der *Stylonychia mytilus* und *Paramecium caudatum. Arch. Protistenk.* 53, 243–52.

115. INABA, F. (1965) Conjugation between two strains of *Blepharisma. J. Protozool.* 12, 146–51.

116. INABA, F., IMAMOTO, K., and SUGANUMA, Y. (1966) Electron-microscopic observations on nuclear exchange during conjugation in *Paramecium multimicronucleatum. Proc. Japan Acad.* 42, 394–8.

117. ITO, S. (1963) (Cytological observations of nuclear behavior in the conjugation of *Diophrys scutum*). *Zool. Mag. (Jap.)* 72, 230–4 (in Japanese with English summary).

118. IVANIĆ, M. (1933) Die Conjugation von *Chilodon cucullulus. Arch. Protistenk.* 79, 313–48.

119. JAMESON, A. (1927) The behaviour of *Balantidium coli* Malm. in cultures. *Parasitology* 19, 411–19.

120. JANKOWSKI, A. V. (1960) (The conjugation processes in *Paramecium trichium* Stokes. I. Amphimixis and autogamy). *Cytology (USSR)*, 2, 581–8 (in Russian).

121. JANKOWSKI, A. V. (1961) (The process of conjugation in the rare salt water *Paramecium, P. woodruffi*). *Doklady Acad. Nauk SSSR* 137, 989–92 (in Russian).

122. JANKOWSKI, A. V. (1962) (The conjugation processes in *Paramecium putrinum* Clap. et Lachm. II. Apomictic reorganization cycles and the system of mixotypes). *Cytology (USSR)* 4, 435–44 (in Russian).

123. JANKOWSKI, A. V. (1962) (The conjugation processes in *Paramecium putrinum* Clap. et Lachm. III. A multiple system of mating types in *P. putrinum*). *Zhurn. Obsh. Biologii* 23, 276–82 (in Russian).

124. JANKOWSKI, A. V. (1962) (Nuclear reorganization of the endomixis type in clones of *Cyclidium glaucoma* O.F.M.). *Nauchnye Doklady Vysshej Shkoly, Biol. Nauki*, No. 4, 14–19 (in Russian).

125. JANKOWSKI, A. V. (1965) (The conjugation processes in *Paramecium putrinum* Clap. et Lachm. V. Return to amphimixis in mixotype B). *Doklady Akad. Nauk SSSR* 163, 523–5 (in Russian).

126. JANKOWSKI, A. V. (1965) (The conjugation processes in *Paramecium putrinum* Clap. et Lachm. VII. Nuclear phenomena during autogamy in singles, induced with a new method of "multiple mating"). *Acta Protozool.* 3, 239–63 (in Russian with English summary).

127. JANKOWSKI, A. V. (1966) (The conjugation processes in *Paramecium putrinum* Clap. et Lachm. VI. Induction and cytological study of triple conjugation). *Cytology (USSR)* 8, 70–79 (in Russian with English summary).

128. JANKOWSKI, A. V. (1966) (The conjugation processes in *Paramecium putrinum* Clap. et Lachm. VIII. Nuclear phenomena during re-conjugation). *Zool. Zhurn.* 45, 818–29 (in Russian with English summary).

128a. JAREÑO, M. A., ALONSO, P., and PÉREZ-SILVA, J. (1969) Autogamy in *Stylonychia*. In: *Progress in Protozoology* (Abstr. *Third Internat. Congr. Protozool. Leningrad*), 29–30.

129. JENNINGS, H. S. (1911) Assortative mating, variability, and inheritance of size in the conjugation of *Paramecium. J. Exper. Zool.* 11, 1–134.

130. JENNINGS, H. S. (1944) *Paramecium bursaria*: life history. I. Immaturity, maturity, and age. *Biol. Bull.* 86, 131–45.

131. JENNINGS, H. S. (1944) *Paramecium bursaria*: life history. II. Age and death of clones in relation to the results of conjugation. *J. Exper. Zool.* 96, 27–52.

131a. JURAND, A., and SELMAN, G. G. (1969) *Anatomy of Paramecium aurelia*. Macmillan & Co., London.

132. KALTENBACH, R. (1915) Die Conjugation von *Ophrydium versatile*. *Arch. Protistenk.* 36, 67–71.

133. KASANZEFF, W. (1910) Zur Kenntnis von *Loxodes rostrum*. *Arch. Protistenk.* 20, 79–96.

134. KATASHIMA, R. (1952) Studies on *Euplotes*. I. Conjugation and cytogamy induced by split pair method in *Euplotes harpa*. *J. Sci. Hiroshima Univ.*, ser. B, div. 1, 13, 111–20.

135. KATASHIMA, R. (1953) Studies on *Euplotes*. II. Macronuclear reorganization process, double and giant animals from two-united exconjugants. *J. Sci. Hiroshima Univ.*, ser. B, div. 1, 14, 57–71.

136. KATASHIMA, R. (1959) A correlation between morphogenesis and old macronucleus during sexual reproduction in *Euplotes eurystomus*. *J. Sci. Hiroshima Univ.*, ser. B, div. 1, 18, 99–107.

137. KATASHIMA, R. (1959) The intimacy of union between the two members of the conjugating pairs in *Euplotes eurystomus*. *Jap. J. Zool.* 12, 329–43.

138. KATASHIMA, R. (1960) Correlations between nuclear behavior and disposition of the nuclei in cytoplasm during conjugation of *Euplotes patella*. *J. Sci. Hiroshima Univ.*, ser. B, div. 1, 18, 239–63.

139. KAY, M. W. (1946) Studies on *Oxytricha bifaria* Stokes. III. Conjugation. *Trans. Amer. Microscop. Soc.* 65, 132–48.

140. KIDDER, G. W. (1933) Studies on *Conchophthirius mytili*. II. Conjugation and nuclear reorganization. *Arch. Protistenk.* 79, 25–49.

141. KIDDER, G. W. (1933) On the genus *Ancistruma* Strand (= *Ancistrum* Maupas). II. The conjugation and nuclear reorganization of *A. isseli* Kahl. *Arch. Protistenk.* 81, 1–18.

142. KIMBALL, R. F. (1941) Double animals and amicronucleate animals in *Euplotes patella*, with special reference to their conjugation. *J. Exper. Zool.* 86, 1–32.

143. KIMBALL, R. F. (1942) The nature and inheritance of mating types in *Euplotes patella*. *Genetics* 27, 269–85.

144. KIMBALL, R. F., and GAITHER, N. (1955) Behavior of nuclei at conjugation in *Paramecium aurelia*. I. Effect of incomplete chromosome sets and competition between complete and incomplete nuclei. *Genetics* 40, 878–89.

145. KING, R. L. (1962) Photomicrographs of reproduction in a suctorian. *J. Protozool.* 9 (Suppl.), 7–8.

146. KLEE, E. E. (1925) Der Formwechsel im Lebenskreis reiner Linien von *Euplotes longipes*. *Zool. Jahrb., Abt. Allgem. Zool. Physiol.* 42, 307–65.

147. KLEIN, B. (1927) Silberliniensystem der Ciliaten. Ihr Verhalten während Teilung und Conjugation, neue Silberbilder, Nachträge. *Arch. Protistenk.* 58, 55–142.

148. KLITZKE, M. (1914) Über Wiederconjuganten bei *Paramecium caudatum*. *Arch. Protistenk.* 33, 1–20.

149. KLITZKE, M. (1915) Ein Beitrag zur Kenntnis der Kernentwicklung bei den Ciliaten. *Arch. Protistenk.* 36, 215–35.

150. KOFOID, C. A., and BUSH, M. (1936) The life cycle of *Parachaenia myae* gen. nov., sp. nov., a ciliate parasitic in *Mya arenaria* Linn. from San Francisco Bay, California. *Bull. Mus. Roy. Hist. Nat. Belg.* 12, No. 22, 1–15.

151. KORMOS, J. (1935) (Sexueller Dimorphismus und Conjugation von *Prodiscophrya collini* [Root]). *Állat. Közlem.* 32, 152–68 (in Hungarian with German summary).

152. KORMOS, J. (1940) (Über einige Probleme der Konjugation bei Infusorien). *Állat. Közlem.* 37, 39–58 (in Hungarian with German summary).

153. KORMOS, J., and KORMOS, K. (1957) Neue Untersuchungen über den Geschlechtsdimorphismus der Prodiscophryen. *Acta Biol. Acad. Sci. Hung.* 7, 109–25.

154. KORMOS, J., and KORMOS, K. (1958) Äußere und innere Konjugation. *Acta Biol. Acad. Sci. Hung.* 8, 103–26.

155. KORMOS, J., and KORMOS, K. (1960) Direkte Beobachtung der Kernveränderungen der Konjugation von *Cyclophrya katharinae* (Ciliata, Protozoa). *Acta Biol. Acad. Sci. Hung.* **10**, 373–94.

156. KOŚCIUSZKO, H. (1965) Karyologic and genetic investigations in syngen 1 of *Paramecium aurelia*. *Folia Biol.* **13**, 339–70.

157. KOVALEVA, N. E., and JANKOWSKI, A. V. (1965) (Effect of ionizing radiation on nuclear reorganization processes in *Paramecium putrinum*. I. Nuclear processes in irradiated conjugants). *Zhurn. Obsh. Biol.* **26**, 176–89 (in Russian).

158. KOVALEVA, N. E., and JANKOWSKI, A. V. (1966) (Effect of ionizing radiation on nuclear reorganization processes in *Paramecium putrinum*. II. Nuclear processes in exconjugants). *Acta Protozool.* **4**, 25–39 (in Russian with English summary).

158a. KOZLOFF, E. N. (1966) *Phalacrocleptes verruciformis* gen. nov., sp. nov., an unciliated ciliate from the sabellid polychaete *Schizobranchia insignis* Bush. *Biol. Bull.* **130**, 202–10.

159. LANDIS, E. (1925) Conjugation of *Paramecium multimicronucleatum* Powers and Mitchell. *J. Morphol.* **40**, 111–67.

160. MACDOUGALL, M. S. (1925) Cytological observations on gymnostomatous ciliates, with a description of the maturation phenomena in diploid and tetraploid forms of *Chilodon uncinatus*. *Quart. J. Microscop. Sci.* **69**, 361–84.

161. MACDOUGALL, M. S. (1936) Étude cytologique de trois espèces du genre *Chilodonella* Strand. Morphologie, conjugaison, réorganisation. *Bull. Biol. France Belg.* **70**, 308–31.

162. MANWELL, R. D. (1928) Conjugation, division, and encystment in *Pleurotricha lanceolata*. *Biol. Bull.* **54**, 417–54.

163. MARTIN, C.-H. (1909) Some observations on Acinetaria. I. The "Tinctin-Körper" of Acinetaria and the conjugation of *Acineta papillifera*. *Quart. J. Microscop. Sci.* **53**, 351–377.

164. MAUPAS, E. (1888) Recherches expérimentales sur la multiplication des Infusoires ciliés. *Arch. Zool. Expér. Génér.*, sér. 2, **6**, 165–277.

165. MAUPAS, E. (1889) La rajeunissement karyogamique chez les Ciliés. *Arch. Zool. Expér. Génér.*, sér. 2, **7**, 149–517.

166. MCDONALD, B. B. (1966) The exchange of RNA and protein during conjugation in *Tetrahymena*. *J. Protozool.* **13**, 277–85.

167. MESSJATZEV, I. (1924) (Konjugation in *Lionotus lamella* Ehrbg.). *Russkij Arch. Protistol.* **3**, 33–43 (in Russian).

168. MINKIEWICZ, R. (1912) Un cas de reproduction extraordinaire chez un Protiste, *Polyspira delagei* Minkiew. *C.R. Acad. Sci. Paris* **155**, 733–7.

169. MITCHELL, J. B. (1962) Nuclear reorganization in *Paramecium jenningsi*. *J. Protozool.* **9** (Suppl.), 26.

170. MIYASHITA, Y. (1928) On a new parasitic ciliate *Lada tanishi* n. sp., with preliminary notes on its heterogamic conjugation. *Jap. J. Zool.* **1**, 205–18.

171. MOLDENHAUER, D. (1964) Zytologische Untersuchungen zum Austausch der Wanderkerne bei konjugierenden *Paramecium caudatum*. *Arch. Protistenk.* **107**, 163–78.

172. MOLDENHAUER, D. (1965) Zytologische Untersuchungen zur Konjugation von *Stylonychia mytilus* und *Urostyla polymicronucleata*. *Arch. Protistenk.* **108**, 63–90.

173. MÜGGE, E. (1957) Die Konjugation von *Vorticella campanula* (Ehrbg.). *Arch. Protisenk.* **102**, 165–208.

174. MÜLLER, W. (1932) Cytologische und vergleichend-physiologische Untersuchungen über *Paramecium multimicronucleatum* und *Paramecium caudatum*, zugleich ein Versuch zur Kreuzung beider Arten. *Arch. Protistenk.* **78**, 361–462.

175. MULSOW, W. (1913) Die Conjugation von *Stentor coeruleus* und *Stentor polymorphus*. *Arch. Protistenk.* **28**, 363–88.

176. NAKATA, A. (1956) Micronuclear behavior during the conjugation of *Paramecium calkinsi*. *Zool. Mag.* (*Jap.*) **65**, 306–10.

177. NANNEY, D. L. (1953) Nucleo-cytoplasmic interaction during conjugation in *Tetrahymena. Biol. Bull.* **105**, 133–48.
178. NANNEY, D. L. (1956) Inbreeding deterioration in *Tetrahymena pyriformis. Genetics* **41**, 655.
178a. NANNEY, D. L., and NAGEL, M. J. (1964) Nuclear misbehavior in an aberrant inbred *Tetrahymena. J. Protozool.* **11**, 465–73.
179. NELSON, E. C. (1934) Observations and experiments on conjugation of the *Balantidium* from the chimpanzee. *Am. J. Hyg.* **20**, 106–34.
180. NOBILI, R. (1965) La riproduzione sessuale nei ciliati. *Boll. Zool.* **32**, 93–131.
181. NOBILI, R. (1967) Ultrastructure of the fusion region of conjugating *Euplotes* (Ciliata Hypotrichida). *Monit. Zool. Ital.*, n. ser., **1**, 73–89.
181a. NOBILI, R., and LUPORINI, P. (1967) Maintenance of heterozygosity at the mt locus after autogamy in *Euplotes minuta* (Ciliata Hypotrichida). *Genet. Res. (Cambridge)* **10**, 35–43.
182. NOBLE, E. A. (1932) On *Tokophrya lemnarum* Stein (Suctoria), with an account of its budding and conjugation. *Univ. Calif. Publ. Zool.* **37**, 477–520.
183. NOIROT-TIMOTHÉE, C. (1960) Étude d'une famille de Ciliés: les Ophryoscolecidae. *Ann. Sci. Nat., Zool.*, Sér. 12, **2**, 527–718.
184. NOLAND, L. E. (1927) Conjugation in the ciliate *Metopus sigmoides* C. and L. *J. Morphol.* **44**, 341–61.
185. OSSIPOV, D. V. (1966) (Methods for obtaining homozygous clones of *Paramecium caudatum*). *Genetika (Moscow)* **2**, 41–48 (in Russian with English summary).
185a. OSSIPOV, D. V., and SKOBLO, I. I. (1968) The autogamy during conjugation in *Paramecium caudatum* Ehrbg. II. The exautogamont stages of the nuclear reorganization. *Acta Protozool.* **6**, 33–52.
186. PADNOS, M. (1939) The morphology and life history of *Trichodina* infecting the Puffer. *Anat. Rec.* **75** (Suppl.), 157.
187. PADNOS, M., and NIGRELLI, R. F. (1942) *Trichodina spheroidesi* and *Trichodina halli* spp. nov., parasitic on the gills and skin of marine fishes, with special reference to the life-history of *T. spheroidesi. Zoologica* **27**, 65–72.
188. PATTEN, M. W. (1921) The life history of an amicronucleate race of *Didinium nasutum. Proc. Soc. Exper. Biol. Med.* **18**, 188–9.
189. PEARL, R. (1907) A biometrical study of conjugation in *Paramecium. Biometrika* **5**, 213–97.
190. PENARD, E. (1922) *Études sur les Infusoires d'eau douce.* Georg & Cie, Genève.
191. PENN, A. B. K. (1937) Reinvestigation into the cytology of conjugation in *Paramecium caudatum. Arch. Protistenk.* **89**, 45–54.
192. PÉREZ-SILVA, J. (1965) Conjugation in *Frontonia acuminata* Ehrenberg. In: *Progress in Protozoology* (Abstr. *2nd Internat. Conf. Protozool., London*), 216–17.
192a. PÉREZ-SILVA, J., and ALONSO, P. (1966) Demonstration of polytene chromosomes in the macronuclear anlage of oxytrichous ciliates. *Arch. Protistenk.* **109**, 65–70.
192b. PÉREZ-SILVA, J., ALONSO, P., GIL, R., and JAREÑO, M. A. (1969) Puffing in the polytene chromosomes of *Stylonychia mytilus* Ehrenberg. In: *Progress in Protozoology* (Abstr. *Third Internat. Congr. Protozool., Leningrad*), 36.
193. PESHKOVSKAJA, L. S. (1936) (Changes of the nuclear apparatus of *Climacostomum virens* during conjugation). *Biol. Zhurnal (Moscow)* **5**, 207–20 (in Russian with English summary).
194. PESHKOVSKAJA, L. S. (1941) (Changes of the nuclear apparatus during conjugation in some holotrichous ciliates). *Trudy Inst. Cytol., Histol. i Embryol.* **1**, 19–27 (in Russian).
195. PESHKOVSKAJA, L. S. (1948) (Metamorphosis of the nuclear apparatus during the sexual process in two species of hypotrichous ciliates). *Trudy Inst. Cytol., Histol. i Embryol.* **3**, 201–8 (in Russian).

196. PIERI, J. (1965) Interprétation cytophotométrique des phénomènes nucléaires au cours de la conjugaison chez *Stylonychia pustulata*. *C.R. Acad. Sci. Paris* **261**, 2742–4.

197. PIERI, J. (1966) Phénomènes nucléaires au cours de la conjugaison chez *Stylonychia pustulata*. *Protistologica* **2**, No. 1, 53–58.

197a. PIERI, J., and VAUGHIEN, C. (1967) Analyse microspectrographique dans l'ultraviolet d'une processus de biosynthèse nucléaire chez *Chilodonella*. *C.R. Acad. Sci. Paris*, D, **265**, 256–9.

197b. PIERI, J., VAUGHIEN, C., and TROUILLER, M. (1968) Interprétation cytophotométrique des phénomènes micronucléaires au cours de la division binaire et des divisions prégamiques chez *Paramecium trichium*. *J. Cell Biol.* **36**, 664–8.

198. POLJANSKY, G. (1926) Die Conjugation von *Dogielella sphaerii* (Infusoria Holotricha, Astomata). *Arch. Protistenk.* **53**, 407–34.

199. POLJANSKY, G. (1934) Geschlechtsprozesse bei *Bursaria truncatella* O. F. Müll. *Arch. Protistenk.* **81**, 420–546.

200. POLJANSKY, G. (1938) (Reconstruction of the nuclear apparatus in *Bursaria truncatella* O. F. Müll. after experimental separation of conjugating pairs). *Biol. Zhurnal* (*Moscow*) **7**, 123–30 (in Russian).

201. POLJANSKY, G., and STRELKOV, A. A. (1938) (Sexual processes in *Entodinium caudatum* Stein). *Zool. Zhurnal* (*Moscow*) **17**, 75–80 (in Russian).

202. POPOFF, M. (1908) Die Gametenbildung und die Conjugation von *Carchesium polypinum* L. *Zeitschr. Wiss. Zool.* **89**, 478–524.

203. PORTER, E. D. (1960) The buccal organelles in *Paramecium aurelia* during fission and conjugation, with special reference to the kinetosomes. *J. Protozool.* **7**, 211–17.

204. PRANDTL, H. (1906) Die Konjugation von *Didinium nasutum* O. F. M. *Arch. Protistenk.* **7**, 229–58.

204a. PUYTORAC, P. DE, and BLANC, J. (1967) Observations sur les modifications ultrastructurales des micronoyaux au cours de leurs transformations en macronoyaux chez *Paramecium caudatum*. *C.R. Soc. Biol. Paris* **161**, 297–9.

205. PROWAZEK, S. von (1899) Protozoenstudien. 1. *Bursaria truncatella*. 2. *Styloynchia pustulata*. *Arb. Zool. Inst. Univ. Wien* **11**, 195–268.

206. PROWAZEK, S. von (1909) Conjugation von *Lionotus*. *Zool. Anz.* **34**, 626–8.

207. PROWAZEK, S. von (1915) Zur Morphologie und Biologie von *Colpidium colpoda*. *Arch. Protistenk.* **36**, 72.

207a. RADZIKOWSKI, S. (1967) Nuclear behavior during conjugation and polytene chromosomes in the exconjugants of *Chilodonella cucullulus* (O. F. MÜLLER). *Bull. Acad. Polon. Sci.*, ser. biol., Cl. II, **15**, 749–51.

208. RAIKOV, I. B. (1958) (Conjugation in the holotrichous ciliate, *Trachelocerca phoenicopterus* Cohn). *Zool. Zhurn.* (*Moscow*) **37**, 781–800 (in Russian with English summary).

209. RAIKOV, I. B. (1958) Der Formwechsel des Kernapparates einiger niederer Ciliaten. I. Die Gattung *Trachelocerca*. *Arch. Protistenk.* **103**, 129–92.

210. RAIKOV, I. B. (1959) Der Formwechsel des Kernapparates einiger niederer Ciliaten. II. Die Gattung *Loxodes*. *Arch. Protistenk.* **104**, 1–42.

211. RAIKOV, I. B. (1962) Der Kernapparat von *Nassula ornata* Ehrbg. (Ciliata, Holotricha). Zur Frage über den Chromosomenaufbau des Makronucleus. *Arch. Protistenk.* **105**, 463–88.

212. RAIKOV, I. B. (1963) (Some stages of conjugation in the holotrichous ciliate *Trachelocerca coluber*). *Cytology* (*USSR*) **5**, 685–9 (in Russian).

213. RAIKOV, I. B. (1967) (*Karyology of Protozoa*). Publ. House "Nauka", Leningrad (in Russian).

214. RAO, M. V. N. (1964) Nuclear behavior of *Euplotes woodruffi* during conjugation. *J. Protozool.* **11**, 296–304.

215. RAO, M. V. N. (1966) Conjugation in *Kahlia* sp. with special reference to meiosis and endomitosis. *J. Protozool.* **13**, 565–73.

216. RAO, M. V. N. (1968) Nuclear behavior of *Spirostomum ambiguum* during conjugation with special reference to macronuclear development. *J. Protozool.* **15**, 748–52.

216a. RAO, M. V. N. (1968) Macronuclear development in *Euplotes woodruffi* following conjugation. *Exper. Cell Res.* **49**, 411–19.

217. RAY, CH. (1955) Microscopic observation on reciprocal interchange of pronuclei in *Tetrahymena pyriformis*. *Biol. Bull.* **109**, 367.

218. RAY, CH. (1955) Irregularities during meiosis in variety 9 of *Tetrahymena pyriformis*. *Biol. Bull.* **109**, 367.

219. RAY, CH. (1956) Meiosis and nuclear behavior in *Tetrahymena pyriformis*. *J. Protozool.* **3**, 88–96.

220. RAY, CH. (1956) Nuclear aberrations associated with lethal conjugation in *Tetrahymena pyriformis*. *J. Protozool.* **3** (Suppl.), 3.

220a. REIFF, I. (1968) Die genetische Determination multipler Paarungstypen bei dem Ciliaten *Uronychia transfuga* (Hypotricha, Euplotidae). *Arch. Protistenk.* **110**, 372–97.

221. ROQUE, M. (1956) L'évolution de la ciliature buccale pendant l'autogamie et la conjugaison chez *Paramecium aurelia*. *C.R. Acad. Sci. Paris* **242**, 2592–5.

222. ROQUE, M. (1956) La stomatogénèse pendant l'autogamie, la conjugaison et la division chez *Paramecium aurelia*. *C.R. Acad. Sci. Paris* **243**, 1564–5.

223. ROSENBERG, L. E. (1940) Conjugation in *Opisthonecta henneguyi*, a free swimming vorticellid. *Proc. Amer. Philosoph. Soc.* **82**, 437–48.

224. SCHNEIDER, L. (1963) Elektronenmikroskopische Untersuchungen der Konjugation von *Paramecium*. I. Die Auflösung und Neubildung der Zellmembran bei den Konjuganten. *Protoplasma* **56**, 109–40.

225. SCHOOLEY, C. N. (1958) An autoradiographic study of cytoplasmic exchange during conjugation in *Tetrahymena pyriformis*. *J. Protozool.* **5** (Suppl.), 24.

226. SCHWARTZ, V. (1952) Die Sexualität der Infusorien. *Fortschr. Zool.* **9**, 605–19.

227. SCOTT, M. J. (1927) Studies on the *Balantidium* from the guinea pig. *J. Morphol.* **44**, 417–65.

228. SESHACHAR, B. R. (1965) Conjugation in *Spirostomum ambiguum* Ehrbg. In: *Progress in Protozoology* (Abstr. *2nd Internat. Conf. Protozool., London*), 20.

228a. SESHACHAR, B. R., and BAI, A. R. K. (1968) Conjugation in *Spirostomum ambiguum* Ehrbg. *Acta Protozool.* **6**, 51–56.

229. SESHACHAR, B. R., and BHANDARY, A. V. (1962) Observations on the life-cycle of a new race of *Blepharisma undulans* from India. *J. Protozool.* **9**, 265–70.

230. SESHACHAR, B. R., and DASS, C. M. S. (1951) The macronucleus of *Vorticella convallaria* (Linn.) during conjugation. *J. Morphol.* **89**, 187–97.

231. SHUBNIKOVA, E. (1947) (Ribonucleic acid in the life cycle of a protozoan cell). *Doklady Acad. Nauk SSSR* **55**, 521–4 (in Russian).

232. SIEGEL, R. W., and HECKMANN, K. (1966) Inheritance of autogamy and the killer trait in *Euplotes minuta*. *J. Protozool.* **13**, 34–38.

232a. SKOBLO, I. I., and OSSIPOV, D. V. (1968) The autogamy during conjugation in *Paramecium caudatum* Ehrbg. I. Study on the nuclear reorganization up to the stage of the third synkaryon division. *Acta Protozool.* **5**, 273–90.

233. SONNEBORN, T. M. (1936) Factors determining conjugation in *Paramecium aurelia*. I. The cyclical factor: the recency of nuclear reorganization. *Genetics* **21**, 503–14.

234. SONNEBORN, T. M. (1939) Genetic evidence of autogamy in *Paramecium aurelia*. *Anat. Rec.* **75** (Suppl.), 85.

235. SONNEBORN, T. M. (1939) *Paramecium aurelia*: mating types and groups; lethal interactions; determination and inheritance. *Amer. Nat.* **73**, 390–413.

236. SONNEBORN, T. M. (1941) The occurrence, frequency, and causes of failure to undergo reciprocal cross-fertilization during mating in *Paramecium aurelia*, variety 1. *Anat. Rec.* **81** (Suppl.), 66–67.
237. SONNEBORN, T. M. (1941) Sexuality in unicellular organisms. In: *Protozoa in Biological Research* (G. CALKINS and F. SUMMERS, eds.), Columbia Univ. Press, New York, 606–709.
238. SONNEBORN, T. M. (1943) Gene and cytoplasm. I. The determination and inheritance of the killer character in variety 4 of *Paramecium aurelia*. *Proc. Nat. Acad. Sci. U.S.A.* **29**, 329–38.
239. SONNEBORN, T. M. (1943) Gene and cytoplasm. II. The bearing of the determination and inheritance of characters in *Paramecium aurelia* on the problems of cytoplasmic inheritance. *Proc. Nat. Acad. Sci. U.S.A.* **29**, 338–43.
240. SONNEBORN, T. M. (1944) Exchange of cytoplasm at conjugation in *Paramecium aurelia*, variety 4. *Anat. Rec.* **89** (Suppl.), 49.
241. SONNEBORN, T. M. (1947) Recent advances in the genetics of *Paramecium* and *Euplotes*. *Adv. Genet.* **1**, 263–358.
242. SONNEBORN, T. M. (1949) Ciliated Protozoa: cytogenetics, genetics, and evolution. *Ann. Rev. Microbiol.* **3**, 55–80.
243. SONNEBORN, T. M. (1951) Some current problems of genetics in the light of investigation on *Chlamydomonas* and *Paramecium*. *Cold Spring Harbor Symp. Quant. Biol.* **16**, 483–503.
244. SONNEBORN, T. M. (1953) Patterns of nucleocytoplasmic integration in *Paramecium*. *Proceed. 9th Internat. Congr. Genet.*, part 1 (Caryologia, Suppl.), 307–25.
245. SONNEBORN, T. M. (1954) The relation of autogamy to senescence and rejuvenescence in *Paramecium aurelia*. *J. Protozool.* **1**, 38–53.
246. SONNEBORN, T. M. (1957) Breeding systems, reproductive methods, and species problems in Protozoa. In: *The Species Problem*, Amer. Assoc. Adv. Sci., Washington, D.C., 155–324.
247. STEIN, F. (1867) *Der Organismus der Infusionsthiere. II. Naturgeschichte der heterotrichen Infusorien*, Leipzig.
248. SUMMERS, F. M. (1938) Some aspects of normal development in the colonial ciliate *Zoothamnium alternans*. *Biol. Bull.* **74**, 117–29.
249. SUMMERS, F. M., and KIDDER, G. W. (1936) Taxonomic and cytological studies on the ciliates associated with the amphipod family Orchestiidae from the Woods Hole district. II. The coelozoic astomatous parasites. *Arch. Protistenk.* **86**, 379–403.
250. SUZUKI, S. (1957) Parthenogenetic conjugation in *Blepharisma undulans japonicus* Suzuki. *Bull. Yamagata Univ. Nat. Sci.* **4**, 69–84.
250a. TAKAYANAGI, T., INOKI, S., and YOSHIKAWA, K. (1966) Unusual conjugation of *Paramecium polycaryum*. *Jap. J. Genet.* **41**, 241–6.
251. TANNREUTHER, G. W. (1926) Life history of *Prorodon griseus*. *Biol. Bull.* **51**, 303–20.
252. TUFFRAU, M. (1953) Les processus cytologiques de la conjugaison chez *Spirochona gemmipara* Stein. *Bull. Biol. France Belg.* **87**, 314–22.
253. TURNER, J. P. (1930) Division and conjugation in *Euplotes patella* Ehrbg. with special reference to the nuclear phenomena. *Univ. Calif. Publ. Zool.* **33**, 193–258.
254. TURNER, J. P. (1941) Fertilization in Protozoa. In: *Protozoa in Biological Research* (G. CALKINS and F. SUMMERS, eds.), Columbia Univ. Press, New York, 583–645.
255. VISSCHER, J. P. (1927) Conjugation in the ciliated protozoon, *Dileptus gigas*, with special reference to the nuclear phenomena. *J. Morphol.* **44**, 383–415.
256. VIVIER, E. (1960) Contribution à l'étude de la conjugaison chez *Paramecium caudatum*. *Ann. Sci. Nat., Zool.* **2**, 387–506.
257. VIVIER, E. (1962) Démonstration, à l'aide de la microscopie électronique, des échanges cytoplasmiques lors de la conjugaison chez *Paramecium caudatum* Ehrb. *C.R. Soc. Biol. Paris* **156**, 1115–16.

258. VIVIER, E. (1965) Sexualité et conjugaison chez la Paramécie. *Ann. Fac. Sci. Clermont-Ferrand* **26**, 101–14.

259. VIVIER, E., and ANDRÉ, J. (1961) Données structurales et ultrastructurales nouvelles sur la conjugaison de *Paramecium caudatum. J. Protozool.* **8**, 416–26.

260. VIVIER, E., and ANDRÉ, J. (1963) Étude au microscope électronique de quelques problèmes ultrastructuraux relatifs à la conjugaison de *Paramecium caudatum.* In: *Progress in Protozoology (Proc. First Internat. Congr. Protozool. Prague, 1961)*, 422.

261. WATTERS, F. A. (1912) Size relationships between conjugants and non-conjugants in *Blepharisma undulans. Biol. Bull.* **23**, 195–213.

261a. WEBB, T. L., and FRANCIS, D. W. (1968) On the biology of conjugation in *Stentor. J. Protozool.* **15** (Suppl.), 8–9.

262. WEISZ, P. B. (1950) Multiconjugation in *Blepharisma. Biol. Bull.* **98**, 242–6.

263. WELLS, C. (1965) Age associated nuclear anomalies in *Tetrahymena. J. Protozool.* **12**, 561–3.

264. WENRICH, D. H. (1954) Sex in Protozoa. A comparative review. In: *Sex in Microorganisms* (D. H. WENRICH ed.), Amer. Assoc. Adv. Sci., Washington, D.C., 134–265.

265. WICHTERMAN, R. (1937) Division and conjugation in *Nyctotherus cordiformis* (Ehr.) Stein (Protozoa, Ciliata) with special reference to the nuclear phenomena. *J. Morphol.* **60**, 563–611.

266. WICHTERMAN, R. (1937) Conjugation in *Paramecium trichium* Stokes (Protozoa, Ciliata), with special reference to the nuclear phenomena. *Biol. Bull.* **73**, 397–8.

267. WICHTERMAN, R. (1940) Cytogamy. A sexual process occurring in living joined pairs of *Paramecium caudatum*, and its relation to other sexual phenomena. *J. Morphol.* **66**, 423–51.

268. WICHTERMAN, R. (1946) Direct observation of the transfer of pronuclei in living conjugants of *Paramecium bursaria. Science* **104**, 505–6.

269. WICHTERMAN, R. (1946) Time relationships of the nuclear behavior in the conjugation of green and colorless *Paramecium bursaria. Anat. Rec.* **94**, 39–40.

270. WICHTERMAN, R. (1946) Further evidence of polyploidy in the conjugation of green and colorless *Paramecium bursaria. Biol. Bull.* **91**, 234.

271. WICHTERMAN, R. (1946) The behavior of cytoplasmic structures in living conjugants of *Paramecium bursaria. Anat. Rec.* **94**, 93–94.

272. WICHTERMAN, R. (1948) The time schedule of mating and nuclear events of the conjugation of *Paramecium bursaria. Turtox News* **26**, 2–10.

273. WICHTERMAN, R. (1967) Mating types, breeding system, conjugation, and nuclear phenomena in the marine ciliate *Euplotes cristatus* Kahl from the Gulf of Naples. *J. Protozool.* **14**, 49–58.

274. WILLIS, A. G. (1948) Studies on *Lagenophrys tattersalli* (Ciliata Peritricha). II. Observations on bionomics, conjugation, and apparent endomixis. *Quart. J. Microscop. Sci.* **89**, 385–400.

275. WOODARD, J., WOODARD, M., GELBER, B., and SWIFT, H. (1966) Cytochemical studies of conjugation of *Paramecium aurelia. Exper. Cell Res.* **41**, 55–63.

276. WOODRUFF, L. L. (1921) Micronucleate and amicronucleate races of Infusoria. *J. Exper. Zool.* **34**, 329–37.

277. WOODRUFF, L. L., and ERDMANN, R. (1914) A normal periodic reorganization process without cell fusion in *Paramecium. J. Exper. Zool.* **17**, 425–518.

278. YAGIU, R. (1940) The division, conjugation, and nuclear reorganization of *Entorhipidium echini* Lynch. *J. Sci. Hiroshima Univ.*, ser. B, div. 1, **7**, 125–56.

279. ZWEIBAUM, J. (1922) Ricerche sperimentali sulla coniugazione degli Infusori. II. Influenza della coniugazione sulla produzione dei materiali di riserva nel *Paramecium caudatum. Arch. Protistenk.* **44**, 375–96.

RELATIONSHIP BETWEEN CERTAIN
PROTOZOA AND OTHER ANIMALS

NORMAN D. LEVINE †

† *College of Veterinary Medicine, Agricultural Experiment Station and Dept. of Zoology, University of Illinois, Urbana.* Published with the assistance of National Science Foundation Research Grant GB-5667X.

10*

CONTENTS

INTRODUCTION

Kirby's[96] chapter, of which the present chapter is more a replacement than a revision, was a seminal influence in protozoology. It contains 410 references; together with Wenrich,[151] which contains 89 references, it is an excellent source of information on the earlier literature on relationships between certain protozoa and other animals. The literature has burgeoned since. Even at the time Kirby wrote, he said that it was far too large for concise treatment, and that statement is even truer today. Indeed, a single person cannot grasp it all.

Kirby recognized two possible paths of approach.

> Either he could attempt to make a comprehensive survey of the entire expanse of pertinent information, or he could explore in as much detail as possible certain chosen fields of inquiry. The former approach would lead to a generalized account, with selected and perhaps original illustrations; and it would in large part reiterate existing, readily accessible, sometimes commonplace concepts. The latter course, although less exhaustive, permits selection for more detailed consideration of certain representative topics; that is the course which has in the main been followed here.

It is also the course that I, too, have followed in the main.

Kirby discussed in depth the ciliates of sea urchins, the protozoa of termites and the cockroach *Cryptocercus*, and the ciliates of ruminants. I shall discuss only the first of these. Wenrich's[152] review of Cleveland's[26-33] work on the relation of sexual cycles of *Cryptocercus* flagellates to the hormones of their hosts, Cleveland's later work[34-40] and that of Cleveland, Burke and Karlon[41] on the same subject, Kirby's[91-97] papers on termite protozoa, Grassé's[71] sections on termite and cockroach protozoa, and the review of the subject by Dogiel, Poljanskiy and Chejsin[55] make the second unnecessary. And there are enough recent reviews of the ciliates of ruminants to make another review of them redundant.[87, 118, 102, 45] I shall therefore omit these two subjects and add discussions of the Trypanosomatidae and the Haemospororina, both of which are important groups of parasites.

TYPES OF RELATIONSHIP

Protozoan relationships with other animals may be classified in three ways. One is by the size of the parties concerned, the second by their interactions with each other, and the third by the parasite's host range. Using the first classification, we may establish three groups: (1) those in which protozoa interact with larger organisms, (2) those in which they interact with smaller ones, and (3) those in which they interact with organisms of a more or less similar size. The third category could be subdivided into (a) interactions with organisms of the same species, such as occur in colonial forms or in those which stay near each other after dividing, and (b) interactions with organisms of a different species, such as occur among bacteria in which one species breaks down a substrate to a certain stage and another species breaks that stage down still further. Predator-prey relationships occur in all three categories, but I shall not deal with them here. Neither shall I deal with that part of the second category which has to do with parasites of protozoa; these are discussed by Dr. Gordon H. Ball in a separate chapter. And so little is known of the third category as it relates to protozoa that it would be premature to say much about it.

This chapter, then, is concerned primarily with relationships between protozoa and larger organisms. This is the so-called host–parasite relationship. Many terms are used to define the relationships between parasites and hosts, to group the associations discussed under definite categories.

> The categories are defined, and it is shown in what manner and to what extent each separate association can be referred to its proper position. A reader of this literature soon becomes sensible of the lack of agreement in almost every major particular. Unlike names are given to the categories, definitions are dissimilar, there is difference of opinion or lack of exact information on the nature of the relationship itself, and the impossibility of making unequivocal distinctions is apparent in many instances.[96]

Man is a classifying creature, but nature defies classification.

Despite the inadequacy of most definitions, and despite the fact that different people define certain words differently, it is necessary to use definitions and to standardize them. *Parasitism* is defined as an association between two specifically distinct organisms in which one lives on or within the other in order to obtain sustenance. The word is derived from the Greek *parasitos*, meaning "situated beside". It originally referred to sycophants who ate beside or at the tables of the wealthy, and also to priests who collected grain for their temples. A parasite had a definite function in ancient Greece and Rome. He got free meals, it is true, but he also contributed his wit to the conversation, he augmented his host's ego, and he gave his host an appreciative audience; there were advantages to both sides.

While parasitism still has a social meaning, the term has been given a new connotation by biologists. *Parasites* are defined as organisms which live on or within some other living organism, which is known as the *host*. Parasites may be either animals or plants of many phyla, but in this chapter we are concerned only with parasitic protozoa. Parasites are found in many different habitats on or within the host—on its outer surface, its gills or other appendages, or in the lumen of its gut, in its skin, its blood plasma, in various tissues, inside different types of cells, even inside cell nuclei. Indeed, parasitology may be defined as that branch of ecology in which one organism is the habitat or environment of another.

There are several types of parasitism. *Symbiosis* is defined as the permanent association between two specifically distinct organisms so dependent upon each other that life apart is impossible under natural conditions. The relation between many termites and their intestinal protozoa is symbiotic. The termites eat wood, but they cannot digest it; the protozoa can digest wood, turning it into glucose, but they have no way of obtaining it. Working together, the termites ingest wood, breaking it down into particles which the protozoa in their hindguts can take in, and the protozoa break down the cellulose to glucose, which they release for the termites.

Another example of symbiosis—and the original one—is furnished by lichens. They are composed of certain species of algae and fungi living together. Many insects, ticks and mites have symbiotic bacteria and rickettssiae[20, 21, 99]. And a growing number of organisms are known to contain symbiotic algae or dinoflagellates. These include *Paramecium* and other protozoa containing *Chlorella*, the green hydra *Chlorohydra viridissima* with algae in its gastrodermal cells (cf. ref. 115) and *Convoluta* and coelenterate medusae containing dinoflagellates.

Droop[57] tabulated eighty-seven genera of invertebrates in which symbiosis with algae has been recorded, including seven of ciliates, thirteen of sarcodines, four of flagellates, three of Porifera, six of Turbellaria (including the famous *Convoluta paradoxa* and *C. roscoffensis*—cf. refs. 89, 90), seven of hydrozoan coelenterates, five of scyphozoan coelenterates, twenty-two of actinozoan coelenterates, five of lamellibranch molluscs, eight of nudibranch molluscs, one ctenophore, one rotifer, one pulmonate mollusc, one echinoderm, and three ascidians. He also mentioned that Boschma[13] and Yonge and Nicholls[155] had found symbiotic algae in forty genera of corals besides the seven he listed. Coral tissues contain as many as 30,000 zooxanthellae per cubic millimeter. Their principal effect is said to be to increase the rate of calcification, i.e. they have to do with carbon dioxide metabolism. The extent of their activity is indicated by the fact that coral reefs grow about 2.5 cm per year, and that 41 cm per year has even been recorded.

Most of these symbiotic algae have been named *Zoochlorella*, *Zooxanthella* or *Cyanella*. These names are based upon the symbiote's color— zoochlorellae are green, zooxanthellae yellow, and cyanellae blue-green—but the true nature of most of them is unknown. Axenic cultivation is necessary to elucidate this, and a good start has been made. McLaughlin and Zahl[112] cultivated the zooxanthellae of the scyphozoon *Cassiopeia* axenically, and Freudenthal[65] named it *Symbiodinium microadriaticum*, working out its life cycle. It is a dinoflagellate, as can be seen from the appearance of its gymnodinioid zoospore. A further step was taken when McLaughlin and Zahl[113] found that the zooxanthella of the coral *Cladocora stellaria* was a dinoflagellate and cultivated it axenically. McLaughlin and Zahl[113a] reviewed the subject.

Some parasitologists have preferred to use the term symbiosis to replace the customary broad definition of parasitism. They justify this usage by referring to the original Greek meaning of the word symbiosis: "life together" or "organisms living together" (cf. refs. 96, 128, 86). They do not, however, use this argument to extend the term to the whole field of ecology, although to be consistent they should.

When the term "symbiosis" was originally introduced by de Bary,[7] it was used in the broad sense. However, de Bary's work was largely concerned with the association between fungi and algae to form lichens, and later workers misinterpreted him, using the term solely for such associations. Our present narrow definition is the consequence (see Hertig et al.[76]).

Some people define parasitism as involving injury to the host, and the use of the term in this narrow sense and also in the broad one is a source of confusion. This was another reason for the proposal to broaden the meaning of symbiosis. However, this action introduced new confusion of its own. We now have two terms defined both narrowly and broadly, instead of one.

Defining a word solely in terms of its etymology cannot be justified. If it were, the correct definition of "therapy" would be "slavery", since it is derived from *theraps*, the Greek word for slave. Furthermore, an attempt to revert to the old historical meaning of a word once it has been superseded will succeed only in decreasing the word's usefulness by interfering with our clear understanding of its meaning.

Pearse[120] attempted to solve the problem by introducing a new word, "consors" (plural, "consortes"), for parasites in the broad sense, but this term has failed to catch on.

Even when an attempt is made to restrict the definition of parasitism, it fails. For instance, in their report on the principles of parasitism, Huff et al.[86] began by using the broad definition of symbiosis and the narrow one of parasitism, but later on reverted repeatedly, using the terms in their customary, opposite senses.

One reason, I believe, why parasitologists have rejected the idea of using only the narrow definition of parasitism is that it would be too artificial. By this usage, *Entamoeba histolytica* would be a parasite, but *E. coli* would not. Indeed, even the small nonpathogenic races of *E. histolytica* would have to be ejected from the parasitic domain; if one were to extend the usage to pure infections in bacteria-free hosts, the large races would have to be removed as well, for Phillips et al.[123] found it impossible to infect bacteria-free guinea pigs with *E. histolytica* at all, although normal guinea pigs or those infected with *Escherichia coli* or *Aerobacter aerogenes* could be readily infected and subsequently developed intestinal lesions.

The impossibility of separating these various aspects of parasitism is recognized, though perhaps unconsciously, even by those who advocate retaining only its narrow definition and using "symbiosis" as the broad term. They still call themselves parasitologists and not symbiotologists, and I have yet to see a proposal that the *Journal of Parasitology* change its name to the *Journal of Symbiotology*.

Mutualism is defined as an association between two specifically distinct organisms by which both are benefited. It differs from symbiosis in that it is not obligatory for both partners, although it may be for one. An example might be the protozoa which swarm in the rumen and reticulum of ruminants. The protozoa are benefited by having a warm home with plenty of proper food and all the conditions necessary for the abundant life. The ruminants are benefited by being presented with a nutritionally superior protein when the protozoa die and are digested, by obtaining volatile fatty acids from the protozoa, and by the fact that the protozoa help smooth out the fermentation process by withholding carbohydrates from the rumen contents for a while.[87, 118, 45] However, the ruminants can get along perfectly well without their protozoa, and the protozoa can be cultivated in artificial culture media.

Commensalism is defined as an association between host and parasite in which one partner is benefited and the other is neither benefited nor harmed. The term comes from the Latin *com* and *mensa*, meaning together at the same table. Many intestinal protozoa, including *Entamoeba coli*, *Endolimax nana* and various trichomonads, are commensals. They neither harm nor help their hosts, but they need its gut in which to live.

The next two terms both refer to potentially pathogenic parasites. *Parasitosis* is the association between two organisms in which one injures the other, causing signs and lesions of disease; *parasitiasis* is the association between two organisms in which one is potentially pathogenic but does not cause signs of disease.

The difference between parasitosis and parasitiasis is quantitative. In parasitiasis, whatever damage the parasite may do is repaired by the host without discernible effects. In parasitosis, so many parasites are present that the damage they cause is more than the host can cope with. Whether a particular species of pathogenic parasite actually injures its host depends upon many factors, including the number of parasites present, the virulence of the particular strain involved, the host's natural resistance, nutritional state and age, its previous history of exposure which may have led it to develop active immunity against the parasite, the amount of stress to which the host is subjected, the species of concomitant bacteria and other parasites present, etc. One condition grades into the other, and the student of parasites must deal with both.

We often fail to recognize that infections are dynamic equilibria. Interaction between host and parasite is continuous. The moment the host lets down its guard, the parasite surges up and may overwhelm it; when parasites lose their vigor, the host kills or expels them. In parasitiasis, the parasite is at a low ebb; in parasitosis it dominates. Higher animals live in a sea of microscopic invaders; those which have survived to the present have developed the ability to cope with their parasites in the course of their evolution. The *carrier state* is a good example of parasitiasis. Carriers are animals which are infected with a pathogenic parasite but are not harmed by it, usually due to immunity resulting from previous exposure. Carriers are a source of infection for susceptible animals. *Entamoeba histolytica* is transmitted largely by carriers, and so are *Histomonas meleagridis* and most coccidia.

Parasites do not live alone in their hosts. Many species, and many million individuals of some of them, may fill the gut. Each species is in balance not only with its host but also with all the other parasite species. This parasite mix along with its habitat is called a *parasitocenose*.[119] Each species may affect any or all of the others—by producing antibiotics which act against them, by producing metabolic products which they can use or need, by preempting their food or place of residence, by making the host more receptive to invasion or injury, or by making it less so. The situation is so complex that it defies analysis. This is one reason that human actions sometimes produce unexpected results. We can use an antibiotic to eliminate a particular bacterium, but it may also destroy other bacteria and allow yeasts to multiply in the intestine and cause disease.

Different parasites have solved their problems of living in different ways, and these solutions differ in satisfactoriness. Symbiosis is a highly specialized type of association, known to occur only in certain groups. Mutualism is a much looser association, also fairly uncommon; perhaps it is a step on the road to symbiosis. The commonest types of parasitism are commensalism, parasitosis and parasitiasis. Commensalism is clearly the most desirable of these for both host and parasite. In parasitosis the host is harmed, and through it he parasite may be also. Parasitiasis is intermediate between parasitosis and commensalism; harmless in itself, it carries a threat of damage.

The above classification of types of parasitism has to do with the type of interaction between the parasite and the host. Another classification, which cuts across the previous one, is based on the number and types of hosts which a parasite may have. Three pairs of term, all independent of each other, are used in this connection.

A parasite's life cycle may be simple and asexual, involving binary or multiple fission, or it may be complex and involve alternation of sexual and asexual multiplication. A *monogenetic* parasite is one in which there is no alternation of generations; its reproduction is always either sexual or asexual. Asexual *monogenetic* protozoa include the flagellates *Trichomonas*, *Giardia*, *Hexamita*, *Leptomonas* and *Streblomastix*, the sarcodines *Entamoeba*, *Endolimax* and *Iodamoeba*, the sporozoon *Toxoplasma*,† and the entodiniomorphid ciliates *Entodinium*, *Ophryoscolex* and *Cycloposthium*.

† Since this paper was written, *Toxoplasma* has been found to have a sexual phase in the cat but not in other hosts.

In all these cases, fission results in the formation of individuals which look like their parents. Another group of asexual monogenetic protozoa includes genera in which there may be more than one form in the life cycle. This group includes the flagellates *Trypanosoma* and *Leishmania*, and the Sporozoa *Babesia* and *Theileria*. *Sexual monogenetic* protozoa (in which multiplication occurs only following zygote formation) include *Monocystis*, *Gregarina* and other eugregarine Sporozoa.

Heterogenetic protozoa are those in which there is alternation of sexual and asexual generations. Examples include *Eimeria, Isospora*, and other coccidia, and *Plasmodium, Haemoproteus, Leucocytozoon* and other haemospororin blood protozoa. Included in this category might also be such ciliates as *Balantidium*, in which multiplication is generally by binary fission but in which conjugation takes place under unknown circumstances.

A second pair of terms has to do with the number of different types of host a parasite lives in during the course of its normal life cycle. A *monoxenous* parasite is one which has only one type of host—the definitive host. Examples are the flagellates *Trichomonas* and *Giardia*, the sarcodines *Entamoeba* and *Endolimax*, the Sporozoa *Eimeria, Isospora* and some but not all other coccidia, and the ciliates *Balantidium, Entodinium* and *Ophryoscolex*. A *heteroxenous* parasite has two or more types of host in its life cycle. Examples are the flagellates *Leishmania* and *Trypanosoma* (except for *T. equiperdum*, which is monoxenous), and the Sporozoa *Plasmodium, Haemoproteus, Leucocytozoon, Babesia* and *Theileria*.

A third set of terms has to do with the parasite's host range, i.e. with the number of host species in which it may occur. A *stenoxenous* parasite is one with a narrow host range. *Trypanosoma lewisi*, which occurs only in rats, and *T. duttoni*, which occurs only in mice, are stenoxenous, as is the sarcodine *Dientamoeba fragilis*. Coccidia of the genus *Eimeria* are generally stenoxenous, as are most human malaria parasites of the genus *Plasmodium*. The enteric ciliate *Cyathodinium* of the guinea pig and *Cycloposthium* of *Equus* are also stenoxenous.

An *euryxenous* parasite is one with a broad host range. *Trypanosoma avium*, which occurs in many species and families of birds, is euryxenous. *T. rhodesiense*, which occurs in both man and wild game, is euryxenous. Most of the species of *Plasmodium* in birds are euryxenous. *Balantidium coli* is quite euryxenous, being known from the pig, man, various monkeys and other primates, and possibly the guinea pig.

A third term which I should like to introduce is *mesoxenous*. A mesoxenous parasite is one with an intermediate host range. For purposes of separation—but with the understanding that it is arbitrary—I should like to define stenoxenous parasites as those which occur within hosts of a single genus (or in some cases closely related genera), mesoxenous parasites as those which occur within hosts of a single order, and euryxenous parasites as those which occur within hosts of more than one order or class.

There may or may not be a difference in host range between the different types of hosts among heteroxenous parasites. Thus, *Leishmania tropica* may occur in man, the dog or various wild rodents among vertebrates, but its vector is always a species of *Phlebotomus*; it is therefore euryxenous in the vertebrate host and stenoxenous in the invertebrate one. *Plasmodium falciparum*, which occurs only in man so far as is known, and is transmitted by one of several species of *Anopheles*, is stenoxenous in the vertebrate host and essentially stenoxenous in the invertebrate one. *Plasmodium gallinaceum* is stenoxenous in the vertebrate host, occurring only in *Gallus*, and quite euryxenous in the invertebrate host; Huff[83] listed twenty-nine susceptible and one questionable mosquito vectors, of which nineteen were *Aedes*, five *Armigeres*, two (possibly three) *Culex*, one *Anopheles*, one *Culiseta* and one *Mansonia*. *Trypanosoma cruzi* is euryxenous in the vertebrate host, occurring in man, the raccoon, skunk, opossum, dog, cat, pig, woodrat, armadillo and many other mammalian species, and quite mesoxenous in the invertebrate host; its vectors include members of the reduviid genera *Panstrongylus*, *Triatoma*, *Eutriatoma*, *Rhodnius*, and *Eratyrus*; in addition, it may be transmitted through the placenta.

Euryxenous parasites are the causes of zoonoses—diseases transmissible between man and other animals; stenoxenous parasites are not. Chagas' disease (due to *T. cruzi*) is a zoonosis, whereas falciparum malaria is not.

Use of the terms stenoxenous and euryxenous may be deceptive. All intergrades between them exist in nature, and all I have done is name the two extremes of a continuum. This is one reason why I have introduced the term mesoxenous.

Actually, the host range of most parasites is broader than generally supposed. Most animal species have not been examined for parasites, so we really don't know the complete host range of any of the latter.

As I have said, the three sets of terms—monogenetic and heterogenetic; monoxenous and heteroxenous; stenoxenous, mesoxenous and euryxenous—are independent. Twelve different combinations are possible. *Hexamita meleagridis* is monogenetic, monoxenous and stenoxenous. *Eimeria tenella* is heterogenetic, monoxenous and stenoxenous. *Plasmodium falciparum* is heterogenetic, heteroxenous and stenoxenous. *Trypanosoma cruzi* is monogenetic, heteroxenous and euryxenous. Examples of the other eight combinations could also be given.

NUMBER OF PROTOZOAN SPECIES

About 44,250 species of protozoa had been named between 1758 and 1958. Of these, 20,182 were fossil, 17,293 free-living and 6775 parasitic. Among the last, 212 were sarcodines, 1300 flagellates, 3513 "Sporozoa," 1700 ciliates and 50 incertae sedis.[103]

The average number of new species of parasitic protozoa named each year between 1955 and 1958 was ninety-five. If we assume that the average has remained the same since 1958 and that the distribution of species in the various groups has remained the same, we can estimate that by the end of 1966 about 7500 species of parasitic protozoa had been named, including about 220 sarcodines, 1400 flagellates, 4000 "Sporozoa" and 1800 ciliates. This is an extremely rough guess. In these figures the subphylum Cnidospora has been left under the "Sporozoa";[81] the reason is that I have not counted the named species in this group separately and therefore do not know how many to shift.

This is far from the total number of protozoan species, either parasitic or free-living. For instance, Levine and Ivens[108] reported that at the time their monograph on the coccidia of rodents went to press, 225 valid species of coccidia had been named from rodents, including 204 of *Eimeria*, ten of *Isospora*, three of *Wenyonella*, two each of *Cryptosporidium* and *Klossiella*, and one each of *Dorisiella*, *Caryospora*, *Tyzzeria* and *Klossia*. Large as this number may be, it is only a small percentage of the number of species which probably occur in rodents. *Eimeria*, the largest genus, had been described from only 15 per cent of the 337 genera and 4 per cent of the 2688 species of rodents. Most potential hosts have simply not been examined for coccidia or any other parasitic protozoa.

According to Muller and Campbell,[114] there are 33,640 known living species of chordates and 3552 of mammals. At present, approximately 575 species of *Eimeria* have been described from chordates and 362 from mammals. Approximately 751 species of coccidia of all genera have been described from chordates and 420 from mammals (Table 1). According to Pellérdy[121] coccidia have been described from 616 species of chordates and 231 of mammals. In other words, coccidia have been described from about 1.8 per cent of the world's chordates and 6.5 per cent of the world's mammals; and about 1.2 coccidian species have been described from each host chordate species, and about 1.8 coccidian species from each host mammalian species. I estimated[105] that there might actually be a total of 45,000 species of coccidia—roughly equivalent to the total of all living and fossil protozoa so far named.

The observations of Levine and Kantor[109] and Levine[104] on blood protozoa of birds of the order Columborida give further information on the state of our knowledge of this group of parasites. They analyzed 174 papers, of which twenty-two were from Europe, twenty-two from Asia, thirty-two from Africa, sixty-four from North America, thirty-two from South and Central America, and two from Australia. This number of papers might lead one to assume that the blood parasites of these birds are rather well known, but the facts are otherwise. Peters[122] listed 61 genera and 320 species of birds in the order. Levine and Kantor[109] found that the commonest genus, *Haemoproteus*, had been reported from only 31 per cent of the

TABLE 1. NUMBERS OF SPECIES OF EIMERIORINA IN VARIOUS HOST GROUPS†

Genus of coccidium	Host group											Total
	Mammals	Birds	Reptiles	Amphibians	Teleost fish	Elasmobranchs	Lower chordates	Arthropods	Annelids	Molluscs	Others	
Barrouxia								6		1	1	8
Caryospora	1	5	10									16
Cryptosporidium	2	1										3
Cyclospora	1		6									7
Dorisiella	1	2	1						1			5
Echinospora								1				1
Eimeria	362	81	51	25	49	5	2	8	1			584
Isospora	48	37	21	10						1		117
Lankesterella		6	1	5								12
Mantonella								2				2
Octosporella			1									1
Pfeifferinella										2		2
Pythonella			1									1
Schellackia			3									3
Tyzzeria	1	3	1									5
Wenyonella	3	3	1									7
Yakimovella	1											1
Total	420	138	97	40	49	5	2	17	2	4	1	775

† Data from Levine and Ivens[108] for rodent *Eimeria*, and from Pellérdy[121] for remainder.

known host genera and 14 per cent of the known host species, *Plasmodium* from 20 per cent of the genera and 7 per cent of the species, *Leucocytozoon* from 11 per cent of the genera and 5 per cent of the species, *Trypanosoma* from 8 per cent of the genera and 2 per cent of the species, and *Toxoplasma*, *Lankesterella* or something similar from 5 per cent of the genera and 1 per cent of the species. No parasites at all had been reported from the genera *Oreopelia* and *Gallicolumba*, which contain fifteen and eighteen species, respectively, and only two cases from the genus *Ducula*, which contains thirty-seven species. (More recently, Coatney *et al.*[44] found *Haemoproteus* in a *Ducula badia* in Thailand.)

This compilation included birds in zoos. If one omits them, then *Haemoproteus* has been found in only 8 per cent of the known species of columborids, *Plasmodium* in 3 per cent, *Leucocytozoon* in 4 per cent, *Trypanosoma* in 2 per cent, and *Toxoplasma, Lankesterella* or something similar in 1 per cent.

Not only this, but most of these records were more or less casual reports by workers who examined blood smears from a miscellany of birds and did not try to identify the parasites beyond genus. Levine and Kantor[109] found that (omitting reports from zoos) the name of the parasite species had been given in only 58 per cent of 138 reports of *Haemoproteus*, 62 per cent of twenty-six reports of *Plasmodium*, 35 per cent of twenty-six reports of *Leucocytozoon*, 60 per cent of ten reports of *Trypanosoma*, and 77 per cent of thirteen reports of *Toxoplasma, Lankesterella,* or something similar. In 43 per cent of the total of 213 records, the species name of the parasite was not given. Furthermore, in only 24 per cent of the 174 papers analyzed by Levine[104] were ten or more birds of a single species examined and the prevalence of infection given, and in only 7 per cent of them were 100 or more birds examined and the prevalence of infection given.

Beyond this, the number of birds examined for these 174 reports was pitifully small. In quite a few papers, the number of birds examined was not given but it was probably not large. The remainder were based on the examination of blood smears from 2515 birds, of which 296 were from Europe, 356 from Asia, 53 from Africa, 1391 from North America, and 419 from South and Central America. In comparison with the world population of these birds, this is an extremely small number.

Whether the parasite species identifications were correct is a matter of opinion. Some authors have felt that all forms of similar structure in hosts of the whole order belong to the same species, and have named them on that basis. Other authors have preferred to use a different name for similar parasites from birds of different families or genera. Very few, however, have carried out the cross transmission studies necessary to establish the true situation.

Haemoproteus is a case in point. Two morphologic types occur in columborid birds. The less common one is *H. sacharovi*, which has large, ellipsoidal gamonts which replace practically all the host erythrocyte cytoplasm, enlarge and distort the host cell, and often push the host cell nucleus far to one side. It has been reported from the mourning dove *Zenaidura macroura*, the band-tailed pigeon *Columba fasciata* and the domestic pigeon *Columba livia* in North America (cf. refs. 104, 145). What may have been the same species was found by Franchini[64] in the turtledove *Streptopelia turtur* in Italy, although he called it *Leucocytozoon* sp.

The commoner form of *Haemoproteus* in columborid birds has gamonts which do not enlarge the host cell or displace its nucleus but extend along one side of the host cell nucleus and curve around its ends; these are the so-called halter-shaped gamonts. This form was first found in the domestic pigeon by Celli and Sanfelice[22, 23] and named by them *H. columbae*. Some indistinguishable forms from other hosts have been given separate names:

H. maccallumi by Novy and MacNeal[117] from *Zenaidura macroura*, *H. melopeliae* by Laveran and Petit[100] from the white-winged dove *Zenaida asiatica*, *H. turtur* by Covaleda Ortega and Gallego Berenguer[48a] from *Streptopelia turtur*, and *H. vilhenai* by Santos Dias[134] from *Plectopterus gambiensis*. Huff[82] transmitted *H. maccallumi* from the mourning dove to the domestic pigeon, but Coatney[42] was unable to transmit *H. columbae* from the domestic pigeon to the mourning dove. Both used the hippoboscid fly *Pseudolynchia canariensis* as the vector. Levine[104] concluded that there might be strain differences between different hosts but that until greater differences than these were brought out it was probably best to use the name *H. columbae* for all forms with halter-shaped gamonts from columborid birds.

More recently, however, Baker[5] compared the *columbae*-like *Haemoproteus* forms in the domestic pigeon *Columba livia* and the wood pigeon *C. palumbus*. The latter is the only common pigeon in England except in cities, where *C. livia* also lives. *H. columbae* occurs in *C. livia* and is transmitted by *Pseudolynchia canariensis*, which does not occur in the United Kingdom. Baker was unable to infect English wood pigeons with *H. columbae* from Egyptian domestic pigeons, or domestic pigeons with the *columbae*-like *Haemoproteus* from wood pigeons. He also found[2] that the normal vector of the wood pigeon form was not *P. canariensis* but *Ornithomyia avicularia*, another hippoboscid fly. On the basis of these host differences, Baker[6] named the wood pigeon form *Haemoproteus palumbis*.

Baker's findings cast doubt on the validity of the view that all morphologically similar forms, even from the same host genus, are necessarily the same parasite species. This doubt applies not only to *Haemoproteus* in columborids but also in other birds. Herman's[75] view, hesitantly suggested by Levine and Hanson,[107] that all *Haemoproteus* forms with halter-shaped gamonts in ducks are *H. nettionis* becomes questionable.

A further reservation should be kept in mind regarding reports of *columbae*-like *Haemoproteus* from birds. This is that in some cases these protozoa may not be *Haemoproteus* at all, but a species of *Plasmodium*, such as *P. fallax* or *P. circumflexum* with halter-shaped or elongate gamonts which does not happen to have schizonts in the blood at the time of examination.

How many species of parasitic protozoa are there? If only 5.8 per cent of the species of coccidia have been named, does this relationship hold true for all parasitic protozoa? If so, then one could guess that there might actually be about 129,000 species of parasitic protozoa, including 4000 sarcodines, 25,000 flagellates, 69,000 "Sporozoa", and 31,000 ciliates. This assumes, of course, that no new species are found in previously examined hosts. Such a guess is subject to considerable error, but it nevertheless probably represents the right order of magnitude.

One other factor might be mentioned. The above numbers refer to named species, but in many cases the descriptions are poor and next to nothing is known about the life cycles or bionomics of the species. For instance, of the 204 species of rodent *Eimeria* in Levine and Ivens'[108] compilation, the location in the host was known for only forty-five, the endogenous stages for only twenty-five, and complete life cycles had been worked out for only four.

If we assume that only about 6 per cent of the total number of parasitic protozoa have been named, what surprises await us as we continue to study the known species and investigate the unknown 94 per cent? It is impossible to predict. At the turn of the century the physicists believed that little remained for them except to add a few decimals to their measurements. Then their whole neat concept fell apart. Perhaps the same thing will happen in protozoology, but somehow I doubt it.

ORIGIN OF PARASITISM

Parasites have arisen independently from free-living ancestors in all the major groups of protozoa. In the subphyla Apicomplexa and Cnidospora, indeed, no free-living species are known. The Apicomplexa are presumed to have arisen from flagellate ancestors and the Cnidospora perhaps from non-protozoan ancestors, but both of these speculations, and especially the latter, are hazardous. The trypanosomatids probably originated from euglenids, the trichomonads and other enteric flagellates from several different lines of free-living flagellates, the Hypermastigorida perhaps from the trichomonads, the enteric amoebae from limax-like and other free-living amoebae (except for *Dientamoeba*, which probably arose from a *Histomonas*-like flagellate), and the parasitic ciliates from several stocks of related free-living forms. All this is speculation and perhaps not particularly profitable. Further details and references are given by Baker[4] and Mattingly.[111]

RELATIONS BETWEEN PARASITE AND HOST GROUPS

Parasites have evolved along with their hosts, and the relationships between parasites of different host may therefore often (but not always) give clues to the evolution of the hosts. Certain groups of parasites are confined to certain groups of hosts. The Hypermastigorida, for instance, are found only in termites (and primitive cockroaches, from which the termites are thought to have arisen). The present parasite species are not the same that first established themselves as parasites, but their descendants. Once a particular parasite–host relationship was established, both host and parasite evolved together. The resultant species of both may be quite different from their starting-points. This is why their origins generally cannot be traced. One can guess that the Apicomplexa arose from marine dinoflagellates and one can point to *Oodinium* as representing a possible line of development, but one can also find reasons for refusing to accept this guess.

Going briefly through the major groups of protozoa (and using the classification of Honigberg *et al.*[81] and the uniform endings of Levine[101]), the lower Trypanosomatorina occur primarily in insects, and the higher ones in an insect vector and a vertebrate. The Retortamonadorida, Diplomonadorida and Trichomonadorida occur primarily in the enteron of insects and vertebrates; some have some rather close free-living relatives. The Oxymonadorida and Hypermastigorida occur in termites and primitive roaches, the Opalinasica in the intestine of Amphibia, and the parasitic Rhizopodasida in the intestine of vertebrates and insects. The Piroplasmasida are found in the blood cells of ruminants and a few birds, and are transmitted by ticks. The Archigregarinorida are parasites of annelids, the Eugregarinorida of annelids and arthropods, and the Neogregarinorida of insects. Among the coccidia, the Protococcidiorida are parasites of marine annelids, the Adeleorina primarily of marine invertebrates, the Eimeriorina primarily of vertebrates and especially of mammals and birds, and the Haemospororina of reptiles, birds and mammals; the last always have a dipteran vector. The Toxoplasmasida are found in vertebrates and the Haplosporasida in invertebrates. Among the Cnidospora, the Myxospororida occur in coldblooded vertebrates, the Actinomyxorida in annelids and other vertebrates, the Helicospororida in insects, and the Microspororida in arthropods and some other invertebrates, in lower vertebrates, and rarely in higher vertebrates. Among ciliates, the parasitic Gymnostomatorida and Trichostomatorida occur mostly in mammals, the Apostomatorida in marine crustaceans, the Astomatorida mostly in oligochaetes, the Thigmotrichorida usually in or on bivalve molluscs, and the Entodiniomorphorida primarily in ruminants, perissodactylids and elephants.

One can make even smaller host–parasite groups. Among the Eimeriorina, for instance, members of the genus *Aggregata* occur in crustacea and cephalopods (schizogony being in the crustacean and sporogony in the cephalopod), members of the genera *Pseudoklossia* and *Hyaloklossia* in marine mussels, of *Myriospora* in marine snails, of *Merocystis* in a whelk, and of *Angeiocystis* in polychaetes. These protozoa are all members of the relatively primitive family Aggregatidae. In the more advanced family Eimeriidae, members of the genera *Caryospora*, *Tyzzeria* and *Wenyonella* occur in mammals, birds and reptiles, of *Isospora* in all vertebrates except fish (and in one species of mollusc), and of *Eimeria* in all vertebrates including teleost and elasmobranch fish (and in a few lower chordates, arthropods and an annelid).

One can carry this even further. The coccidia of gallinaceous birds are mostly *Eimeria*, of passerine birds mostly *Isospora*. The coccidia of ruminants are mostly *Eimeria*, of carnivores mostly *Isospora*. This certainly does not reflect any phylogenetic relationship between chickens and cattle as opposed to sparrows and dogs, it merely reflects the element of chance in the original host–parasite combinations in the ancestors of our modern host orders. In contrast, the evolution of the genera of Haemospororina seems to have paralleled somewhat that of their hosts (cf. refs. 18, 68, 4, 60, 61). *Haemoproteus* is found only in reptiles and birds, and *Leucocytozoon* only in birds. The former is transmitted by *Culicoides* and hippoboscid flies, and the latter by *Simulium* and *Culicoides*. *Nycteria* and *Polychromophilus* are found in bats, and *Hepatocystis* in a few orders of mammals. The last is transmitted by *Culicoides*;[69] the vectors of the first two are unknown. *Plasmodium* occurs in reptiles, birds and mammals. The species in reptiles and birds are transmitted by culicine mosquitoes and those in mammals by anopheline mosquitoes. Subgeneric names have been introduced for the different types;[48] each subgenus is confined to a fairly restricted host range.

Anomalies do occur, and not all are readily explained. There appears to be no great physiologic difference between the cecum of the domestic rabbit and that of the guinea pig, yet the guinea pig cecum contains many species of protozoa, while the rabbit cecum has very few. I have found records of twenty-six species of protozoa in the lumen of the guinea pig cecum, and only five in that of the rabbit. How has this come about? Conversely, eleven species of coccidia (*Eimeria*) have been described from the domestic rabbit, and only one from the guinea pig. Trichomonads are almost ubiquitous, being found in all classes of vertebrates and even in some invertebrates, yet none are known from rabbits or other lagomorphs. Coccidia are rare in primates, ducks and shorebirds. Ticks carry *Theileria*, they bite both ruminants and many small mammals, yet the latter do not become infected. How have all these things come about? Questions such as these remain unanswered in the present state of our knowledge. It appears, however, to be simply a matter of random evolution and of chance.

FATE OF PARASITES IN FOREIGN HOSTS

When a parasitic protozoon enters a foreign host, one of three things may happen:

(1) it may be killed almost at once or be passed on out without change;
(2) it may survive for a longer or shorter time but not multiply; or
(3) it may develop to maturity; if so, it may or may not cause disease.

The commonest fate of a parasitic protozoon in a foreign host is death. Each protozoan species has a more or less restricted host range, and it cannot establish itself in a new host too distant phylogenetically from its normal one. Higher animals continually ingest lower ones, yet none of the latter's protozoa can establish themselves in the former (except for those such as trypanosomes, which have an invertebrate for one host and a vertebrate for another in their normal life cycle). The protozoa which parasitize fish do not parasitize higher vertebrates, nor do those of higher vertebrates parasitize fish. Birds and mammals each have their own protozoa, and so do reptiles, amphibia, fish and the various invertebrate groups. The association is even stricter than this; few protozoa which parasitize hosts of one order are able to parasitize hosts of another, and in many (perhaps most) cases this statement can be extended to genera. It can be stated as a general rule that each host genus (or sometimes order) has its own protozoan parasites, or conversely that each protozoan species can parasitize only one host genus (or sometimes family or order).

There are some exceptions to the above. There are a few protozoa which have an extremely broad host range. One such is *Toxoplasma gondii*, which has been found in a very large number of mammals and some birds, and which can even infect reptiles experimentally.[88] Some of the trypanosomatids are also exceptions, although their host range is not as broad as that of *Toxoplasma*. *Leishmania tropica*, for instance, occurs naturally not only in man but also in the dog and a number of wild rodents and other wild mammals. *Trypanosoma avium* infects a large number of birds but no mammals.[10] *Trypanosoma cruzi* infects a wide variety of mammals, from man to the fox, skunk, raccoon, dog, cat, pig, woodrat, opossum and armadillo. *T. evansi* occurs in camels, horses, cattle, deer, goats, pigs, dogs, elephants, capybaras and tapirs. *T. brucei* occurs in horses, cattle, camels, sheep, goats, pigs, dogs and many wild game animals. The closely related *T. rhodesiense* occurs in man and wild game; it has been found in the bushbuck *Tragelaphus scriptus*.[74] However, *T. gambiense*, which is morphologically indistinguishable from *T. brucei* and *T. rhodesiense*, occurs only in man (and possibly in the pig—cf. ref. 147). All these trypanosomes can infect laboratory mice and rats experimentally.

But trypanosomes of the lewisi group are more restricted, each species occurring only in a single rodent host genus. *T. lewisi* of the genus *Rattus* will not normally infect *Mus*, and *T. duttoni* of *Mus* will not normally infect *Rattus*.

A second exception is the chick embryo. Its defenses are not well developed, and it can be infected with many foreign protozoa, including some of mammals. *Tritrichomonas foetus* of the ox, for instance, can be propagated in the chick embryo.[116, 106, 79] Trypanosomes, too, can be cultivated in chick embryos (cf. refs. 124–5). And Richardson[130] was able to infect newly hatched chicks with mammalian trichomonads.

Another general rule which I state with much less confidence than the previous one, because less information is available to sustain it, is that the lower the host group in the animal kingdom, the broader are its parasites' host ranges. In general, a protozoan species parasitizing fish is found in more genera, families and higher groups than a protozoan species parasitizing mammals. The protozoa which parasitize insects seem seldom limited to a single host genus and sometimes not even to a single host family. *Crithidia fasciculata* occurs in mosquitoes of the genera *Aedes, Anopheles, Armigeres, Culex, Culiseta* and *Mansonia.*[150] Becker[9] found *Herpetomonas muscarum* in flies of the genera *Phormia, Phaenicia, Calliphora, Sarcophaga* and *Musca*, and infected flies of the genera *Musca, Phaenicia, Calliphora, Callitroga, Sarcophaga, Drosophila* and *Phormia* with it. This species has also been reported from flies of the genera *Coelopa, Neuroctena, Limosina, Scatophaga, Fannia, Chrysomya, Lucilia* and *Pollenia.*[150] These fly genera occur in seven families. On the other hand, the gregarines of insects are generally considered to be relatively stenoxenous, each species being usually limited to a single host genus. This statement may, however, be more an indication of ignorance than of fact. We have examined far too few insects for parasites, and we have made far too few cross transmission studies among them to be able to draw solid conclusions.

The second possible fate of a parasitic protozoon in a foreign host is that it may survive for a time without multiplying. Little is actually known about this possibility among the protozoa. It is common among nematodes and is responsible for cutaneous and visceral larva migrans. It is also the cause of sparganosis among cestodes. In the protozoa, however, the normal life span is ordinarily short, and prolonged survival is unusual. A possible exception is *Trypanosoma melophagium*, which occurs in the blood of sheep and the gut of the sheep ked *Melophagus ovinus*. Multiplication is known to occur in the ked, but the protozoa are so sparse in sheep that it has been suggested that no multiplication occurs in this host.

The third possible fate of a parasitic protozoon in a foreign host is that it may multiply in it more or less well, with or without causing disease. Experimental infections of laboratory animals with parasites which they would never normally acquire are examples of this category.

The fact that parasites may multiply in a foreign host is the basis for using laboratory animals at all. *Entamoeba histolytica* has never been found naturally in the guinea pig, rabbit or mouse, and only rarely in rats, dogs and cats, yet these hosts are often infected experimentally. *Trypanosoma brucei* and many other ruminant and primate trypanosomes never occur normally in the rat and mouse, yet these animals can be easily infected and are the usual laboratory hosts. The normal host of *Plasmodium berghei* is the tree rat *Thamnomys surdaster*, but almost all the research done on it has been in the laboratory mouse *Mus musculus* and laboratory rat *Rattus norvegicus*.

Abortive infections may occur in foreign hosts. It is not at all uncommon for protozoa to multiply or survive a short time in foreign hosts and then to die out. This has been observed in many species, including *Trichomonas, Trypanosoma, Entamoeba, Eimeria* and *Plasmodium*.

It has often been said that protozoa in new hosts are generally more pathogenic than in their accustomed ones. This statement is based on two presumptions: (1) that in evolutionarily long-standing associations the host and parasite must have worked out some sort of mutual toleration, i.e. commensalism; and (2) that when the parasite enters a new host, the latter is more susceptible to it than the normal one and therefore that it causes disease in the former. This is true of some protozoa but not of most. *Plasmodium berghei* is much more pathogenic in the laboratory mouse than in its normal host *Thamnomys surdaster*, but it is not so pathogenic in the laboratory rat.

Much of the evidence for this notion comes from the study of African trypanosomes, but even here analysis of the data would have shown how wrong it is. *Trypanosoma brucei*, *T. vivax* and *T. congolense* are common in African large game animals, but cause no apparent disease in them. However, they cause a fatal disease in cattle. *T. rhodesiense* occurs in the bushbuck and presumably other game animals without harming them, but it causes a rapidly fatal disease in man. These trypanosomes can cause a fatal disease in laboratory rats also.

On the other hand, *T. brucei*, *T. vivax* and *T. congolense* are unable to infect man at all, even though they must have been introduced into him thousands of times by biting tsetse flies, and *T. rhodesiense* and *T. gambiense* cannot infect cattle at all. One must select the right new host and ignore all the others in order to make this notion seem true.

Indeed, the whole idea of increased pathogenicity in a new host is contrary to the concept of producing vaccines which are avirulent in the host for which they are designed by repeated passage in another host in which they are not very pathogenic in the first place, e.g. yellow fever, smallpox, etc.

For the great majority of protozoa, new hosts are less susceptible than old ones.

How, then, can one explain high pathogenicity? In unusual cases it may be due to infection with a new or foreign parasite. In some cases it may be due to infection of the normal host species in an area where the parasite never occurred before and where the host had therefore not become adapted to the parasite. This is what happened when smallpox, measles, tuberculosis, etc., were introduced into North America by the European colonists, and it was what happened when these diseases were introduced into the Pacific islands by Europeans.

More usually, however, it is probably due to mutation of the parasite itself. Chickens have eight species of *Eimeria*, all their oocysts look more or less alike, but only three are very pathogenic. The most pathogenic species is *E. tenella*. It cannot have been introduced from some other host, since no other host is known for it. Most likely, it arose by mutation from a previously relatively innocuous species. The same thing can be said for most pathogenic species of *Eimeria* and other protozoa in various hosts.

MORPHOLOGIC ADAPTATIONS TO PARASITISM

Adaptation to a parasitic existence has involved many modifications, both morphologic and physiologic; the latter are the more important. Among morphologic modifications are restriction of locomotion, development of organelles of fixation, reduction of mouthparts, elimination of contractile vacuoles, and increase in reproduction rate.

Restriction in locomotion once a parasite has entered a host may have occurred in some cases because it was advantageous not to move around so actively that the parasite was accidentally evicted from its habitat. In other cases, loss of a locomotory organelle may have resulted from a chance mutation, and the parasite may have survived because the loss did not affect its success. Blindness in cave denizens probably arose in this way. Two whole subphyla of protozoa, all parasitic—the Apicomplexa and Cnidospora—have no external organelles of locomotion except for flagella in some microgametes. Even the pseudopods of *Plasmodium* trophozoites are used for ingestion rather than locomotion. The symbiotic algae and dinoflagellates have no flagella when in their hosts, although some may develop flagella for passage from one host to another. On the other hand, loss of locomotory power is not a necessary concomitant of parasitism. The Hypermastigorida, all of which are parasites of insect guts, have many more flagella than nonparasitic flagellates, and there has been no reduction in the number of cilia of parasitic genera such as *Balantidium, Buetschlia, Isotricha, Dasytricha* and *Blepharozoum*. Again, the cilia have been reduced to cirri in a somewhat similar way in the free-living Hypotrichorida and the parasitic Entodiniomorphorida. One can only conclude that restriction of locomotion is not necessary to parasitism, although it may occur in some groups.

A second concomitant of parasitism is the development of organelles of fixation. Most of the ventral surface of *Giardia* is a sucking disc which presumably helps it to retain its position in the upper small intestine. *Streblomastix strix* is often attached to the gut wall of its termite host by a holdfast, as are the Oxymonadinae. The peritrich *Ellobiophrya donacis*, which lives in the gills of the wedge shell mollusc *Donax vittatus*, is attached to the gill grid of its host by a ring formed by a doubled posterior stalk. The gregarines are attached to the intestinal wall of their host, often by a special organelle, the epimerite. Some astomatid ciliates have holdfasts. One might even think of intracellular parasites, such as the coccidia, piroplasmids, malaria parasites, symbiotic dinoflagellates and Cnidospora, as cases of advanced (or internal) fixation. On the other hand, organelles of fixation are not necessary for parasitism, even in the gut. Trichomonads have none, nor do most other parasitic flagellates, ciliates or amoebae.

A third concomitant of parasitism may be reduction in mouthparts. This may result from a change in type of nutrition from holozoic to saprozoic. None of the Apicomplexa or Cnidospora have permanent mouths, nor do the great majority of parasitic flagellates, sarcodines and many of the parasitic ciliates. This does not prevent them from taking in large particles through the body wall. The Hypermastigorida in the termite gut may ingest pieces of wood as big as themselves, and the malaria parasites and enteric amoebae ingest food by means of pseudopods. On the other hand, many free-living protozoa also have saprozoic nutrition, so reduction or absence of a mouth is no indication of parasitism.

A fourth concomitant of parasitism is the elimination of contractile vacuoles. The presence of contractile vacuoles is said to be an adaptation to the freshwater habitat, where the low osmotic pressure and consequent intake of water through the body wall makes some type of osmoregulation necessary. Parasitic amoebae lack contractile vacuoles, whereas their freshwater relatives have them. Marine amoebae, too, generally lack contractile vacuoles. This comparison cannot be pushed too far, however. It is true that contractile vacuoles are absent in many parasitic protozoa, including the Apicomplexa and Cnidospora, but they are present in others. Most of the enteric ciliates of ruminants and equids, both holotrichs, gymnostomes and entodiniomorphids, have contractile vacuoles. The presence or absence of this structure, therefore, has more to do with the phylogeny of the particular parasite group or species than with the parasitic mode of life.

The fifth concomitant of parasitism is increase in rate of reproduction. The production of large numbers of merozoites by schizogony or of large numbers of sporozoites by a similar process is characteristic of many parasitic protozoa, and especially of the Cnidospora and Apicomplexa. Other parasitic protozoa produce great numbers of daughter cells by repeated binary fissions. The daughter cells may be diffused through the whole body of the host, as are *Trypanosoma* or *Babesia* in the blood, or they may be concentrated in certain areas, as are *Leishmania* in reticuloendothelial cells or *Babesia* and *Theileria* in the tick's salivary glands. However, in other parasites such as *Trypanosoma melophagium* in the sheep, reproduction may be absent or so slow as to be undetectable. When the reproduction rate is increased, it is more likely to be a response to increased temperature and especially to increase in the concentration of available nutrients than to parasitism per se.

No single structural feature is characteristic of a parasitic existence. Parasites closely resemble their free-living relatives and very few of their structures can be considered adaptations for parasitism. The essential differences between a parasite and a non-parasite are physiologic and biochemical, not morphologic.

PHYSIOLOGIC ADAPTATIONS TO PARASITISM

Some of the features I have mentioned above may be considered physiologic. One involves reproduction, and another osmoregulation.

The primary characteristic of a parasite is that it is able to survive in its host without destruction. Structural modifications have little to do with this; biochemical and immunologic ones are all-important.

What keeps parasites which live in their host's intestine from being digested? This question has been answered by saying that the same mechanism operates which prevents the hosts from digesting themselves, that the parasites protect themselves by producing mucus, or that mucoproteins in their integument protect them, or that they secrete antienzymes, or that the surface membrane of living organisms is impermeable to proteolytic enzymes. One or more of these answers may be correct, but much more research must be done before we are sure which are.

Why are anatomically similar species restricted to different hosts which may themselves be anatomically quite similar? Why, for instance, should *Trypanosoma lewisi* occur only in *Rattus*, and the indistinguishable *T. duttoni* only in *Mus*? Why is it that some protozoa such as *Colpoda steinii* can be facultative parasites (cf. refs. 129, 148), while others are not? Why does the ox have one series of sporozoan parasites, and the sheep another, with little overlap between them?

Such questions are at present unanswerable. Compatibility of host and parasite protoplasm is invoked, as is some form of immunologic tolerance. However, these terms merely name the phenomenon; they do not explain it. The question of how this compatibility is brought about remains unanswered, and a great deal of biochemical and immunochemical research must be done before it can be answered (cf. refs. 9, 127, 14, 16, 144).

The subject of immunologic tolerance deserves at least brief mention, for it may hold the key to the question. Under this view (cf. refs. 144, 51, 52, 49) parasites have developed in the course of evolution which have antigens resembling those of their hosts; this was called molecular mimicry by Damian[49]. The host is then unable to produce antigens effective against its parasites. Dineen[51, 52] suggested that this immunologic tolerance would continue until the parasite population reached a critical level of abundance. At that time, the host would react immunologically against the parasite and eliminate most or all of its population. Working with the nematodes *Nematodirus spathiger* and *Haemonchus contortus* in sheep, Donald *et al.*[56] and Dineen *et al.*[53, 54] obtained experimental evidence which might be interpreted in this way.

However, the concept of molecular mimicry alone does not explain all the facts. It is well known that each species of parasite contains or produces several different antigens against which the host may produce antibodies. Such humoral antibodies in a "normal" host may not interfere with the parasite's development; there may be no correlation between the presence of humoral antibody and resistance to infection. A corollary of the above hypothesis, therefore, is that the antibodies elicited by certain parasite antigens are not effective against the parasite species that elicited them. Why this should be is unknown; indeed, it is not clear that such a statement is justified without qualification. Nevertheless, it has considerable value.

Schad[135] has gone a step further in relating this type of antibody response to competition between parasite species within a particular host, introducing the concept of nonreciprocal cross immunity. In his view, an antibody that has been elicited in the host by a particular parasite species may not be effective against that parasite species, but may be highly effective against another. This type of immunity he called nonreciprocal cross immunity. In the course of natural selection, any mechanism which favors one parasite species over another in a particular host species would be selected for. Nonreciprocal cross immunity is one such mechanism; Schad suggested that it was one of several mechanisms involved in the adaptation of a parasite to its host environment, and that it limited the populations of competing parasite species. Nonreciprocal cross immunity, he said, might "be viewed as a special kind of competitive interaction between two species, which is mediated by a third species, the host".

While the above ideas have been developed as a consequence of studies on helminth, and especially nematode, parasites (and on viruses, other non-animal parasites and mammalian grafts), it is quite possible that they operate in protozoan infections as well.

HOST SPECIFICITY

Much of what I have already said deals with host specificity, and some of the examples I shall give later do so also. At this point I should like to summarize some of this information and some of these views in seven bald statements.

First, parasites arose from free-living ancestors. Different groups of parasites arose independently from different free-living ancestors; even within a group, more than one free-living species adopted the parasitic habit independently.

Second, host specificity or host restriction depends upon compatibility between the host and parasite.

Third, during the millions of years of evolutionary history, the directions and nature of change of the hosts were more or less random, but they increased the animals' adaptability to their environment and therefore their survivability.

Fourth, the parasites also evolved along with their hosts; again, the directions and nature of their changes were more or less random and fortuitous but again they tended to increase compatibility and transmissibility, and therefore to increase survivability.

Fifth, as a consequence of the two previous developments, as the hosts became more and more widely separated phylogenetically and in evolutionary time, the resemblance between their parasites also decreased progressively. However, the parasites often appeared quite similar.

Sixth, structural and developmental adaptations to parasitism are of importance, but they do not explain why a particular parasite occurs in one host and not in another which does not appear to be too greatly different.

Seventh, the basic answer to the problem of host specificity is to be sought in the field of biochemistry and immunochemistry—it is a matter of immunologic tolerance or whatever one may wish to call it.

SELECTED HOST–PARASITE RELATIONSHIPS

In this section I should like to describe in some detail some selected host–parasite relationships which illustrate some of the above principles. Other selections which could have been made would be just as useful, but these are satisfactory and representative. There will necessarily be some duplication with what has gone before, but I shall avoid it as much as I can.

Ciliates of Sea Urchins

Kirby[96] discussed the ciliates of sea urchins (echinoids) as an example of distributional host relationships and host specificity. He discussed the observations of seven investigators as presented in fourteen papers. He did not attempt to tabulate the ciliate or host species, but he concluded that (1) given species of sea urchins have characteristic faunules and a few sea urchins have no faunules, facts which suggest that host relationships should be investigated experimentally; (2) the same host species may have different parasite species in different localities; (3) the number of each species of ciliate in a given host varies; some ciliates are found in abundance in most hosts, whereas others are found less commonly, and still others are rare; (4) host specificity is not strong among these ciliates; the same species of ciliate may occur in several host species or genera (Kirby mentioned specifically *Entorhipidium fukuii*, which Uyemura[149] had found in five hosts of four genera); however, some ciliates are more host-specific than others (he said, for instance, that *Metopus histophagus* occurred only in one species of echinoid in Tortugas, whereas *M. circumlabens* occurred in a number of hosts at Bermuda, Tortugas, Amoy and Japan); (5) some genera of these ciliates are restricted to sea urchins; (6) cross infection should be easy to accomplish, although no experimental work had been done on it. He thought that the average number of ciliate species per host was probably four or five; (7) there is evidence that the occurrence of some of these ciliates is based more on geographic factors than on host specificity; for instance,

Yagiu[154] found two species of ciliates in all but one of the echinoid species he examined at Yaku Island, Japan, that contained any ciliates at all, and Powers[126] found two species of ciliates in all the sea urchins at Tortugas that were infected; (8) there is no evidence that any of the sea urchin ciliates injures its host; (9) there is no evidence in the literature that the ciliate parasites of sea urchins are accidentally introduced free-living forms, although some species belong to genera most of whose members are free-living.

The most recent review of echinoid ciliates is that of Berger.[12] He gave 180 references; even though not all of them dealt with the ciliates themselves, the difference between this number and the fourteen of Kirby gives some indication of how our knowledge has grown. Nevertheless, it is far from complete. According to Berger,[12] on whose monographic thesis I am leaning heavily in what follows, sixty-three species of echinoids have been examined for ciliates, but these represent only 8 per cent of the total number of echinoid species. A total of fifty-three named species of ciliates has been found in forty-six (73 per cent) of these host species.

Table 2 gives the species of ciliates which have been found in sea urchins and the number of host genera and species in which each has been found. It lists fifty named and six unnamed ("sp.") ciliate species in three subclasses and six orders. By far the largest number of species are hymenostomes; thirty-six (64 per cent) of the species belong to this order. The other holotrich order, the trichostomes, includes nine species (18 per cent). There are three species of peritrichs (5 per cent), four of heterotrichs (7 per cent), one of oligotrichs (2 per cent) and three of hypotrichs (5 per cent).

The number of host genera for each parasite species ranges from 1 (14 species) to 16, with a mean of 3.9. The number of host species for each parasite species ranges from 1 (10 species) to 26, with a mean of 5.9.

Information on the hosts of the ciliates by subclass and order is summarized in Table 3. The holotrichs average 4.1 host genera and 6.2 host species per parasite species. The peritrichs average 1.7 host genera and 4.0 host species per parasite species. The spirotrichs average 3.5 host genera and 5.5 host species per parasite species. There is no good evidence of any greater host restriction in any one ciliate order than in any other, especially in view of the fact that only 8 per cent of the potential hosts have been examined. One can safely conclude, therefore, that the great majority of echinoid ciliates are euryxenous.

Berger[12] also listed the echinoid species which have been examined for intestinal ciliates. I have summarized his list in Tables 4 and 5; it includes some dubious records which he apparently did not include in his statement referred to above. Ciliates have been found only in the subclass Euechinoidea and not in the subclass Perischoechinoidea. In the former subclass, they have been found in 70 per cent of the seventy-one species examined: in 100 per cent of eight species of the superorder Diadematacea, in 88 per cent of forty-three species of the superorder Echinacea, in 33 per cent of nine

TABLE 2. SPECIES OF CILIATES OCCURRING IN SEA URCHINS AND NUMBER OF HOST GENERA AND SPECIES IN WHICH EACH IS FOUND [†]

Ciliate species	No. of host genera	No. of host species
Subclass HOLOTRICHASINA		
Order TRICHOSTOMATORIDA		
Family COLPODIDAE		
Genus *Colpoda*		
fragilis	1	1
Family PLAGIOPYLIDAE		
Genus *Lechriopyla*		
mystax	1	4
Genus *Plagiopyla*		
minuta	1	1
nyctotherus	1	1
Genus *Plagiopyliella*		
pacifica	1	3
striata	7	9
Genus *Schizocaryum*		
dogieli	1	4
Family THYROPHYLACIDAE		
Genus *Thyrophylax*		
strongylocentroti	2	5
vorax	1	1
Order HYMENOSTOMATORIDA		
Suborder TETRAHYMENORINA		
Family COHNILEMBIDAE		
Genus *Anophrys*		
aglycus	4	4
dogieli	2	2
echini	2	2
elongata	16	26
vermiformis	5	5
sp.	3	3
Genus *Cohnilembus*		
caeci	5	6
Genus *Madsenia*		
indomita	6	12
INCERTAE SEDIS		
Genus "*Colpidium*"		
echini	2	2
Family CRYPTOCHILIDAE		
Genus *Biggaria*		
bermudensis	10	12
sp.	5	6

[†] Based on data given by Berger;[12] the classification of higher taxa is that of Honigberg *et al.*[81]

TABLE 2 (*cont.*)

Ciliate species	No. of host genera	No. of host species
Genus *Cryptochilum*		
caudatum	2	7
echini	15	16
fukuii	6	6
minor	3	7
sigmoides	4	7
sp.	8	10
Genus *Metaxystomium*		
echinometricum	8	12
ozakii	4	4
polynucleatum	6	6
Genus *Tanystomium*		
gracile	3	7
Family ENTORHIPIDIIDAE		
Genus *Entorhipidium*		
echini	7	13
multimicronucleatum	1	1
pilatum	2	5
tenue	4	10
triangularis	2	7
Family ENTODISCIDAE		
Genus *Entodiscus*		
borealis	10	17
powersi	1	2
sabulonis	3	4
Suborder PLEURONEMATORINA		
Family PLEURONEMATIDAE		
Genus *Cyclidium*		
amoyensis	1	1
echinophilum	2	2
ozakii	6	6
rhabdodectum	4	4
stercoris	2	7
sp.	3	5
Genus *Mesogymnus*		
interruptus	2	2
Subclass PERITRICHASINA		
Order PERITRICHORIDA		
Suborder MOBILORINA		
Family URCEOLARIIDAE		
Genus *Trichodina*		
sp.	2	5
Genus *Urceolaria*		
lytechini	1	1
strongylocentroti	2	6

TABLE 2 (cont.)

Ciliate species	No. of host genera	No. of host species
Subclass SPIROTRICHASINA		
Order HETEROTRICHORIDA		
Suborder HETEROTRICHORINA		
Family GYROCORYTHIDAE		
Genus *Metopus*		
brevicristatus	1	1
circumlabens	12	19
histophagus	1	1
rotundus	6	9
Order OLIGOTRICHORIDA		
Family STROBILIDIIDAE		
Genus *Strobilidium*		
rapulum	3	5
Order HYPOTRICHORIDA		
Suborder SPORADOTRICHORINA		
Family EUPLOTIDAE		
Genus *Euplotes*		
balteatus	2	5
charon	1	1
sp.	2	3

TABLE 3. NUMBER OF HOST GENERA AND SPECIES IN WHICH EACH HIGHER TAXON OF SEA URCHIN CILIATE OCCURS†

Ciliate taxon	No. of species	No. of host genera per parasite species	No. of host species per parasite species
Subclass HOLOTRICHASINA	45	4.1	6.2
Order TRICHOSTOMATORIDA	9	1.8	3.2
Order HYMENOSTOMATORIDA	36	4.7	6.9
Subclass PERITRICHASINA	3	1.7	4.0
Order PERITRICHORIDA	3	1.7	4.0
Subclass SPIROTRICHASINA	8	3.5	5.5
Order HETEROTRICHORIDA	4	5.0	7.5
Order OLIGOTRICHORIDA	1	3.0	5.0
Order HYPOTRICHORIDA	3	1.7	3.0
Total	56	3.9	5.9

† Based on data given by Berger;[12] the classification is that of Honigberg et al. [81].

TABLE 4. NUMBER OF INTESTINAL CILIATE GENERA AND SPECIES OCCURRING
IN SEA URCHINS †

Sea Urchin Taxon	No. of parasite genera	No. of parasite species
Subclass PERISCHOECHINOIDEA		
Order CIDAROIDEA		
Genus *Goniocidaris*		
biserialis	0	0
Genus *cidaris*		
rugosa	0	0
Genus *Eucidaris*		
metularia	0	0
thouarsii	0	0
tribuloides	0	0
Genus *Chondrocidaris*		
gigantea	0	0
Subclass EUECHINOIDEA		
Superorder DIADEMATACEA		
Order DIADEMATOIDA		
Genus *Astropyga*		
magnifica	2	2
Genus *Diadema*		
setosum	5	5
savignyi	3	3
antillarum	6	9
paucispinum	3	3
Genus *Echinothrix*		
calamaris	5	6
diadema	5	6
Order ECHINOTHURIOIDA		
Genus *Calveriosoma*		
gracile	1	2
Superorder ECHINACEA		
Order PHYMOSOMATOIDA		
Genus *Glyptocidaris*		
crenularis	4	6
Genus *Stopneustes*		
variolaris	‡	‡
Order ARBACIOIDA		
Genus *Arbacia*		
lixula	0	0
punctulata	1	1
Order TEMNOPLEUROIDA		
Genus *Temnopleurus*		
toreumaticus	4	5

† Based on data given by Berger.[12]
‡ Unidentified ciliates found.

TABLE 4 (*cont.*)

Sea Urchin Taxon	No. of parasite genera	No. of parasite species
Genus *Salmacis*		
bicolor	0	0
virgulata		
Genus *Mespilia*		
globulus	7	7
Genus *Lytechinus*		
variegatus	6	10
pictus	1	1
anamesus	4	4
Genus *Toxopneustes*		
pileolus	2	3
roseus	6	7
Genus *Tripneustes*		
ventricosus	7	9
gratilla	5	6
Genus *Sphaerechinus*		
granularis	2	2
Genus *Pseudobolita*		
indiana	1	1
Genus *Pseudocentrotus*		
depressus	6	7
Order ECHINOIDA		
Genus *Echinus*		
esculentus	2	2
acutus	1	1
Genus *Sterechinus*		
neumayeri	1	1
Genus *Psammechinus*		
miliaris	2	2
Genus *Paracentrotus*		
lividus	6	6
Genus *Strongylocentrotus*		
droebachiensis	14	22
echinoides	14	22
intermedius	10	15
pulchellus	6	9
nudus	7	11
purpuratus	12	17
franciscanus	17	26
Genus *Hemicentrotus*		
pulcherrimus	9	13
Genus *Allocentrotus*		
fragilis	10	16
Genus *Parasalenia*		
gratiosa	0	0

Table 4 (*cont.*)

Sea Urchin Taxon	No. of parasite genera	No. of parasite species
Genus *Evechinus*		
chloroticus	1	1
Genus *Echinostrephus*		
aciculatus	3	4
Genus *Anthocidaris*		
crassispina	11	17
Genus *Echinometra*		
lucunter	8	12
vanbrunti	6	7
mathaei	8	11
Genus *Heterocentrotus*		
mammilatus	4	6
trigonarius	2	2
Genus *Colobocentrotus*		
mertensii	0	0
atratus	4	4
Superorder GNATHOSTOMATA		
Order HOLECTYPOIDA		
Genus *Echinoneus*		
cyclostomus	0	0
Order CLYPEASTEROIDA		
Genus *Clypeaster*		
rosaceus	5	6
subdepressus	4	5
Genus *Fibularia*		
ovulum	0	0
Genus *Echinarachinius*		
parma	0	0
Genus *Dendraster*		
excentricus	0	0
Genus *Mellita*		
quinquiesperforata	0	0
sexiesperforata	0	0
Genus *Encope*?		
sp.	1	1
Superorder ATELOSTOMATA		
Order SPATANGOIDA		
Genus *Maretia*		
planulata	0	0
Genus *Echinocardium*		
cordatum	0	0
Genus *Brisaster*		
townsendi	0	0

Table 4 (*cont.*)

Sea Urchin Taxon	No. of parasite genera	No. of parasite species
Genus *Moira*		
atropos	0	0
Genus *Rhinobrissus*		
hemiasteroides	0	0
Genus *Plagiobrissus*		
grandis	0	0
Genus *Brissus*		
unicolor	‡	‡
latecarinatus	0	0
agassizi	0	0
Genus *Meoma*		
ventricosa	0	0
Genus *Metalia*		
spatagus	0	0

TABLE 5. NUMBER OF CILIATE GENERA AND SPECIES FOUND IN HIGHER TAXA OF SEA URCHINS

Sea Urchin Taxon	No. of species examined	No. of parasite genera per host species†	No. of parasites pecies per host species†
Subclass PERISCHOECHINOIDEA	6	0	0
Order CIDAROIDEA	6	0	0
Subclass EUECHINOIDEA	71	5.1	6.9
Superorder DIADEMATACEA	8	3.8	4.5
Order DIADEMATOIDA	7	4.1	4.9
Order ECHINOTHURIOIDA	1	1.0	2.0
Superorder ECHINACEA	43	5.7	7.8
Order PHYMOSOMATOIDA	2	2.5	3.5
Order ARBACIOIDA	2	1.0	1.0
Order TEMNOPLEUROIDA	14	4.2	5.2
Order ECHINOIDA	25	6.9	9.9
Superorder GNATHOSTOMATA	9	3.3	4.0
Order HOLECTYPOIDA	1	0	0
Order CLYPEASTEROIDA	8	3.3	4.0
Superorder ATELOSTOMATA	11	1.0	1.0
Order SPATANGOIDA	11	1.0	1.0
Total	77	5.1	6.9

† Only parasitized species included; + given value of 1 in calculations.

11*

species of the superorder Gnathostomata, and in 9 per cent of eleven species of the superorder Atelostomata. In other words, they are not evenly distributed throughout the whole group, but are found predominantly in the superorders Diadematacea and Echinacea. Among the parasitized sea urchin species, the number of ciliate genera ranged from 1 to 17, with a mean of 5.1, and the number of ciliate species ranged from 1 to 26, with a mean of 6.9. This last figure is somewhat higher than Kirby's[96] estimate of four or five ciliate species per host, but many more sea urchins have been examined since that time.

Despite the fact that the echinoid ciliates are euryxenous, their hosts are often quite closely related, so that similarities between the fauna of the hosts can often be used as indicators of host relationships. If this is done, however, it should be done with great caution. Indeed the structural characters of the sea urchins themselves are far better criteria of taxonomic relationships than are the relationships of their parasites.

Berger[12] said (p. 358) that "a very strong argument can be made for zoogeographical restriction being more important in the evolution of these entocommensals than their restriction to particular host taxa". However, any statements about this problem "are rendered somewhat illogical due to the circular argument: Do ciliates inhabit a particular host because it is that particular species of echinoid, or are they present in that host because they can only survive in a certain geographic locality due to the interaction of the innumerable biological and physical factors present in the marine habitat?"

Table 6 summarizes the data given by Berger[12] on the geographic distribution of sea urchin ciliates in the Northern Hemisphere. He divided the Atlantic and Pacific Oceans into northern and southern parts at 33–34° N. latitude. Twenty-three per cent of the species occurred only in the Atlantic Ocean, 8 per cent being in the northern part, 13 per cent in the southern part, and 2 per cent in both. Forty-seven per cent of the species occurred only in the Pacific Ocean, 32 per cent being in the northern part, 8 per cent in the southern part, and 7 per cent in both. Fifty-five per cent of the species occurred only in the northern area, of which 8 per cent were in the Atlantic, 32 per cent in the Pacific and 15 per cent in both. Twenty-eight per cent of the species occurred only in the southern area, of which 13 per cent were in the Atlantic, 8 per cent in the Pacific and 7 per cent in both. Only 8 per cent of the species occurred in a combination of three of the four areas, and none in all four. These findings bear out the view that geographic area is an important factor in the distribution of sea urchin ciliates.

Few cross transmission studies have been made. In Hawaii, Heckman[73] tried unsuccessfully to infect *Echinometra mathei* with unidentified ciliates from *Pseudobolita indiana*, and to infect both of these hosts with *Metopus* sp. and other ciliates from *Echinothrix diadema*. Berger[12] was unable to infect *Dendraster excentricus* (which normally has no enteric ciliates) with

TABLE 6. GEOGRAPHIC DISTRIBUTION OF ECHINOID CILIATES IN THE NORTHERN HEMI-
SPHERE†

Locality	Number of ciliate species	
	Number	Per cent
Northern Atlantic alone	5	8
Southern Atlantic alone	8	13
Northern Pacific alone	19	32
Southern Pacific alone	5	8
Northern Atlantic and southern Atlantic	1	2
Northern Atlantic and northern Pacific	9	15
Northern Atlantic and southern Pacific		
Southern Atlantic and northern Pacific		
Southern Atlantic and southern Pacific	4	7
Northern Pacific and southern Pacific	4	7
Northern Atlantic, southern Atlantic and northern Pacific	1	2
Northern Atlantic, southern Atlantic and southern Pacific		
Northern Atlantic, northern Pacific and southern Pacific	1	2
Southern Atlantic, northern Pacific and southern Pacific	3	5
All four areas		
Atlantic alone	14	23
Pacific alone	28	47
Northern area alone	33	55
Southern area alone	17	28
Combination of three areas	5	8
Total number of species	60	100

† Based on data given by Berger;[12] 0–33° N. lat. considered southern, 34–90° N. lat.
considered northern.

the ciliates *Entodiscus borealis*, *Madsenia indomita*, *Cryptochilum sigmoides*,
Tanystomium gracile and *Cyclidium stercoris* from the sea urchins *Strongylo-
centrotus droebachiensis*, *S. echinoides* and *S. franciscanus* at Friday Harbor,
Washington. These findings suggest that there is a certain degree of host
specificity among the protozoa. However, no controls were run to determine
whether the infection techniques they used would have succeeded even in the
normal hosts.

There is a question whether the sea urchin ciliates may be marine species
which have simply found a snug harborage in the sea urchin intestine. Presumably the ocean was the original source from which they came. However,
none of the present species is known to live free in the ocean, although many
are congeners of others which do. According to Berger,[12] if one omits

questionable identifications and dubious host records, nine genera (*Entodiscus, Entorphipidium, Lechriopyla, Mesogymnus, Metaxystomium, Plagiopyliella, Schizocaryum, Tanystomium* and *Thyrophylax*) parasitize echinoids and related forms alone, five genera (*Biggaria, Cryptochilum, Madsenia, Trichodina* and *Urceolaria*) parasitize both echinoids and other invertebrates, while seven genera (*Anophrys, Cohnilembus, Cyclidium, Euplotes, Metopus, Plagiopyla* and *Strombilidium*) contain parasites of echinoids and free-living species.

Thus, although we now know a great deal more about sea urchin ciliates than Kirby did, most of his conclusions remain essentially unchanged. Cross transmission experiments still need to be done, but in the main our newer data simply flesh out Kirby's statements.

The Trypanosomatidae

This family is the only one in the suborder Trypanosomatorina, order Kinetoplastorida. This order was established only recently by Honigberg[80] for zooflagellates with one to four flagella and an argentophile, Feulgen-positive kinetoplast; the kinetoplast is self-replicating and has mitochondrial affinities. The order contains two suborders, of which members of the suborder Bodonorina typically have two flagella, while members of the suborder Trypanosomatorina have one flagellum.

No complete study of the family Trypanosomatidae has been attempted since Wenyon's[153] review, although Grassé[70] discussed all groups without giving the names of most species. Other writers, such as Baker[3, 4] Hoare,[77, 78] Stephen[147] and Wallace,[150] have discussed limited groups or aspects. My own discussion will necessarily be far from complete, but members of this family form an excellent evolutionary series and their host-parasite relations and life cycles, which vary with their group, are of endless interest.

There are nine genera in the family: *Leptomonas, Phytomonas, Herpetomonas, Crithidia, Blastocrithidia, Rhynchoidomonas, Trypanosoma, Endotrypanum* and *Leishmania*. The second is heteroxenous and occurs in plants and insects, the last three are heteroxenous and occur in vertebrates and invertebrates, while the remainder are monoxenous and occur only in invertebrates.

The whole family probably arose from some free-living, euglenorid, phytomastigophorean ancestor, but this is speculation and will probably never be either substantiated or refuted. The trypanosomatids are usually elongate and slender, but some species or forms may be short and broad or even round. They all have the following structures: (1) a vesicular nucleus; (2) a single flagellum (which may not extend out of the body and which may possibly be absent in a few species) composed of the standard 9 + 2 fibrils and which arises from (3) a basal granule (basal body, centriole) composed of a cylinder formed by the continuation of the flagellum's nine peripheral fibrils; (4) a reservoir just above the basal granule through which the flagellum passes; it is an invagination of the outer cell membrane; it is at the anterior end of the organism in *Leptomonas, Herpetomonas* (usually) and *Crithidia,* and along the side in *Blastocrithidia* and *Trypanosoma*; it can be seen with the phase contrast microscope as a clear space around the base of the flagellum (cf. ref. 25); (5) a kinetoplast just posterior to the basal granule; it is a modified mitochondrion which contains a band of anteroposterior fibers or a loose meshwork of fibers that are believed to be DNA; and (6) one or more contractile vacuoles that open into the side of the reservoir.

The genera of Trypanosomatidae are differentiated on the basis of the location of the above structures and also on the basis of their hosts. In the course of their life cycles, some of them pass through stages resembling other genera. A source of confusion is the fact that *Leptomonas* and certain other "lower" genera occur only in insects, while insects are also vectors of *Trypanosoma* and *Leishmania* and contain leptomonad (and, in the case of *Trypanosoma*, other "lower") forms in their guts; hence a person who sees a leptomonad in an insect gut cannot be sure whether it is a parasite of the insect alone or the leptomonad stage of a trypanosome or *Leishmania* of vertebrates.†

The leptomonad form, with its kinetoplast at the anterior end, is considered to be the most primitive member of the family. *Leptomonas* is a monoxenous parasite of the intestine of many insects and also of some nematodes and molluscs.

There are two theories as to the origin of the various trypanosomatids from a primitive leptomonad ancestor. One, espoused by Baker[4] and a large number of earlier workers, is that *Trypanosoma* and *Leishmania* of vertebrates arose from the intestinal leptomonads of invertebrates when the latter sucked blood from the former. Figure 1 gives Baker's proposed phylogenetic history of the family; note that it contains two questionmarks and three "hypotheticals". The second view, espoused most recently by Wallace,[150] is that the family originated as blood parasites of vertebrates which secondarily came to be transmitted by insects.

According to the first theory, a primitive leptomonad stock gave rise to *Leptomonas, Leishmania, Phytomonas* and presumably *Endotrypanum* and the subgenus *Schizotrypanum* of the genus *Trypanosoma*; the primitive leptomonad stock also gave rise to a hypothetical crithidial stock which in turn gave rise to our modern genera *Crithidia, Blastocrithidia, Rhynchoidomonas* and all *Trypanosoma* except members of the subgenus *Schizotrypanum*. According to the second theory, *Leishmania* may have given rise to *Leptomonas, Herpetomonas* and *Crithidia*, while *Trypanosoma* may have given rise to *Blastocrithidia* and *Rhynchoidomonas*. I prefer the first theory,

† Since this chapter was written Hoare and Wallace (1966, *Nature* **212**, 1385–6) introduced a new terminology to describe the stages. The former trypanosome form is now known as the trypomastigote form, and the herpetomonad, crithidial and leptomonad forms as the opisthomastigote, epimastigote and leptomonad forms, respectively

mostly because I can see a logical connection between a free-living euglenorid and a monoxenous intestinal parasite, whereas I can see no such connection between a blood parasite of vertebrates and any free-living ancestral form.

In Table 7 are given the morphologic characteristics of each genus (position of kinetoplast, position of reservoir, presence or absence of an undulating membrane, number of forms in the life cycle), their host–parasite relationship (whether monoxenous or heteroxenous), the host groups of their various stages, and the approximate numbers of species which have been named from each host group.

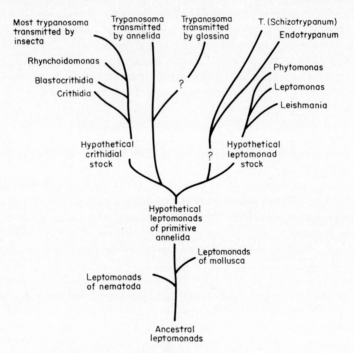

FIG. 1. Baker's[4] proposed phylogenetic history of the suborder Trypanosomatorina.

As will be seen, the table includes fifty-eight species of *Leptomonas*, of which 93 per cent are from insects; nine species of *Phytomonas* from plants and insects; fifteen species of *Herpetomonas*, of which 60 per cent are from insects; fifteen species of *Crithidia*, all from insects; thirty species of *Blastocrithidia*, all from arthropods (insects and acarines); six species of *Rhynchoidomonas*, all from insects; 190 species of *Trypanosoma*, of which 98 per cent are from vertebrates, including 42 per cent from mammals, 1 per cent from birds, 11 per cent from reptiles, 11 per cent from amphibia and 34 per cent from fish; one species of *Endotrypanum* from a mammal; and seven species of *Leishmania*, of which 43 per cent are from mammals, 43 per cent from reptiles and 14 per cent from protozoa. Arthropods (the great majority being insects) and vertebrates are by far the commonest hosts; indeed, the vectors of the mammalian and avian *Trypanosoma* and of all the *Phytomonas* are insects. Of the 331 species of Trypanosomatidae in this listing, 61 per cent occur in vertebrates and 63 per cent in insects and a few acarines.

I am not as sure about the validity of some species as of others. As will be seen from the footnotes, some of the "species" (especially of *Trypanosoma*) listed in the *Zoological Record* are probably synonyms, while other species not mentioned in that source are not included. The validity of some species of mammalian, reptilian, amphibian and piscine *Trypanosoma* is especially dubious; forty-eight species of avian *Trypanosoma* are listed in the *Zoological Record*, but Bennett,[10] whom I am following, doubted the validity of all but two of them. He found that the trypanosomes of twenty-four species of birds from eleven families were morphologically similar, and he was able to transmit trypanosomes from nine species of birds to birds of different species, families and even orders, using *Simulium* spp., *Prosimulium decemarticulatum* and *Aedes aegypti* as vectors. He considered that all the trypanosomes that he was dealing with (except one) were *Trypanosoma avium*, and thought that the avium complex might contain many physiologically distinct strains or species. (The exception was *T. paddae*, which he found in a chipping sparrow *Spizella passerina*; this species had been originally described by Laveran and Mesnil, 1904, from the Java sparrow *Padda oryzivora*, and differed morphologically from the others.) Stabler, Holt and Kitzmiller[146] also considered that practically all avian trypanosomes belonged to one species. They reported *T. avium* in the bone marrow of ninety-six species of wild birds, of which forty-three were now host records.

In addition, it is likely that many of the nine named species of *Phytomonas* may be synonyms of other species. Finally, modern methods should be used to determine whether the eight species of *Leptomonas* and *Herpetomonas* reported from reptiles, frogs and bats between 1921 and 1931 actually belong to these genera.

This table should not be considered definitive. It gives my best estimate of the situation and illustrates the present state of our knowledge. I hope that it can be revised and corrected in the future as the result of definitive studies and analyses of some of the groups.

Hoare[77, 78] revised the classification of mammalian *Trypanosoma*, but without tackling the members of this genus in other hosts. His classification is based on morphologic and biological characters. He divided the genus into two sections, Stercoraria and Salivaria, based on whether they developed in the posterior station (i.e. hindgut) of the vector and were transmitted by contamination with feces (Stercoraria) or on whether they developed in the anterior station (i.e. salivary glands and/or proboscis) of the vector and were

TABLE 7. CHARACTERISTICS OF THE TRYPANOSOMATIDAE

Genus	Position of kinetoplast	Position of reservoir	Undulating membrane	No. of forms in life cycle	Host-parasite relationship	Hosts — Of intermediate stages	Hosts — Of final stages	Approx. no. of species named
Leptomonas	Anterior	Anterior	Absent	1	Monoxenous	—	Protozoa Nematodes Molluscs Insects Amphibia Bat	1[1] 1[a] 54[b] 1[e] 1[e] 9[d]
Phytomonas	Anterior	?	Absent	1	Heteroxenous	Insects	Plants	9[b]
Herpetomonas	From one-fifth of body length from ant. end to post. tip	one-fifth to nine-tenths of body length from ant. end	Absent	1	Monoxenous	—	Insects Reptiles	6[e]
Crithidia	Posterior	Anterior	Absent (body truncate anteriorly)	1	Monoxenous	—	Insects	15[b]
Blastocrithidia	Near nucleus	Near nucleus	Present	1(?)	Monoxenous	—	Arthropods	30[b]
Rhynchoidomonas	Behind nucleus	?	Absent	1	Monoxenous	—	Insects	6[b]
Trypanosoma	Posterior	Posterior	Present	2–4	Heteroxenous	Insects Insects Invertebrates Invertebrates Invertebrates ?	Mammals Birds Reptiles Amphibia Fish Other	79[a] 2[c] 20[a] 26+[k] 64[a] 4[a]

TABLE 7 (cont.)

Endotrypanum	Anterior or posterior to nucleus	?	Present	3(?)	Heteroxenous	Insect[j]	Mammal	1[f]
Leishmania	Anterior	?	Absent	2	Heteroxenous	Insects	Mammals Reptiles Protozoa	3[g] 3[a] 1[h]

[a] From the *Zoological Record*; some of these "species" are probably synonyms; in addition, some species are not to be found in this source.

[b] From Wallace.[150]

[c] Cf. Bennett;[10] forty-eight names are given in the *Zoological Record*.

[d] This number of species or more has been named; however, Hanson, McGhee and Blake[72] thought it possible that "many, if not all, existing species of *Phytomonas* may in reality be one".

[e] These numbers of names are listed in the *Zoological Record*; whether the species of *Leptomonas* and *Herpetomonas* reported from reptiles, frogs and bats between 1921 and 1931 really belong to these genera should be determined by modern studies.

[f] Cf. Shaw.[137–38]

[g] At least seven species of *Leishmania* have been named from mammals; I am tentatively accepting as valid only *L. donovani*, *L. enriettii* and *L. tropica*.

[h] *L. esocis* Georgevitch, 1936 on a *Myxidium* from a pike.

[i] *L. karyophilum* Gillies and Hanson, 1963 in the macronucleus of *Paramecium trichium*.

[j] Shaw[137] thought the vector to be *Phlebotumus*.

[k] Diamond (1964) recognized twenty-six species from the Anura alone. The *Zoological Record* gave twenty-one species for all amphibians.

11a*

transmitted by inoculation (Salivaria). He then divided each section into subgenera, recognizing three in the Stercoraria and four in the Salivaria. Since, however, the type species of *Trypanosoma* is *T. rotatorium* (Mayer, 1843) Gruby, 1843 of frogs, he had to accept an eighth subgenus as the nominative subgenus (i.e. the subgenus *Trypanosoma*, which is not one of the seven in mammals).

Hoare realized that his new classification might not be favorably received by medical and veterinary workers, but pointed out that the subgeneric names need not be used. In addition, his classification includes only species from mammals. He felt that revision of the taxonomy of trypanosomes of other vertebrates was long overdue, and suggested that the subgeneric names *Trypanosoma* Gruby, 1843 might be used for the trypanosomes of amphibia, *Haematomonas* Mitrophanow, 1883 for those of fish and *Trypanomorpha* Woodcock, 1906 for those of birds. He said, too, that his nomenclature could not be regarded as firmly stabilized, especially in the Stercoraria. The life cycles and host–parasite relations of most species in this group are not known, so they have been assigned more or less arbitrarily to the various subgenera.

Another point made by Hoare which should be re-emphasized here is that the sections Stercoraria and Salivaria have no taxonomic status, but are merely convenient groupings for the subgenera.

One might think that all these subgenera should be elevated to genus rank, and indeed some authors have done so in the past. However, retention of the single generic name *Trypanosoma* makes it possible for non-taxonomists to ignore the subgeneric names if they wish. In addition, all members of the group are so similar morphologically that splitting them into several genera is not justified.

In the following classification, I have accepted Hoare's[78] subgenera of mammalian trypanosomes and have quoted their diagnoses and those of his sections, but I have refrained from adding subgenera for the trypanosomes of other vertebrates. This should be done in order to make the classification complete, but I feel that too little is known about them to justify it at this time. In addition, I have added information on metabolism based on Ryley,[133] von Brand,[15] and von Brand and Tobie[17] and quoted from Levine.[102]

Section STERCORARIA

Trypanosomes in which a free flagellum is always present; kinetoplast large and not terminal; posterior end of body pointed. Multiplication in the mammalian host is discontinuous, typically taking place in the crithidial or leishmanial stages. Development in the vector is completed in the posterior station and transmission is contaminative (in *T. rangeli* also in the anterior station, with inoculative transmission). Trypanosomes typically not pathogenic. *Metabolism*: blood forms have high respiratory quotient and low sugar consumption, producing acetic, succinic and lactic acids aerobically and succinic, lactic, acetic and pyruvic acids anaerobically. Cyanide markedly inhibits oxygen consumption. Sulfhydryl antagonists moderately inhibit oxygen consumption. Culture forms have high respiratory quotient and moderately high sugar consumption, producing acetic and succinic acids aerobically and succinic and acetic acids anaerobically; cyanide markedly inhibits oxygen consumption; sulfhydryl antagonists moderately to markedly inhibit oxygen consumption. Cytochrome pigments, cytochrome and succinic oxidase activity are present in *T. cruzi* and *T. lewisi*.

Subgenus *Megatrypanum* Hoare, 1964

Large mammalian trypanosomes, with kinetoplast typically situated near the nucleus and far from the posterior end of the body. These trypanosomes have affinities with some corresponding parasites of amphibians and reptiles. Multiplication known only for *T. theileri*.

Type species: *Trypanosoma* (*Megatrypanum*) *theileri* Laveran, 1902 of bovines and antelopes.

Other selected species: *T. tragelaphi* Kinghorn *et al.*, 1913 of antelopes; *T. melophagium* Flu, 1908 of sheep; *T. mazamarum* Mazza *et al.*, 1932 of deer; *T. cephalophi* Bruce *et al.*, 1913 of antelopes; *T. ingens* Bruce *et al.*, 1910 of cattle, antelopes and chevrotains.

Subgenus *Herpetosoma* Doflein, 1901

Trypanosomes of medium size, with subterminal kinetoplast, lying at some distance from the pointed posterior end of the body. Reproduction typically in the crithidial stage.

Type species: *Trypanosoma* (*Herpetosoma*) *lewisi* (Kent, 1880) Laveran and Mesnil, 1901 of rats.

Other selected species: *T. duttoni* Thiroux, 1900 (syn. *T. musculi* Kendall, 1906) of house mice; *T. grosi* Laveran and Pettit, 1909 of wood mice; *T. rabinowitschae* Brumpt, 1906 of hamsters; *T. primatum* (Reichenow, 1917) of anthropoid apes; *T. nabiasi* Railliet, 1895 of rabbits; *T. zapi* Davis, 1952 of jumping mice; *T. rangeli* Tejera, 1920 of man, dogs, opossums and monkeys; *T. otospermophili* Wellman and Wherry, 1910 (syn. *T. spermophili* Laveran, 1911) of susliks and ground squirrels.

Subgenus *Schizotrypanum* Chagas, 1909; emend. Nöller, 1931. Trypanosomes relatively small (range of lengths 12–29 µ, means *ca.* 15–24 µ) and typically C-shaped; voluminous kinetoplast situated very near the short pointed posterior end of the body. Multiplication intracellular, typically in leishmanial stage. Homogeneous assemblage of morphologically indistinguishable species.

Type species: *Trypanosoma* (*Schizotrypanum*) *cruzi* Chagas, 1909, of man, dogs, cats, armadillos, opossums, etc.
Other selected species: *T. vespertilionis* Battaglia, 1904 of bats; *T. pipistrelli* Chatton and Courrier, 1921 of bats; *T. phyllostomae* Cartaya, 1910 of bats; *T. hipposideri* Mackerras, 1959 of bats; *T. prowazeki* Berenberg-Gossler, 1908 of uakari monkeys; *T. lesourdi* Leger and Porry, 1918 of spider monkeys; *T. sanmartini* Garnham and Gonzales-Mugaburu, 1962 of squirrel monkeys (possibly syn. of *T. cruzi*).

Section SALIVARIA

Subgenus Duttonella Chalmers, 1918

Trypanosomes of the former *vivax* group, represented by monomorphic forms in which a free flagellum is always present; posterior end of the body typically rounded; kinetoplast large and usually terminal. Development in vector (*Glossina*) takes place in proboscis exclusively. *Metabolism*: blood forms have high respiratory quotient and high sugar consumption, producing pyruvic, acetic, lactic acids and glycerol aerobically and glycerol, pyruvic, lactic and acetic acids anaerobically. Cyanide does not inhibit oxygen consumption. Sulfhydryl antagonists markedly inhibit oxygen consumption. Cytochrome and succinic oxidase activity present.
Type species: *Trypanosoma* (*Duttonella*) *vivax* Ziemann, 1905 of ruminants and equids; long forms with mean lengths of 21.0–25.4 μ.
Only other species: *T. uniforme* Bruce *et al.*, 1911 of ruminants; short forms with mean lengths of 14.6–16.5 μ.

Subgenus *Nannomonas* Hoare, 1964

Trypanosomes of former congolense group, represented by small forms in which a free flagellum is usually absent; kinetoplast of medium size, typically marginal. Development in vector (*Glossina*) takes place in midgut and proboscis. *Metabolism*: blood forms have high respiratory quotient and high sugar consumption, producing acetic acid, succinic acid, glycerol and pyruvic acid anaerobically. Cyanide and sulfhydryl antagonists moderately inhibit oxygen consumption. Cytochrome pigments absent, but cytochrome and succinic oxidase activity present. Culture forms of *T. congolense* have R. Q. of 0.9, produce pyruvate, acetate and smaller amounts of lactate, succinate and glycerol aerobically, and pyruvate, acetate and succinate with small amounts of glycerol and no carbon dioxide anaerobically. Cyanide and sulfhydryl antagonists (iodoacetate and sodium arsenite) inhibit oxygen consumption, but Krebs cycle inhibitors (fluoroacetate and malonate) do so only slightly.
Type species: *Trypanosoma* (*Nannomonas*) *congolense* Broden, 1904 of ruminants, equids, pigs and dogs.

Other species: *T. dimorphon* Laveran and Mesnil, 1904 of equids, ruminants and pigs (monomorphic long forms with mean lengths of 15.3–17.6 μ); *T. simiae* Bruce *et al.*, 1912 of pigs, warthogs, monkeys and other animals (polymorphic [short, long and slender] forms with mean lengths of 17.0–18.2 μ).

Subgenus *Pycnomonas* Hoare, 1964

Trypanosomes of former *brucei* group—*suis* subgroup, represented by stout monomorphic forms with short free flagellum and small subterminal kinetoplast. Development in vector (*Glossina*) takes place in midgut and salivary glands. Monotypical.

Type and only species: *Trypanosoma* (*Pycnomonas*) *suis* Ochmann, 1905 of pigs.

Subgenus *Trypanozoon* Lühe, 1906

Trypanosomes of former *brucei* group—*brucei* and *evansi* subgroups, represented by polymorphic forms (slender, intermediate and stumpy), with or without free flagellum; kinetoplast small, subterminal (absent in *T. equinum*). Development in vector (*Glossina*) takes place in midgut and salivary glands (except *T. evansi*, *T. equinum* and *T. equiperdum*). *Metabolism*: blood forms have very low respiratory quotient and very high sugar consumption, producing pyruvic acid and sometimes glycerol aerobically and pyruvic acid and glycerol anaerobically. Cyanide does not inhibit oxygen consumption. Sulfhydryl antagonists markedly inhibit oxygen consumption. Culture forms have high respiratory quotient and moderately high sugar consumption, producing acetic, succinic, pyruvic and lactic acids aerobically. Cyanide moderately inhibits oxygen consumption. Sulfhydryl antagonists markedly inhibit oxygen consumption. Cytochrome pigments have not been found in *T. rhodesiense* or *T. equiperdum*, but cytochrome and succinic oxidase activity are present in *T. rhodesiense*.

Type species: *Trypanosoma* (*Trypanozoon*) *brucei* Plimmer and Bradford, 1899 of all domestic mammals, antelopes.

Other species: *T. rhodesiense* Stephens and Fantham, 1910 of man and antelopes; *T. gambiense* Dutton, 1902 of man (and pig? cf. Stephen, 147); *T. evansi* (Steel, 1885) (syns. *T. hippicum* Darling, 1910, etc.) of camels, equids, bovids, dogs, etc.; *T. equinum* Voges, 1901 of equids and bovids; *T. equiperdum* Doflein, 1901 of equids.

In general, the different groups of *Trypanosoma* have different host ranges. Of the mammalian subgenera, *Megatrypanum*, *Herpetomonas*, *Schizotrypanum* and *Pycnomonas* are stenoxenous or at least mesoxenous [with the exceptions of *T. (H.) rangeli* and *T. (S.) cruzi*, which are euryxenous], while *Duttonella*, *Nannomonas* and *Trypanozoon* are euryxenous [with the exceptions of *T. (D.) uniforme*, which is mesoxenous, and *T. (T.) equiperdum*, which is stenoxenous; *T. (T.) gambiense* is generally thought to be stenoxenous, occurring only in man, but it is likely that it also infects the pig (cf. ref. 147)].

Of the species in birds, *T. avium* is clearly euryxenous. Little is known about the host ranges of the species in reptiles, amphibia and fish; cross-infection studies are needed to determine the categories to which they belong.

The Haemospororina

The sporozoan order Eucoccidorida as defined by Honigberg *et al.*[81] includes three suborders, the Adelorina, Eimeriorina and Haemospororina. In the first, the macrogamete and microgametocyte are associated in syzygy during development, the microgametocyte usually produces few microgametes, and the sporozoites are enclosed in an envelope. In the second, the macrogamete and microgametocyte develop independently, syzygy is absent, the microgametocyte typically produces many microgametes, the zygote is non-motile, the oocyst does not increase in size during sporogony, and the sporozoites are typically enclosed in a sporocyst. In the third, the macrogamete and microgametocyte develop independently, syzygy is absent, the microgametocyte produces a moderate number of microgametes, the zygote is motile in some forms, the oocyst increases in size during sporogony, and the sporozoites are naked.

The life cycles of all three suborders include both sexual and asexual phases, and all are parasites of epithelial or blood cells of invertebrates or vertebrates. Members of the first two orders may be monoxenous or heteroxenous. Members of the Haemospororina are always heteroxenous; schizogony takes place in a vertebrate and sporogony in an invertebrate host.

Most Adeleorina are parasites of invertebrates, most Eimeriorina of vertebrates, and most Haemospororina of both. Members of the Eimeriorina are mostly parasites of intestinal epithelial cells, while members of the Haemospororina are mostly parasites of blood cells of vertebrates. The life cycles of the two suborders are similar, except that sporogony takes place in an invertebrate in the Haemospororina and outside a host in the Eimeriorina.

It is generally considered that the Haemospororina arose from the coccidia, but whether from the Adeleorina or the Eimeriorina is uncertain (cf. ref. 4); this is a matter which I shall not discuss. The suborder contains one, two, three or perhaps even four families, depending upon the authority. The older, customary view was to subdivide the suborder into two families, the Plasmodiidae containing the genus *Plasmodium* and the Haemoproteidae containing the genera *Haemoproteus* and *Leucocytozoon*. The principal difference was that in the Plasmodiidae schizogony was thought to take place only in the erythrocytes, while in the Haemoproteidae it takes place in the lungs, liver, spleen, kidney and other internal organs. However, when the complete life cycles of several species of avian and human *Plasmodium* were worked out[84, 85, 142, 141, 67, 2] it was realized that schizogony may occur both within the erythrocytes and exoerythrocytically. The distinction between the two families thus became an artificial one, and some authors considered that there was no point in retaining more than a single family in the suborder (cf. refs. 102, 110). Bray,[18, 19] however, continued to use the two families. There has also been a feeling that more rather than fewer families should be included in the order. Fallis[60] felt that the life cycles of the various genera justified the establishment of the families Plasmodiidae, Haemoproteidae and Leucocytozoidae, and said that a fourth family might possibly be justified. Baker[4] included the Leucocytozoidae with a question mark in his proposed phylogenetic scheme of the suborder.

Some workers have also split the genus *Plasmodium* either into genera or subgenera. Garnham[66] revived the genus *Laverania* Feletti and Grassi, 1890 for mammalian parasites with crescentic gamonts, and Bray[18] recommended revival of the genus *Haemamoeba* Feletti and Grassi, 1889 for the avian and reptilian malaria parasites, the vectors of which are culicine rather than anopheline mosquitoes. Corradetti, Garnham and Laird[48] retained the name *Plasmodium* for the avian species (as did Bray,[18] despite his recommendation), but suggested that the genus be divided into a number of subgenera. Confining themselves to the avian species, they suggested four subgenera: (1) the subgenus *Haemamoeba* for species with large, round erythrocytic schizonts, round gamonts and exoerythrocytic schizogony in the reticulo-endothelial system; (2) the subgenus *Giovannolaia* for species with large erythrocytic schizonts containing plentiful cytoplasm, with elongate gamonts, and with exoerythrocytic schizogony in the reticulo-endothelial system; (3) the subgenus *Novyella* for species with small erythrocytic schizonts containing scanty cytoplasm, with elongate gamonts, and with exoerythrocytic schizogony in the reticulo-endothelial system; and (4) the subgenus *Huffia* for species with small erythrocytic schizonts, elongate gamonts, slight exoerythrocytic schizogony in the reticulo-endothelial system but prolific e–e schizogony in the haemopoietic system.

The above picture is further confused by the idea (cf. refs. 18, 68, 4) that the overall genus *Plasmodium* has arisen from two different lines. One group is thought to have evolved from a karyolysid type of ancestor. This group is based on *Haemoproteus*, from which two lines are thought to have arisen: *Polychromophilus* in bats, and the avian species of *Plasmodium* (the genera or subgenera *Giovannolaia*, *Novyella*, *Huffia* and *Haemamoeba*). The second group is thought to have evolved from a hepatozoid type of ancestor. *Nycteria* (of bats), *Hepatocystis* (of mammals) and the mammalian species of *Plasmodium* (i.e. members of the subgenera *Laverania* and *Plasmodium*) are included here. It is uncertain how *Leucocytozoon* and the reptilian and amphibian species of *Plasmodium* fit into this scheme. Furthermore, even some species of avian and mammalian *Plasmodium* cannot be fitted into it because information on their vectors and exo-erythrocytic schizogony is lacking.

The taxonomy of the Haemospororina, once quite static, has been rendered fluid by recent research. It is impossible to settle on a definitive classification until further research is done, but tentatively I am accepting only a single family, the Plasmodiidae. I do not feel that the size of the schizonts, the number of merozoites produced by them, or their type of host cell can be used to differentiate genera, much less families; all these factors differ within the single sporozoan genus *Eimeria*, yet this genus remains intact.

The only other character which has been used is the type of vector, but here we are in the unfortunate position of knowing both too much and not enough. At one stage in our knowledge, it was simple enough. All the vectors were Diptera, and each genus had its own group of vectors. Mammalian *Plasmodium* species were transmitted by *Anopheles* mosquitoes, avian *Plasmodium* species by culicine mosquitoes, *Haemoproteus* by hippoboscids, and *Leucocytozoon* by *Simulium*.

Now, however, our added knowledge has complicated the picture. Some avian *Plasmodium* species can be transmitted by *Anopheles*, and some mammalian species presumably by culicine mosquitoes.[60, 83a, 110] Indeed, Garnham, Heisch and Minter[69] suggested that *Culicoides* might be a vector of *P. elongatum* and other primitive species. *Culicoides* has been added to the vectors of *Haemoproteus*[62, 63] and of *Leucocytozoon*;[1] it has also been found to be the vector of *Hepatocystis*.[69]

But this picture is based on incomplete information, and we often fail to realize how incomplete it is. To give one example, perhaps twenty-five species of avian *Plasmodium* are generally believed to be valid, but as Manwell[110] said, "It is probable that for no species of avian malaria parasite are all the vector species known, and there are only five for which any vectors at all have been discovered". Huff[83a] listed fifteen susceptible mosquito species for *P. cathemerium*, three for *P. circumflexum*, seven for *P. elongatum*, seven for *P. fallax*, thirty-seven for *P. gallinaceum*, one for *P. heroni*, one for *P. juxtanucleare*, six for *P. lophurae* and twenty-five for *P. relictum*. He listed eight generally accepted species of *Plasmodium* and twelve of questionable validity for which no susceptible mosquitoes were known. (It must be pointed out that mosquitoes found susceptible in the laboratory are not necessarily vectors in nature; further, by far the great majority of mosquito species have not been tested.) And no vector is known for any of the twenty species of *Plasmodium* in reptiles.

In Table 8 I have listed what is known about the host groups of the genera of Plasmodiidae, but this table is based on limited (and in some cases suspect) information, and should not be taken as anything more than a suggestion of the true situation. Undoubtedly many more species remain to be discovered, and the validity of some of the present names is not clear. A question mark could be put after almost any number in the last column. I

TABLE 8. HOSTS OF PLASMODIIDAE

Genus	Vectors (so far as known)	Vertebrate hosts	No. of species in vert. hosts
Plasmodium	Mosquitoes	Frogs	2
		Lizards	18
		Snakes	2
		Birds	25
		Bats	2
		Rodents	5
		Ruminants	1
		Lemurs	2
		Lower primates	8
		Anthropoid apes	7
		Man	4
Hepatocystis	*Culicoides*	Bats	3
		Rodents	3
		Ruminants	2
		Hippopotamus	1
		Lower primates	4
Haemoproteus	*Culicoides*, Hippoboscids	Toads	3
		Reptiles	1
		Birds	84
Leucocytozoon	*Simulium, Culicoides*	Birds	58
Nycteria	?	Bats	1
Polychromophilus	?	Bats	3

am not sure whether the genera *Nycteria* and *Polychromophilus* should be accepted, but am including them nevertheless. And I am uneasy as to the relationship between *Haemoproteus* and *Hepatocystis*.

One fact emerges clearly from a study of this table. The Plasmodiidae are not distributed evenly throughout the vertebrates. Each genus has its own groups of hosts. *Leucocytozoon* occurs only in birds. *Haemoproteus* occurs primarily in birds but also infects a few amphibia and reptiles. *Hepatocystis* occurs only in mammals, and among them only in bats, rodents, lower primates, a few ruminants and the hippopotamus. *Plasmodium* is more widely distributed, occurring in amphibia, reptiles, birds and mammals. In the last, however, it is found in only a few orders—the bats, rodents, primates, and ruminants—and in only one species of the last.

I have listed eighty-four species of *Haemoproteus* among the birds. Whether all are valid is dubious. At one time, I was inclined to lump most of those from a single host family or even host order together. Now, however, Baker's[6] finding that the domestic pigeon *Columba livia* and the English wood pigeon *C. palumbus* have different species of *Haemoproteus* whose gamonts look alike, has made me uncertain about this view. The fact is that most species of avian *Haemoproteus* have been named from their gamonts only and on the basis of their discovery in different hosts. Cross transmission studies are conspicuous by their absence, and without them we cannot be sure of speciation.

Almost the same thing could be said about *Leucocytozoon* as about *Haemoproteus*. The fifty-eight species that I have listed merely indicate an order of magnitude. Again, most species have been named from their gamonts only and from the fact that they were found in different hosts. Few cross transmission studies have been carried out. Future research will have to determine which species are valid. It was once thought that the gamonts of *Leucocytozoon* occurred only in leucocytes, but now we know that this is not true; they may also invade erythrocytes (cf. refs. 47, 50). It has also been thought that each species had only one type of gamont—either circular or elongate. However, it is now clear (cf. refs. 98, 50) that, in *L. simondi* of ducks, at least, both types of gamont may be present at different stages in the life cycle. Our present information is but a prelude to future studies; it has raised questions which they must answer.

Thousands of papers have been written about members of the genus *Plasmodium*, and one would expect that our knowledge of it is approaching completion. This is far from true. We know more about it than about *Haemoproteus* or *Leucocytozoon*, but we have a great deal to learn. It was only in 1948 that the first report was presented that schizogony occurred outside the erythrocytes in the mammalian species.[139-40, 142-3]

The fact that malaria is a zoonosis was not recognized until Rodhain and Dellaert[132] and Rodhain[131] showed that the organism in chimpanzees formerly called *Plasmodium rodhaini* was actually a strain of *P. malariae*, and that man could be infected with the form from the chimpanzee, and the chimpanzee with the form from man. *P. brasilianum* of South American monkeys has also been transmitted to man,[46] and I myself have wondered whether this species might not have actually arisen from a strain of *P. malariae* brought to the New World in an infected man and transmitted to monkeys from him. The vivax-type *Plasmodium*, *P. cynomolgi*, which occurs in macaques, can infect man.[59, 136, 11] The malariae-type *P. inui shortti* of macaques can also infect man,[43] as can the quotidian *P. knowlesi* of macaques and leaf monkeys.[19, 58, 24] Chin *et al.*[24] found a natural human case resulting from exposure in Malaya; this was the first proof that simian malaria is a true natural zoonosis. Even more recently, Deane, Deane and Neto[49a] found a natural human case of *P. simium* infection in Brazil.

The recent resurgence of interest in primate malaria has resulted in the discovery of several new species, and still others may well be discovered. In Table 9 I have given the natural hosts, so far as known, of the named primate *Plasmodium* species. This table reveals immediately how many species have been named recently and also gives some idea of the waning and

TABLE 9. NATURAL HOSTS OF PRIMATE *Plasmodium* SPECIES

Plasmodium species	Natural hosts
brasilianum Gonder and Berenberg-Gossler, 1908	Cacajao (*Cacajao calvus*), howler monkeys (*Alouatta palliata*, *A. villosa*), spider monkeys (*Ateles geoffroyi*, *A. fuscipes*, *A. paniscus*), capuchin monkeys (*Cebus capucinus*, *C. albifrons*, *C. apella*), squirrel monkeys (*Saimiri sciurea*, *S. boliviensis*), Lagothrix lagotricha, *L. infumata*
coatneyi Eyles *et al.*, 1962	Cynomolgus monkey (*Macaca irus*)
cynomolgi Mayer, 1907	Macaques (*Macaca irus*, *M. mulatta*, *M. nemestrina*), leaf monkeys (*Presbytis* spp.), man (accidental)
eylesi Warren *et al.*, 1965	Gibbon (*Hylobates lar*)
falciparum (Welch, 1897)	Man
fieldi Eyles, Laing and Fong, 1962	Pig-tailed macaque (*Macaca nemestrina*)
girardi Bück, Coudurier and Quesnel, 1952	*Lemur fulvus*
gonderi Rodhain and van den Berghe, 1936	Mangabey monkeys (*Cercocebus* spp.), drill (*Mandrillus leucophaeus*)
hylobati Rodhain, 1941	Gibbon (*Hylobates moloch*)
inui Halberstaedter and von Prowazek, 1907	Macaques (*Macaca cyclopis*, *M. irus*, *M. mulatta*, *M. nemestrina*, *M. radiata*), *Cynopithecus niger*
jefferyi Warren, Coatney and Skinner, 1966	Gibbon (*Hylobates lar*)
knowlesi Sinton and Mulligan, 1932	Macaques (*Macaca cyclopis*, *M. irus*, *M. nemestrina*), leaf monkeys (*Presbytis* spp.) man (one case)
lemuris Huff and Hoogstraal, 1963	*Lemur collaris*
malariae (Feletti and Grassi, 1889)	Man, chimpanzee
ovale (Craig, 1900)	Man
pitheci Halberstaedter and von Prowazek, 1907	Orangutan
reichenowi (Sluiter, Swellengrebel and Ihle, 1922)	Chimpanzee, gorilla
schwetzi Brumpt, 1939	Chimpanzee, gorilla
simium da Fonseca, 1951	Howler monkey (*Alouatta fusca*)
vivax (Grassi and Feletti, 1889)	Man
youngi Eyles *et al.*, 1964	Gibbon (*Hylobates lar*)

waxing of interest in primate malaria. Three species were named in the nineteenth century, and all the human species had been named by 1900. Five species were named in the first decade of the twentieth century, none in the second decade, one in the third decade, three in the fourth decade, one in the fifth decade, two in the sixth and six in the first 7 years of the seventh decade.

In Table 10 I have summarized information on the named species of

TABLE 10. SPECIES OF *Plasmodium* IN PRIMATES

Type	Man	Higher anthropoids	Gibbons	Asian monkeys	African monkeys	New World monkeys	Lemurs
Quotidian	(knowlesi)[1]			knowlesi			
Tertian	falciparum	reichenowi	youngi	coatneyi[2]			lemuris
	vivax	schwetzi	eylesi	cynomolgi	gonderi		
		pitheci	jefferyi				
Quartan	ovale	malariae	hylobati	fieldi[3]		simium[4]	girardi
	malariae			inui		brasilianum	

[1] A single natural case known.
[2] Resembles *falciparum*, but gamonts round.
[3] Resemblance to *ovale* superficial.
[4] Resembles *ovale* and also *vivax* in some respects.

primate *Plasmodium* by host group and by type of schizogony or fever. There are only three types of malaria in terms of the interval between paroxysms (which is correlated with the length of time between release of batches of merozoites, i.e. with the length of the schizogonous cycle). In quotidian malaria, there is a paroxysm every day; in tertian malaria, there is a paroxysm every other day; and in quartan malaria, there is a paroxysm every third day.

Note that each group of primates, by and large, has its own special species of *Plasmodium*, yet certain ones look alike and can be related to the well-known species occurring in man. Man has one group of species, gibbons another, the chimpanzee and gorilla another, Asian monkeys another, African monkeys another (I have omitted the commonest species, *Hepatocystis kochi*, from this table, of course), New World monkeys another, and lemurs still another.

Plasmodium knowlesi appears to be separate from the other species; it is the only one which causes quotidian malaria. It occurs only in Asian monkeys, but in a number of them. A single natural human case has been reported, in Malaya.[24]

Among the tertian malaria species, *P. reichenowi* of the chimpanzee and gorilla resembles *P. falciparum* of man, as does *P. coatneyi* of Asian monkeys except that its gamonts are round.

In the group which more or less resembles *P. vivax* of man are *P. schwetzi* of the chimpanzee and gorilla; *P. pitheci* of the orangutan; *P. youngi*, *P. eylesi* and *P. jefferyi* of gibbons; *P. cynomolgi* of Asian monkeys; *P. gonderi* of African monkeys; and *P. lemuris* of lemurs.

In the group which more or less resembles *P. ovale* of man are *P. fieldi* of Asian monkeys and *P. simium* of New World monkeys. However, the resemblance of *P. fieldi* is superficial, and *P. simium* resembles both *P. ovale* and *P. vivax* in some respects. *P. ovale* may be more or less peculiar to man; however, too little is known about it to draw any safe conclusions.

The human quartan parasite, *P. malariae*, also occurs in the chimpanzee. *P. hylobati* of the gibbon, *P. inui* of Asian monkeys, *P. brasilianum* of New World monkeys and *P. girardi* of lemurs are also quartan. As mentioned above, I have speculated that *P. brasilianum* might actually be a strain of *P. malariae* introduced by man into the New World, gone wild in monkeys and adapted to them.

Interestingly, the wild rhesus monkey, *Macaca mulatta*, rarely has malaria, although it is used commonly for laboratory studies of the disease. The great majority of wild rhesus monkeys come from India, and most of the macaque species of *Plasmodium* have been described from eastern and southeastern Asia. Schmidt *et al.*[136] found *P. cynomolgi* in two *M. mulatta* in East Pakistan—the first report of its natural occurrence in this host—and thought that these monkeys may have become infected from a reservoir in other monkeys species further east.

REFERENCES

1. AKIBA, K. (1960) Studies on the *Leucocytozoon* found in the chicken in Japan. II. On the transmission of *L. caulleryi* by *Culicoides arakawae*. *Jap. J. Vet. Sci.* **22**, 309–19.

2. BAKER, J. R. (1957) A new vector of *Haemoproteus columbae* in England. *J. Protozool.* **4**, 204–8.

3. BAKER, J. R. (1963) Speculations on the evolution of the family Trypanosomatidae Doflein, 1901. *Exp. Parasit.*, **13**, 219–33.

4. BAKER, J. R. (1965) The evolution of parasitic protozoa. In *Evolution of Parasites*, pp. 1–27, ed. A. E. R. TAYLOR. Blackwell, Oxford, England.

5. BAKER, J. R. (1966) The host-restriction of *Haemoproteus* sp. indet. of the wood-pigeon *Columba p. palumbus*. *J. Protozool.* **13**, 406–8.

6. BAKER, J. R. (1966) *Haemoproteus palumbis* sp. nov. (Sporozoa, Haemosporina) of the English wood-pigeon *Columba p. palumbus*. *J. Protozool.* **13**, 515–19.

7. DE BARY, A. (1879) *Die Erscheinung der Symbiose.* Trübner, Strassburg, Germany.

8. BECKER, E. R. (1923) Transmission experiments on the specificity of *Herpetomonas muscae-domesticae* in muscoid flies. *J. Parasit.* **10**, 25–34.

9. BECKER, E. R. (1953) How parasites tolerate their hosts. *J. Parasit.* **39**, 467–80.

10. BENNETT, G. F. (1961) On the specificity and transmission of some avian trypanosomes. *Can. J. Zool.* **39**, 17–33.

11. BENNETT, G. F., and WARREN, M. (1965) Transmission of a new strain of *Plasmodium cynomolgi* to man. *J. Parasit.* **51**, 79–80.

12. BERGER, J. (1964) The morphology, systematics, and biology of the entocommensal ciliates of echinoids. Ph.D. thesis, Univ. of Illinois. vii + 534 pp.

13. BOSCHMA, H. (1924) On the food of Madreporaria. *Proc. Acad. Sci. Amst.* **27**, 13 (cited by Droop, 1963).

14. VON BRAND, T. (1952) *Chemical Physiology of Endoparasitic Animals.* Academic Press, New York. x + 339 pp.

15. VON BRAND, T. (1956) Beziehungen zwischen Stoffwechsel und taxonomischer Einteilung der Sägetiertrypanosomen. *Zool. Anz.* **157**, 119–23.

16. VON BRAND, T. (1966) *Biochemistry of Parasites.* Academic Press, New York. x + 429 pp.

17. VON BRAND, T., and TOBIE, E. J. (1959) Observations on the metabolism of the culture form of *Trypanosoma congolense*. *J. Parasit.* **45**, 204–8.

18. BRAY, R. S. (1957) Studies on the exo-erythrocytic cycle in the genus *Plasmodium*. *London Sch. Hyg. Trop. Med. Mem.* No. 12. H. K. Lewis & Co., London. viii + 192 pp.

19. BRAY, R. S. (1963) Malaria infections in primates and their importance to man. *Ergebn. Mikrobiol. Immunitätsfor. Exp. Therap.* **36**, 168–213.

20. BROOKS, M. A. (1963) Symbiosis and aposymbiosis in arthropods. In: *Symbiotic Associations*, ed. P. S. NUTMAN and B. MOSSE, pp. 200–31. Cambridge Univ. Press, Cambridge, England.

21. BUCHNER, P. (1953) *Endosymbiose der Tiere mit pflanzlichen Mikroorganismen.* Verlag Birkhäuser, Basel, Switzerland.

22. CELLI, A., and SANFELICE, F. (1891) Sui parassiti del globulo rosso nell' uomo e negli animali. *Ann. Ist. Sper., Univ. Roma* (N.S.), **1**, 33–63.

23. CELLI, A., and SANFELICE, F. (1891a) Über die Parasiten des rothen Blutkörperchens im Menschen und in Tieren. *Fortschr. Med.* **9**, 499–511, 541–52, 581–6.

24. CHIN, W., CONTACOS, P. G., COATNEY, G. R., and KIMBALL, H. R. (1965) A naturally acquired quotidian-type malaria in man transferable to monkeys. *Science* **149**, 865.

25. CLARK, T. B. (1959) Comparative morphology of four genera of Trypanosomatidae. *J. Protozool.* **6**, 227–32.

26. CLEVELAND, L. R. (1949) Hormone-induced sexual cycles of flagellates. I. Gametogenesis, fertilization and meiosis in *Trichonympha*. *J. Morphol.* **85**, 197–295.

27. CLEVELAND, L. R. (1950) Hormone-induced sexual cycles of flagellates. II. Gameto-genesis, fertilization and one-division meiosis in *Oxymonas*. *J. Morphol.* **86**, 185–214.

28. CLEVELAND, L. R. (1950) Hormone-induced sexual cycles of flagellates. III. Gameto-genesis, fertilization and one-division meiosis in *Saccinobaculus*. *J. Morphol.* **86**, 215–28.

29. CLEVELAND, L. R. (1950) Hormone-induced sexual cycles of flagellates. IV. Meiosis after syngamy and before nuclear fusion in *Notila*. *J. Morphol.* **87**, 317–48.

30. CLEVELAND, L. R. (1950) Hormone-induced sexual cycles of flagellates. V. Fertiliza-tion in *Eucomonympha*. *J. Morphol.* **87**, 349–68.

31. CLEVELAND, L. R. (1951) Hormone-induced sexual cycles of flagellates. VI. Gameto-genesis, fertilization, meiosis, oocysts, and gametocysts in *Leptospironympha*. *J. Morphol.* **88**, 194–244.

32. CLEVELAND, L. R. (1951) Hormone-induced sexual cycles of flagellates. VII. One-division meiosis and autogamy without cell division in *Urinympha*. *J. Morphol.* **88**, 385–440.

33. CLEVELAND, L. R. (1952) Hormone-induced sexual cycles of flagellates. VIII. Meiosis in *Rhynchonympha* in one cytoplasmic and two nuclear divisions followed by auto-gamy. *J. Morphol.* **91**, 269–323.

34. CLEVELAND, L. R. (1953) Hormone-induced sexual cycles of flagellates. IX. Haploid gametogenesis and fertilization in *Barbulanympha*. *J. Morphol.* **93**, 371–403.

35. CLEVELAND, L. R. (1954) Hormone-induced sexual cycles of flagellates. X. Autogamy and endomitosis in *Barbulanympha* resulting from interruption of haploid gametogene-sis. *J. Morphol.* **95**, 189–212.

36. CLEVELAND, L. R. (1954) Hormone-induced sexual cycles of flagellates. XI. Reorgani-zation in the zygote of *Barbulanympha* without nuclear or cytoplasmic division. *J. Morphol.* **95**, 213–35.

37. CLEVELAND, L. R. (1954) Hormone-induced sexual cycles of flagellates. XII. Meiosis in *Barbulanympha* following fertilization, autogamy and endomitosis. *J. Morphol.* **95**, 557–619.

38. CLEVELAND, L. R. (1956) Brief account of the sexual cycles of the flagellates of *Crypto-cercus*. *J. Protozool.* **3**, 161–80.

39. CLEVELAND, L. R. (1956a) Hormone-induced sexual cycles of flagellates. XIV. Gametic meiosis and fertilization in *Macrospironympha*. *Arch. Protistenk.* **101**, 99–170.

40. CLEVELAND, L. R. (1957) Correlation between molting period of *Cryptocercus* and sexuality in its Protozoa. *J. Protozool.* **4**, 168–75.

41. CLEVELAND, L. R., BURKE, A., and KARLON, P. (1960) Ecdysone induced modifications in the sexual cycles of the protozoa of *Cryptocercus*. *J. Protozool.* **7**, 229–39.

42. COATNEY, G. R. (1933) Relapse and associated phenomena in the Haemoproteus infection of the pigeon. *Am. J. Hyg.* **18**, 133–60.

43. COATNEY, G. R., CHIN, W., CONTACOS, P. G., and KING, H. K. (1966) *Plasmodium inui*, a quartan-type malaria parasite of Old World monkeys transmissible to man. *J. Parasit.* **52**, 660–3.

44. COATNEY, G. R., ELBEL, R. E., and KOCHARATANA, P. (1960) Some blood parasites found in birds and mammals from Loei Province, Thailand. *J. Parasit.* **46**, 701–2.

45. COLEMAN, G. S. (1963) The growth and metabolism of rumen ciliate protozoa. In *Symbiotic Associations*, ed. P. S. NUTMAN and B. MOSSE, pp. 298–324. Cambridge Univ. Press, Cambridge, England.

46. CONTACOS, P. G., LUNN, J. S., COATNEY, G. R., KILPATRICK, J. W., and JONES, F. E. (1963) Quartan-type malaria parasite of New World monkeys transmissible to man. *Science* **142**, 676.

47. COOK, A. R. (1954) The gametocyte development of *Leucocytozoon simondi*. *Proc. Helm. Soc. Wash.* **21**, 1–9.

48. CORRADETTI, A., GARNHAM, P. C. C., and LAIRD, M. (1963) New classification of the avian malaria parasites. *Parassitologia* **5**, 1–4.

48a. COVALEDA ORTEGA, J., and GALLEGO BERENGUER, J. (1950) Hemoproteus aviares. *Rev. Iber. Parasit.* **10**, 141–85.

49. DAMIAN, R. T. (1964) Molecular mimicry: antigen sharing by parasite and host and its consequences. *Am. Nat.* **98**, 129–49.

49a. DEANE, L. M., DEANE, M. P., and NETO, J. F. (1966) Studies on transmission of simian malaria and on a natural infection of man with *Plasmodium simium* in Brazil. *Bull. World Health Org.* **35**, 805–8.

50. DESSER, S. S. (1967) Schizogony and gametogony of *Leucocytozoon simondi* and associated reactions in the avian host. *J. Protozool.* **14** (in press).

50a. DIAMOND, L. S. (1965) Trypanosomes of Anura. *Wildlife Dis.* No. **44**, 82 pp.

51. DINEEN, J. K. (1963) Immunological aspects of parasitism. *Nature* **197**, 268–9.

52. DINEEN, J. K. (1963) Antigenic relationship between host and parasite. *Nature* **197**, 471–2.

53. DINEEN, J. K., DONALD, A. D., WAGLAND, B. M., and OFFNER, J. (1965) The dynamics of the host–parasite relationship. III. The response of sheep to primary infection with *Haemonchus contortus. Parasitology* **55**, 515–25.

54. DINEEN, J. K., DONALD, A. D., WAGLAND, B. M., and TURNER, J. H. (1965). The dynamics of the host–parasite relationship. II. The response of sheep to primary and secondary infection with *Nematodirus spathiger. Parasitology* **54**, 163–71.

55. DOGIEL, V. A., POLJANSKY, J. I. and CHEJSIN, E. M. (1965). *General Protozoology.* 2nd ed. Clarendon Press, Oxford, England. xiv + 747 pp.

56. DONALD, A. D., DINEEN, J. K., TURNER, J. H., and WAGLANG, B. M. (1964). The dynamics of the host–parasite ralationship. I. *Nematodirus spathiger* infection in sheep. *Parasitology,* **54**, 527–44.

57. DROOP, M. R. (1963) Algae and invertebrates in symbiosis. In *Symbiotic Associations,* ed. P. S. NUTMAN and B. MOSSE, pp. 171–99. Cambridge Univ. Press, Cambridge, England.

58. EYLES, D. E. (1963) The species of simian malaria: Taxonomy, morphology, life cycle, and geographical distribution of the monkey species. *J. Parasit.* **49**, 866–87.

59. EYLES, D. E., COATNEY, G. R., and GETZ, M. E. (1960) Vivax-type malaria parasite of macaques transmissible to man. *Science* **131**, 1812–13.

60. FALLIS, A. M. (1965) Protozoan life cycles. *Am. Zool.* **5**, 85–94.

61. FALLIS, A. M. (1966) On the development and transmission of *Leucocytozoon* and *Haemoproteus. Proc. First Intern. Cong. Parasit.* **1**, 239–40.

62. FALLIS, A. M., and BENNETT, G. F. (1961) Sporogony of *Leucocytozoon* and *Haemoproteus* in simuliids and ceratopogonids and a revised classification of the Haemosporidiida. *Can. J. Zool.* **39**, 215–28.

63. FALLIS, A. M., and WOOD, D. M. (1957) Biting midges (Diptera: Ceratopogonidae) as intermediate hosts for Haemoproteus of ducks. *Can. J. Zool.* **34**, 425–35.

64. FRANCHINI, G. (1924) Observations sur les hematozoaires des oiseaux d'Italie (2ᵉ Note). *Ann. Inst. Pasteur* **38**, 470–515.

65. FREUDENTHAL, H. D. (1962) *Symbiodinium* gen. nov. and *Symbiodinium microadriaticum* sp. nov., a zooxanthella; taxonomy, life cycle, and morphology. *J. Protozool.* **9**, 45–52.

66. GARNHAM, P. C. C. (1953) Terminology of Haemosporidiidea. *Trans. Fifth Intern. Congr. Trop. Med. Malariol., Istanbul* **2**, 228–31.

67. GARNHAM, P. C. C. (1954) Life history of the malaria parasites. *Ann. Rev. Microbiol.* **8**, 153–66.

68. GARNHAM, P. C. C. (1963) Distribution of simian malaria parasites in various hosts. *J. Parasit.* **49**, 905–11.

69. GARNHAM, P. C. C., HEISCH, R. B., and MINTER, D. M. (1961) The vector of *Hepato-cystis* (= *Plasmodium*) *kochi*; the successful conclusion of observations in many parts of tropical Africa. *Trans. Roy. Soc. Trop. Med. Hyg.* **55**, 497–502.

70. GRASSÉ, P.-P. (1952) Ordre des trypanosomides (Trypanosomidea n. n.). In *Traité de Zoologie*, ed. P.-P. GRASSÉ, **1**(1), 602–68. Masson, Paris.

71. GRASSÉ, P.-P. (1952) Ordre des trichomonadines. Ordre des pyrsonymphines. Ordre des oxymonadines. Ordre des joeniides. Ordre des lophomonadines. Ordre des trichonymphines. Ordre des spirotrichonymphines. La symbiose flagellé-termites. In *Traité de Zoologie*, ed. P.-P. GRASSÉ, **1**(1), 704–823, 836–962. Masson, Paris.

72. HANSON, W. L., McGHEE, R. B., and BLAKE, J. D. (1966) Experimental infection of various latex plants of the family Asclepiadaceae with *Phytomonas elmassiani*. *J. Protozool.* **13**, 324–7.

73. HECKMAN, R. A. (1961) Entocommensal ciliates of sea urchins from Hawaii. Unpub. Res. Progr. Rep., Pauley Fund. 7 pp. (cited by Berger, 1964).

74. HEISCH, R. B., McMAHON, J. P., and MANSON-BAHR, P. E. C. (1958) The isolation of Trypanosoma rhodesiense from a bushbuck. *Brit. Med. J.* **1958** (Nov.), 1203–4.

75. HERMAN, C. M. (1954) *Haemoproteus* infections in waterfowl. *Proc. Helm. Soc. Wash.* **21**, 37–42.

76. HERTIG, M., TALIAFERRO, W. H., and SCHWARTZ, B. (1937) Report of the Committee on Terminology. *J. Parasit.* **23**, 325–9.

77. HOARE, C. A. (1964) Morphological and taxonomic studies on mammalian trypanosomes. X. Revision of the systematics. *J. Protozool.* **11**, 200–7.

78. HOARE, C. A. (1966) The classification of mammalian trypanosomes. *Ergeb. Mikrobiol. Immunitätsf. Exp. Therap.* **39**, 43–57.

79. HOGUE, M. (1939) Infections of Trichomonas foetus in chick embryos and young chicks. *Am. J. Hyg.* **30**, 65–67.

80. HONIGBERG, B. M. (1963) A contribution to the systematics of the non-pigmented flagellates. *Progress in Protozoology. Proc. First Intern. Congr. Protozool., Prague*, 68–69.

81. HONIGBERG, B. M., BALAMUTH, W., BOVEE, E. C., CORLISS, J. O., GOJDICS, M., HALL, R. P., KUDO, R. R., LEVINE, N. D., LOEBLICH, A. R., JR., WEISER, J., and WENRICH, D. H. (1964) A revised classification of the phylum Protozoa. *J. Protozool.* **11**, 7–20.

82. HUFF, C. G. (1932) Studies on Haemoproteus of mourning doves. *Am. J. Hyg.* **16**, 618–23.

83. HUFF, C. G. (1954) A review of the literature on susceptibility of mosquitoes to avain malaria, with some unpublished data on the subject. *Res. Rep. Naval Med. Res. Inst.* **12**, 619–44.

83a. HUFF, C. G. (1965) Susceptibility of mosquitoes to avian malaria. *Exp. Parasit.* **16**, 107–32.

84. HUFF, C. G., and COULSTON, F. (1944) The development of *P. gallinaceum* from sporozoite to erythrocytic trophozoite. *J. Infec. Dis.* **75**, 231–49.

85. HUFF, C. G., and COULSTON, F. (1946) The relation of natural and acquired immunity of various avian hosts to the cryptozoites and metacryptozoites of *P. gallinaceum* and *P. relictum. J. Infec. Dis.* **78**, 99–117.

86. HUFF, C. G., NOLF, L. O., PORTER, R. J., READ, C. P., RICHARDS, A. G., RIKER, A. J., and STAUBER, L. A. (1958) An approach toward a course in the principles of parasitism. *J. Parasit.* **44**, 28–45.

87. HUNGATE, R. E. (1955) Mutualistic intestinal protozoa. In *Biochemistry and Physiology of Protozoa*, eds. S. H. HUTNER and A. LWOFF, **2**, 159–99. Academic Press, New York.

88. JACOBS, L. (1956) Propagation, morphology, and biology of Toxoplasma. *Ann. N.Y. Acad. Sci.* **64**, 154–79.

89. KEEBLE, F. (1908) The yellow-brown cells of *Convoluta paradoxa. Quart. J. Micr. Sci.* **52**, 431–80.

90. KEEBLE, F., and GAMBLE, F. W. (1907) The origin and nature of the green cells of *Convoluta roscoffensis. Quart. J. Micr. Sci.* **51**, 167–220.

91. KIRBY, H., JR. (1930) Trichomonad flagellates from termites. I. *Tricercomitus* gen. nov., and *Hexamastix* Alexeieff. *Univ. Calif. Publ. Zool.* **33**, 393–444.

92. KIRBY, H., JR. (1931) Trichomonad flagellates from termites. II. *Eutrichomastix* and the subfamily Trichomonadinae. *Univ. Calif. Publ. Zool.* **36**, 171–262.

93. KIRBY, H., JR (1932) Flagellates of the genus *Trichonympha* in termites. *Univ. Calif. Publ. Zool.* **37**, 349–476.

94. KIRBY, H., JR. (1932) Protozoa in termites of the genus *Amitermes. Parasitology* **24**, 289–304.

95. KIRBY, H., JR. (1937) Host–parasite relations in the distribution of Protozoa in termites. *Univ. Calif. Publ. Zool.* **41**, 189–212.

96. KIRBY, H., JR. (1941) Relationships between certain protozoa and other animals. Chap. XIX in *Protozoa in Biological Research*, eds. G. N. CALKINS and F. M. SUMMERS, pp. 890–1008. Columbia Univ. Press, New York.

97. KIRBY, H., JR. (1944) The structural characteristics and nuclear parasites of some species of Trichonympha in termites. *Univ. Calif. Publ. Zool.* **49**, 185–282.

98. KOCAN, R. M., and CLARK, D. T. (1966) Anemia in ducks infected with *Leucocytozoon simondi. J. Protozool.* **13**, 465–8.

99. KOCH, A. (1960) Intracellular symbiosis in insects. *Ann. Rev. Microbiol.* **14**, 121–40.

100. LAVERAN, M. A., and PETIT, A. (1909) Sur une hemamibe de *Melopelia leucoptera* L. *C.R. Soc. Biol.* **66**, 952–4.

101. LEVINE, N. D. (1959 (1958)) Uniform endings for the names of higher taxa. *System. Zool.* **7**, 134–5.

102. LEVINE, N. D. (1961) *Protozoan Parasites of Domestic Animals and of Man.* Burgess, Minneapolis. iii + 412 pp.

103. LEVINE, N. D. (1962) Protozoology today. *J. Protozool.* **9**, 1–6.

104. LEVINE, N. D. (1962) Geographic host distribution of blood parasites in columborid birds. *Trans. Ill. State Acad. Sci.* **55**, 92–111.

105. LEVINE, N. D. (1963) Coccidiosis. *Ann. Rev. Microbiol.* **17**, 179–98.

106. LEVINE, N. D., BRANDLY, C. A., and GRAHAM, R. (1939) The cultivation of *Tritrichomonas foetus* in developing chick eggs. *Science* **89**, 161.

107. LEVINE, N. D., and HANSON, H. C. (1953) Blood parasites of the Canada goose, *Branta canadensis interior. J. Wildlife Manag.* **17**, 185–96.

108. LEVINE, N. D., and IVENS, V. (1965) *The Coccidian Parasites (Protozoa, Sporozoa) of Rodents.* Ill. Biol. Monog. No. 33. Univ. Ill. Press, Urbana. 365 pp.

109. LEVINE, N. D., and KANTOR, S. (1959) Checklist of blood parasites of birds of the order Columbiformes. *Wildl. Dis.* **1**(1), 38 pp.

110. MANWELL, R. D. (1965) The lesser Haemosporidina. *J. Protozool.* **12**, 1–9.

111. MATTINGLY, P. F. (1965) The evolution of parasite-arthropod vector systems. In *Evolution of Parasites*, ed. A. E. R. TAYLOR, pp. 29–45. Blackwell, Oxford, England.

112. MCLAUGHLIN, J. J. A., and ZAHL, P. A. (1959) Axenic zooxanthellae from various invertebrate hosts. *Ann. N.Y. Acad. Sci.* **77**, 55–72.

113. MCLAUGHLIN, J. J. A., and ZAHL, P. A. (1962) Axenic cultivation of the dinoflagellate symbiont from the coral Cladocora. *Arch. Mikrobiol.* **42**, 40–41.

113a. MCLAUGHLIN, J. J. A., and ZAHL, P. A. (1966) Endozoic algae. In *Symbiosis*, Vol. 1. *Associations of Microorganisms, Plants, and Marine Organisms*, ed. S. M. HENRY, pp. 257–97. Academic Press, New York.

114. MULLER, S. W., and CAMPBELL, A. (1954) The relative number of living and fossil species of animals. *System. Zool.* **3**, 168–70.

115. MUSCATINE, L., and LENHOFF, H. M. (1963) Symbiosis: on the role of algae symbiotic with Hydra. *Science*, **142**, 956–8.

116. NELSON, P. (1938) Cultivation of *Trichomonas foetus* in the chick embryo. *Proc. Soc. Exp. Biol. Med.* **39**, 258–9.

117. NOVY, E. G., and MACNEAL, W. J. (1905) Trypanosomes and bird malaria. *Proc. Soc. Exp. Biol. Med.* **2**, 23–28.

118. OXFORD, A. E. (1955) The rumen ciliate protozoa: Their chemical composition, metabolism, requirements for maintenance and culture, and physiological significance for the host. *Exp. Parasit.* **4**, 569–605.

119. PAVLOVSKII, E. N. (1963) The present state of knowledge regarding natural foci of human infections. In *Natural Foci of Human Infections*, ed. E. N. PAVLOVSKII, translated from Russian by H. BRACHYAHU, edited by O. THEODOR, pp. 3–22. Israel Program Sci. Transl., Jerusalem.

120. PEARSE, A. S. (1942) *Introduction to Parasitology*, Thomas, Springfield, Ill. ix + 357 pp.

121. PELLÉRDY, L. (1963) *Catalogue of Eimeriidea (Protozoa: Sporozoa)*. Akad. Kiado, Budapest. 160 pp.

122. PETERS, J. L. (1937) *Check List of Birds of the World*, Vol. 3. Harvard Univ. Press, Cambridge, Mass.

123. PHILLIPS, B. P., WOLFE, P. A., REES, C. W., GORDON, H. A., WRIGHT, W. H., and REYNIERS, J. A. (1955) Studies on the ameba–bacteria relationship in amebiasis. Comparative results of the intracaecal inoculation of germfree mono-contaminated, and conventional guinea pigs with *Entamoeba histolytica. Am. J. Trop. Med. Hyg.* **4**, 675–92.

124. PIPKIN, A. C. (1960) Avian embryos and tissue culture in the study of parasitic protozoa. II. Protozoa other than *Plasmodium. Exp. Parasit.* **9**, 167–203.

125. PIPKIN, A. C., and JENSEN, D. V. (1958) Avian embryos and tissue culture in the study of parasitic protozoa. *Exp. Parasit.* **7**, 491–530.

126. POWERS, P. B. A. (1935) Studies on the ciliates of sea urchins. A general survey of the infestations occurring in Tortugas echinoids. *Pap. Tortugas Lab., Carnegie Inst., Washington, D.C.* **29**, 292–326.

127. READ, C. P., JR (1950) The vertebrate small intestine as an environment for parasitic helminths. *Rice Inst. Pamph.* **37**(2), 1–94.

128. READ, C. P. (1958) A science of symbiosis. *A.I.B.S. Bull.* **8**, 16–17.

129. REYNOLDS, B. D. (1936) *Colpoda steini*, a facultative parasite of the land slug, *Agriolimax agrestis. J. Parasit.* **22**, 48–53.

130. RICHARDSON, F. L. (1934) Studies on experimental epidemiology of intestinal protozoan infections in birds. *Am. J. Hyg.* **20**, 373–403.

131. RODHAIN, J. (1948) Contribution a l'étude des Plasmodiums des anthropoïdes africains. Transmission du *Plasmodium malariae* de l'homme au chimpanzé. *An. Soc. Belge Med. Trop.* **28**, 39–49.

132. RODHAIN, J., and DELLAERT, R. (1943) L'infection à *Plasmodium malariae* du chimpanzé chez l'homme. Etude d'une première souche isolée de l'anthropoide *Pan satyrus verus. An. Soc. Belge Med. Trop.* **23**, 19–46.

133. RYLEY, J. F. (1956) Studies on the metabolism of the Protozoa. 7. Comparative carbohydrate metabolism of eleven species of trypanosome. *Biochem. J.*, **62**, 215–22.

134. SANTOS DIAS, J. A. T. (1953) Resultados de um reconhecimento zoologico no Alto Limpopo efectuado pelos Drs. F. ZUMPT e J. A. T. SANTOS DIAS. V. Hematozoarios das aves: Generos *Haemoproteus* Kruse e *Carpanoplasma* n. gen. *Mocambique* **73**, 61–99.

135. SCHAD, G. A. (1966) Immunity, competition, and natural regulation of helminth populations. *Am. Nat.* **100**, 359–64.

136. SCHMIDT, L. H., GREENLAND, R., and GENTHER, C. S. (1961) The transmission of *Plasmodium cynomolgi* to man. *Am. J. Trop. Med. Hyg.* **10**, 679–88.
137. SHAW, J. J. (1964) A possible vector of *Endotrypanum schaudinni* of the sloth *Choloepus hoffmani*, in Panama. *Nature* **201**, 417–18.
138. SHAW, J. J. (1966) The relationship of *Endotrypanum* to other members of the Trypanosomatidae and its possible bearing upon the evolution of certain haemoflagellates of the New World. *Proc. First Intern. Congr. Parasit.* **1**, 332–3.
139. SHORTT, H. E. (1948) The pre-erythrocytic cycle of *P. cynomolgi. Proc. Fourth. Congr. Trop. Med. Malariol.* **1**, 607–17.
140. SHORTT, H. E. (1948) The life cycle of *Plasmodium cynomolgi* in its insect and mammalian hosts. *Trans. Roy. Soc. Trop. Med. Hyg.* **42**, 227–30.
141. SHORTT, H. E., FAIRLEY, N. H., COVELL, G., SHUTE, P. G., and GARNHAM, P. C. C. (1951) The pre-erythrocytic stage of *P. falciparum. Trans. Roy. Soc. Trop. Med. Hyg.* **44**, 405–19.
142. SHORTT, H. E., and GARNHAM, P. C. C. (1948) The pre-erythrocytic development of *P. cynomolgi* and *P. vivax. Trans. Roy. Soc. Trop. Med. Hyg.* **41**, 785–95.
143. SHORTT, H. E., and GARNHAM, P. C. C. (1948) pre-erythrocytic stage in mammalian malaria parasites. *Nature* **161**, 126.
144. SPRENT, J. F. A. (1963) *Parasitism: An Introduction to Parasitology and Immunology for Students of Biology. Veterinary Science, and Medicine.* Ballière, Tindall & Cox, London. x + 145 pp.
145. STABLER, R. M., and HOLT, P. A. (1963) Hematozoa from Colorado birds. I. Pigeons and doves. *J. Parasit.* **49**, 320–2.
146. STABLER, R. M., HOLT, P. A., and KITZMILLER, N. J. (1966) *Trypanosoma avium* in the blood and bone marrow from 677 Colorado birds. *J. Parasit.* **52**, 1141–4.
147. STEPHEN, L. E. (1966) *Pig Trypanosomiasis in Tropical Africa.* Rev. Ser. No. 8, Commonwealth Bur. An. Health, Commonw. Ag. Bur., Farnham Royal, England. ix + 65 pp.
148. TOEPFER, E. W., JR. (1964) *Colpoda steinii* in oral swabbings from mourning doves (*Zenaidura macroura* L.). *J. Parasit.* **50**, 703.
149. UYEMURA, M. (1934) Über einige neue Ciliaten aus dem Darmkanal von japanischen Echinoideen. (I) *Sci. Rep. Tokyo Bunrika Daig.* (B) **1**, 181–91.
150. WALLACE, F. G. (1966) The trypanosomatid parasites of insects and arachnids. *Exp. Parasit.* **18**, 124–93.
151. WENRICH, D. H. (1935) Host–parasite relations between parasitic protozoa and their hosts. *Proc. Am. Philos. Soc.* **75**, 605–50.
152. WENRICH, D. H. (1954) Sex in Protozoa. A comparative review. In *Sex in Microorganisms*, eds. D. H. WENRICH, I. F. LEWIS and J. R. RAPER pp. 134–265. Am. Ass. Adv. Sci., Washington, D.C.
153. WENYON, C. M. (1926) *Protozoology*, 2 vols, Wood, New York. xvi + 1563 pp.
154. YAGIU, R. (1935) Studies on the ciliates from sea urchins of Uaku Island, with a description of a new species, *Cryptochilidium ozakii* sp. nov. *J. Sci. Hiroshima Univ.* (B, 1) **3**, 139–47.
155. YONGE, C. M., and NICHOLLS, A. G. (1931) Studies on the physiology of corals. IV. The structure, distribution and physiology of the zooxanthellae. *Sci. Rep. Great Barrier Reef Exped.* **1**, 135–76.

To the above should be added the following important book, which I did not see until after having written this chapter:

156. GARNHAM, P. C. C. (1966) *Malaria parasites and other Haemosporidia.* Blackwell, Oxford, England. xviii + 1114 pp

CHROMOSOMES AND NUCLEOLI
IN SOME OPALINID PROTOZOA

Tze-Tuan Chen

Department of Biological Sciences, University of Southern California
Los Angeles, California

CONTENTS

I. INTRODUCTION

Opalinids are a group of astomatous Protozoa, most species of which live in the rectum of Anurans. Opalinids have only one type of nucleus, differing from most other ciliates which exhibit dimorphism of the nuclear apparatus.†

Much misinterpretation and confusion has existed concerning the chromosomes in the opalinids. However, beginning with the work of Chen,[4-7] the true nature and behavior of the chromosomes and nucleoli in opalinids were described in detail. The general similarity of mitosis in opalinids to that in Metazoa and Metaphyta is now established.

Pfitzner[35] was the first to record that the behavior of chromosomes in the opalinids is essentially the same as that found in the Metazoa. Unfortunately, *Opalina ranarum* on which he worked has very small nuclei. Some later investigators, including Metcalf,[26] expressed considerable skepticism about Pfitzner's report.

Metcalf[26] reported that in *Protoopalina intestinalis* mitosis is not of the kind generally found in other organisms. According to him, a spireme stage is absent and the "chromosomes" divide transversely. In his subsequent papers Metcalf[27-29] described the following nuclear conditions among the binucleate opalinids: (1) the nuclei "rest" in a mid-mitotic stage before they complete their division and this mid-mitotic stage may be either an anaphase or a metaphase; (2) in each dividing nucleus there are two sets of chromosomes, the "macrochromosomes" which are ribbon-shaped and the "microchromosomes" which are granular and linear; (3) the "macrochromosomes" and the "microchromosomes" are equal in number in some species, probably in all. These conclusions of Metcalf were widely accepted and followed in whole or in part by later investigators; these included Awerinzew,[1] Töniges,[40, 41] Da Cunha and Penido,[15] Lavier,[23] Fantham and Robertson,[17] Fantham,[16] and Nie.[32] Some textbooks, such as those of Hegner and Taliaferro,[19] Wenyon,[43] and Reichenow (Doflein),[38] also accepted Metcalf's conclusions in the main.

Chen,[4-10] Ivanić[20, 21] and Valkanov,[42] however, did not share Metcalf's views.

Sugiyama[39] reported that in *Opalina japonica* irregular chromosomes are formed during mitosis and divide transversely.

---

† Lwoff[24, 25] reported that some other ciliates, *Stephanopogon mesnili*, *Blepharostoma* sp., and *Euchelys* sp., have only one type of nucleus as in the opalinids. Nuclear structure and division in *Spirochona* as described by Swarczewsky[39a] appears to be quite different from conditions in the opalinids and the ciliates described by Lwoff.

Bĕlăr[2] interpreted Metcalf's investigations to mean that mitosis in the opalinids is not of the kind generally found in other organisms and Calkins[3] apparently believed there are no real chromosomes in the opalinids.

The descriptive studies of opalinid chromosomes by Chen[10] were primarily concerned with five species of *Zelleriella* found in the rectum of *Bufo valliceps* secured from New Orleans, Louisiana, and one species of *Protoopalina* in *Microhyla ornata* from Nanking, China. (Incidental references will also be made to some other species of *Zelleriella*.) These six species are listed below. Several hundred preparations of these opalinids, made over a period of 10 years, were used, including: *Zelleriella elliptica* Chen, *Zelleriella louisianensis* Chen, *Zelleriella pfitzneri* Chen, *Zelleriella valliceps* Chen, *Zelleriella intermedia* Metcalf, *Protoopalina caudata microhyla* Nie.

Smear preparations were used exclusively. Since whole animals were studied, all mitotic figures were complete and every chromosome was present in each dividing nucleus.†

Both iron hematoxylin (without a counterstain) and Feulgen's nucleal reaction (counterstained with fast green) were used. Details of these techniques may be found elsewhere.[8] The best preparations were made with iron hematoxylin, destaining in saturated aqueous solution of picric acid. Preparations treated according to the Feulgen technique presented the same history of the chromosomes, but the chromosomes appeared rather homogenous and the nucleolar substance was unstained. The following accounts are based chiefly on the study of the hematoxylin-stained material.

II. RELATIONSHIPS OF THE OPALINIDS

The systematic position of the family Opalinidae is still open to debate. Though Metcalf considered them as the most primitive of all ciliates,[26, 27, 29] in recent years this idea has been challenged by Grassé[18] and Corliss[13, 14] who both consider them zooflagellates. Further, Wessenberg[44] found that it is unjustifiable to class them with either group, since they contain almost equal numbers of features of the flagellates and the cilates.[36] Honigberg et al.[19a] put them in the Sarcomastigophora, as a superclass Opalinata—correlative with Mastigophora. Still later, Kudo[22a] placed the Opalinida under the Peritricha. Thus far, my personal study on the opalinid chromosomes does not throw any light on their taxonomic relationships.

---

† Observations were made with a Zeiss research microscope equipped with a Zeiss 90X, N.A. 1.4 objective, compensating oculars, and a Zeiss Pancratic Condenser. All drawings were made with a camera lucida. Most of the drawings were made by the author.

Four genera, *Protoopalina*, *Zelleriella*, *Cepedea*, and *Opalina*, and many species are recognized. In the genus *Protoopalina*, with the exception of several multinucleate species, all are binucleate. Species of the genus *Zelleriella* are binucleate. Species of the genera *Cepedea* and *Opalina* are multinucleate. *Protoopalina* and *Cepedea* are more or less cylindrical whereas *Zelleriella* and *Opalina* are flattened.

A. Structure of the opalinids

As an introduction to the morphology of the opalinids, we may describe *Zelleriella elliptica* (Fig. 1) and make some comparison with some other opalinids. *Z. elliptica* is binucleate, flat, and somewhat elliptical in outline. When compared with other members of the genus it is fairly large. The average length of fully or nearly fully grown trophozoites is 184 μ; average width, 90.8 μ.

The 2 resting nuclei are of the same shape and size and lie in the midregion of the body, some distance apart. In addition to the 2 nuclei, the endosarc contains a large number of small bodies which have been variously designated. We shall refer to them simply as endosarc bodies.

B. Division of the opalinids

During mitosis the behavior of the chromosomes is fundamentally the same as in the Metazoa and Metaphyta, including: (a) the transformation of the chromatin reticulum into chromosomes which become condensed and shortened in the prophase; (b) arrangement of chromosomes in the equatorial region of the nucleus and the subsequent separation of sister chromatids in the metaphase; (c) movement of daughter chromosomes toward opposite poles in the anaphase; and (d) formation of the chromatin reticulum of the daughter nuclei by the chromosomes in the telophase. The nuclear membrane is persistent throughout mitosis.

No centrosomes or other kinetic centers have been observed either in the resting stage or during mitosis. Events in the 2 nuclei are usually more or less synchronous during mitosis (Figs. 2, 3), although cases have also been observed in which the nuclear events in one nucleus follow the other.

In the division of the body, a groove appears first on the surface of the animal and deepens until it cuts through the entire cytoplasm (Fig. 4). Division of the body is longitudinal and in *Zelleriella elliptica* is completed at the metaphase. After division, each daughter animal contains one metaphase nucleus (Fig. 4). Finally, the nucleus completes its division (Figs. 5, 6, 7), giving rise to 2 daughter nuclei—the normal binucleate condition being thus restored in each daughter animal. The daughter animals are relatively small and narrow (Fig. 7). The resting stage seems to be long, accompanied by considerable growth of both nuclei and cytoplasm.

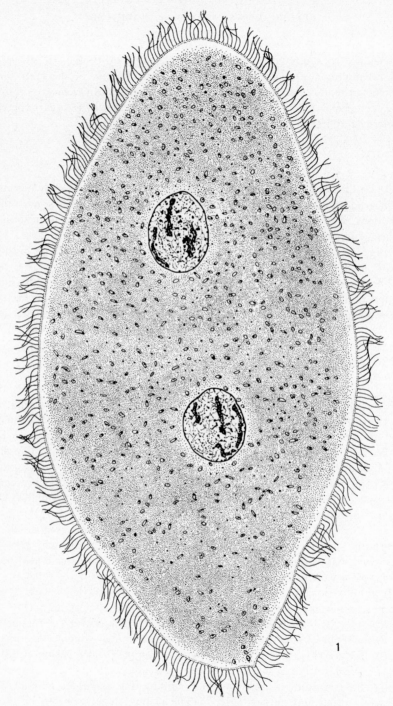

1

FIG. 1. *Zelleriella elliptica*. A typical vegetative animal. ×878.

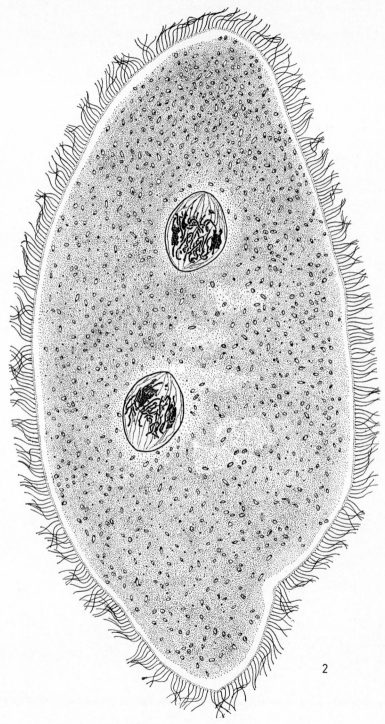

2

FIG. 2. *Zelleriella elliptica*. Late prophase. ×878.

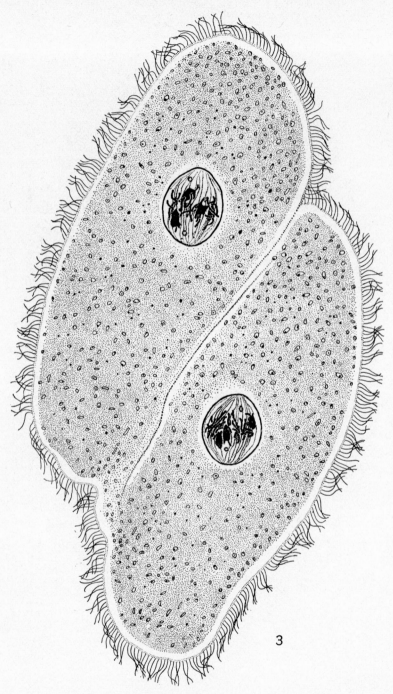

3

FIG. 3. *Zelleriella elliptica*. Early metaphase. Division of the body is almost completed. ×878.

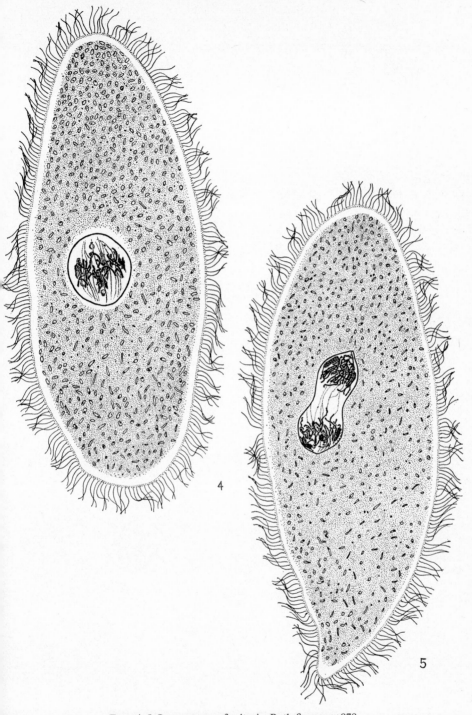

4

5

Figs. 4–5. Later stages of mitosis. Both figures ×878.

Fig. 4. *Zelleriella elliptica*. A daughter animal with a metaphase nucleus.

Fig. 5. *Zelleriella elliptica*. A daughter animal with a late anaphase nucleus.

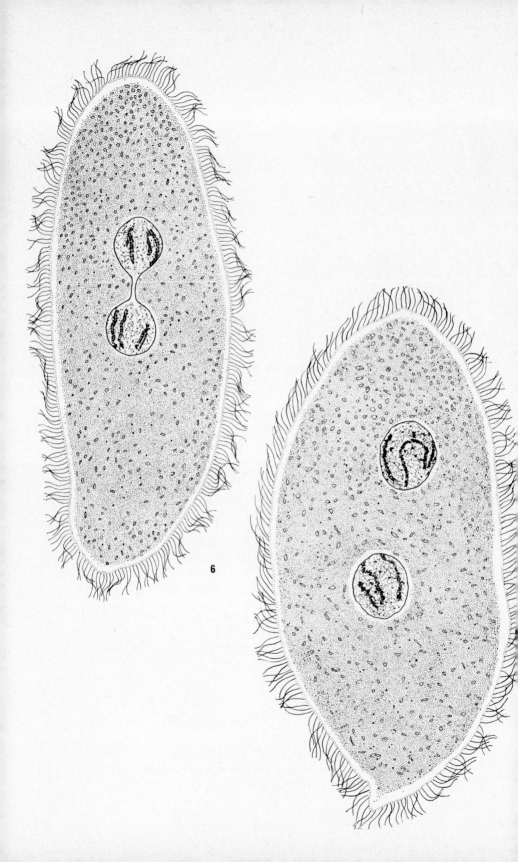

6

Although nuclear division is usually accompanied by the division of the body, there are cases in which the latter division is delayed, thus giving rise to specimens with four nuclei. Animals with eight nuclei have also been observed—they may have been the result of the division of the nuclei in a quadri-nucleate individual, with division of the body further delayed. Animals with three, five, six, seven and nine nuclei were also found.

III. THE RESTING NUCLEUS

The resting nuclei in *Zelleriella elliptica* will be described as an example. The two resting nuclei are either spherical or ovoid, averaging 21.2 μ in diameter. Three principal structures, as seen in the fixed and stained preparations (Fig. 75, left)- are: (1) the nuclear membrane, (2) the chromatin reticulum, and (3) the four elongated, darkly stained structures.

The nuclear membrane has an appreciable thickness and is persistent throughout mitosis, as is the case in most Protozoa. The chromatin reticulum, which is evenly distributed throughout the nucleus, gives rise to chromosomes during the prophase of mitosis. The four elongated, darkly stained structures found in the periphery of the nucleus are the nucleoli. As seen during mitosis, they are constantly associated with certain portions of certain chromosomes.

IV. BEHAVIOR OF CHROMOSOMES DURING MITOSIS

The following account is based chiefly on studies of hematoxylin-stained preparations, as this method gives greater differentiation and more intense coloring of the chromosomes than the Feulgen technique. The behavior of chromosomes during mitosis is essentially the same in these species of *Zelleriella*, and therefore the following description applies to all phases since the two nuclei in each animal are alike.

A. Prophase

In early prophase the fine chromatin reticulum transforms into slender chromosomes which are separate and distinct (Fig. 9). There is no continuous spireme.

FIG. 6. *Zelleriella elliptica*. A daughter animal with a late telophase nucleus.
×878.
FIG. 7. *Zelleriella elliptica*. A daughter animal with two well-developed nuclei.

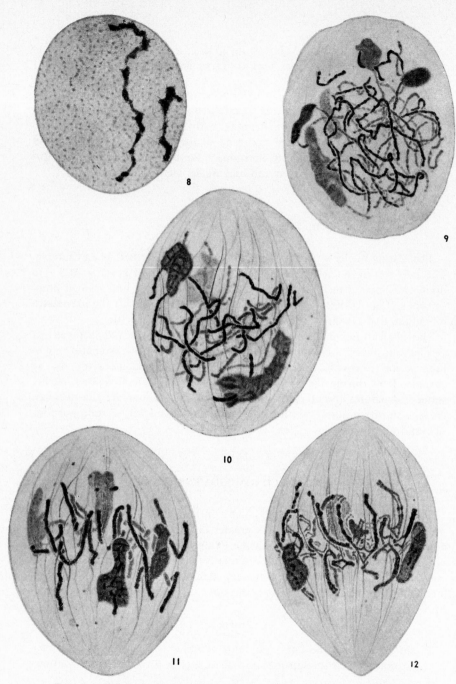

FIGS. 8–12. Nuclei of *Zelleriella* in different stages of mitosis. All figures
×2600.

FIG. 8. A resting nucleus of *Zelleriella elliptica.*

FIG. 9. An early prophase nucleus of *Z. pfitzneri.*

FIG. 10. A late prophase nucleus of *Z. elliptica.*

FIG. 11. An early metaphase nucleus of *Z. elliptica.*

FIG. 12. A metaphase nucleus of *Z. pfitzneri.*

The structure of the chromosomes is more apparent in this stage than in the late prophase and metaphase. In properly stained preparations the chromosomes show a series of chromatic granules, the chromomeres, separated by more lightly stained portions (Fig. 9). In certain places the chromomeres are close together; in other places they are farther apart. No evidence of replication has been observed at this stage, nor as a rule is there any sign of duality—only a single row of chromomeres being visible in each chromosome. The chromosomes have rough contours and each shows considerable variation in diameter. Certain regions are more or less constricted.

The chromosomes become shorter and thicker as their condensation continues (Fig. 10), although the chromomeres are still visible. The spindle fibers and their attachments to the chromosomes are obvious in the late prophase. The chromosomes now begin to congregate at the equatorial region of the nucleus, which in the meantime has become somewhat elongated.

B. Metaphase

In the metaphase (Fig. 11) the chromosomes arrange themselves in the equatorial region of the nucleus. The entire lengths of the shortest chromosomes are found in the equatorial region, while only the spindle fiber attachment portion of the long chromosomes is found near or within the equatorial region, with the arms of the chromosomes stretching more or less axially for a considerable distance. No metaphase plate is formed. This condition is reminiscent of mitosis in some of the higher plants with long chromosomes.

The chromomeres are still visible in the metaphase, although the chromosomes are much condensed and thickened. The chromosomes split longitudinally in the late metaphase (Fig. 12). A careful examination shows the two chromatids of each chromosome to be mirror images of each other with respect to the number, size and position of the chromomeres.

C. Anaphase

There is apparently only a short period between the appearance of the longitudinal split and the movement of the two chromatids toward opposite poles. The separation starts from the fiber attachment region (Fig. 13). The shape of each individual daughter chromosome during the anaphase becomes definite and is correlated with the location of the spindle attachment. Chromosomes with subterminal fiber attachments are J-shaped in the anaphase, with the apex leading. Those with median (or nearly median) attachments are V-shaped, the apex leading. Since all the twenty-four chromosomes in each of these several species of *Zelleriella* are atelomitic, they are either J- or V-shaped in the anaphase; for the most part they are J-shaped. Chromomeres are easily seen.

12a*

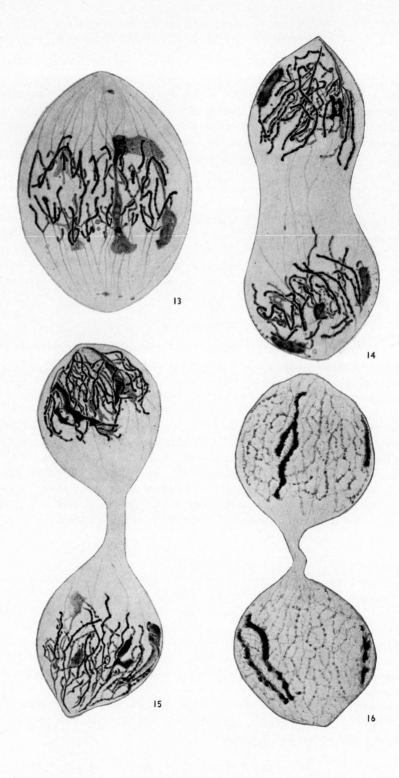

D. Telophase

After the chromosomes reach the polar regions of the elongated nucleus the latter constricts at the middle (Figs. 14, 15). The arms of the chromosomes, with chromomeres still apparent, now swing away from the longitudinal axis of the nucleus and gradually lose their chromaticity as they become greatly elongated. They assume a crooked, zig-zag appearance and their contours appear fuzzy (Fig. 16). Finally, outlines of the individual chromosomes grow indistinct as the resting stage is approached. Meanwhile two separate daughter nuclei are formed.

Most of the animals in division are either in the prophase or telophase, suggesting that these two stages are of comparatively longer duration. The anaphase is seen most infrequently and, therefore, is probably completed in a very short time.

V. MORPHOLOGY OF CHROMOSOMES

The chromosomes in the opalinids behave essentially in the same manner as those in the Metazoa and Metaphyta. There is no question about their exact and longitudinal division. Furthermore, they possess definite individuality, having constant and distinctive characteristics which are not lost during the resting stage, but are retained from one cell division to another. The presence of remarkable structural modifications at certain regions of particular chromosomes is additional proof of the complexity of the chromosomes of these Protozoa. Two types of structural differentiation will be described: the constrictions and the nucleolus-regions. The spindle fibers will also be described.

A. Constrictions

Three kinds of constrictions are present: a primary constriction located at the attachment region; a secondary constriction located somewhere on the arms; and a nucleolar constriction at the nucleolus region. The primary constriction is more pronounced in some chromosomes than in others. Such constrictions are particularly evident in some of the nucleolus-chromosomes (Figs. 40, 42, 43, 44, 45, 49, 51, 54). In most nucleolus-chromosomes the attachment region is so constricted and lightly stained that it is apt to be

---

FIGS. 13–16. Nuclei of *Zelleriella* in different stages of mitosis. All figures × 2600.

FIG. 13. An anaphase nucleus of *Z. intermedia*.

FIG. 14. A late anaphase nucleus of *Z. elliptica*.

FIG. 15. An early telophase nucleus of *Z. elliptica*.

FIG. 16. A late telophase nucleus of *Z. elliptica*.

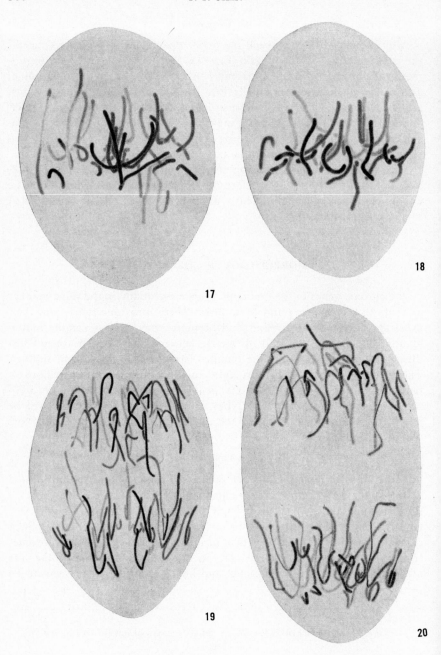

FIGS. 17–20. Nuclei of *Zelleriella elliptica* after Feulgen nucleal reaction. Note that only chromosomes are stained, the nucleoli are not. All figures ×2600.

overlooked, and a single nucleolus-chromosome thus may be mistaken for two separate chromosomes. Some previous investigators apparently made this mistake and described the two arms as two or more separate chromosomes. In addition to these primary constrictions at the fiber attachment region, the arms of certain chromosomes have secondary constrictions, and the nucleolus-chromosomes may also have nucleolar constrictions at the regions where nucleoli are formed.

B. Nucleolus-regions

Among the twenty-four chromosomes in each of these several species of *Zelleriella*, four or six particular chromosomes are associated with nucleoli which are located at specific regions. There are three ways in which the nucleolar substance is associated with the chromosomes. The nucleolar material may be located on the outside of the chromosomes, forming extrachromosomal structures (Figs. 27–45). In some cases the segment on which the nucleolus is located is normal, having the same diameter as other portions of the same chromosome and as other chromosomes in the same nucleus; in other cases this segment may be constricted or precociously split (Figs. 46, 47). In another type of association, the nucleolar substance is apparently located inside the chromosome, with a great increase in the diameter, as well as great changes in the structure of this portion of the chromosome. Instead of the usual chromomeres, chromatic spirals are found at this region (Figs. 57–68). In still another type of association, the nucleolar material is apparently located both inside and outside the chromosome, accompanied by conspicuous changes in the diameter as well as in the structure of the segment involved (Figs. 51–56).

C. Spindle fibers

The chromosomal fibers in *Zelleriella* are unusually thick and conspicuous (Figs. 27–74), more so than those commonly found in other organisms. They differ in size in different species of *Zelleriella*, although the same technique of fixation, staining and destaining was used for all. They are especially thick in *Z. intermedia* and *Z. elliptica*. They are not uniformly thick, but taper gradually toward the poles (Figs. 27–74).

VI. NUMBER OF CHROMOSOMES

As shown in Table 1, all five of the species of *Zelleriella* used in the present investigation have twenty-four chromosomes. However, in an unidentified species of *Zelleriella* found in *Bufo marinus* obtained from Trinidad, British West Indies, the chromosome number is close to thirty-eight.

TABLE 1. CHROMOSOME NUMBER IN THE FIVE SPECIES OF *Zelleriella* FOUND IN *Bufo valliceps* COLLECTED FROM NEW ORLEANS, LOUISIANA

Opalinid	Host	Locality collected	Chromosome number
Zelleriella elliptica	*Bufo valliceps*	New Orleans, Louisiana	24
Zelleriella intermedia	*Bufo valliceps*	New Orleans, Louisiana	24
Zelleriella louisianensis	*Bufo valliceps*	New Orleans, Louisiana	24
Zelleriella pfitzneri	*Bufo valliceps*	New Orleans, Louisiana	24
Zelleriella valliceps	*Bufo valliceps*	New Orleans, Louisiana	24

In this connection it should be noted that Metcalf's conclusion (ref. 29, p. 256) that the number of chromosomes in the genus *Zelleriella* is small (from four to ten) is unconfirmed. In the study of different species of the genus *Zelleriella* secured from different parts of this country, as well as some other countries, including those species previously studied by Metcalf, I have not yet found a single species having fewer than twenty-four chromosomes.

VII. INDIVIDUALITY OF CHROMOSOMES

In three of the five species of *Zelleriella* used in the present investigation (*Z. elliptica, Z. louisianensis, Z. intermedia*) the chromosomes have been studied in considerable detail. As representative of these three species, the chromosomes in *Zelleriella elliptica* will be described in some detail.

The chromosomes are graded in length, there being two of each size. There are three long pairs, three short pairs and six pairs of intermediate lengths. A member of each pair is illustrated in Fig. 21.

In the anaphase each chromosome has a particular shape correlated with the location of the spindle fiber attachment. All the chromosomes in this species are atelomitic, having median, submedian or subterminal fiber attachments. Nine pairs are J-shaped and three V-shaped (one of the latter is distinctly asymmetrical) in the anaphase. The ratio between the lengths of the two arms is peculiar to each chromosome—the members of each chromosome pair are alike.

The three longest chromosomes all have submedian fiber attachments. Chromosome 1 is J-shaped, the short arm being half the length of the long arm. A large nucleolus is always found associated with the mid-region of the long arm. Chromosome 2 is V-shaped, the short arm being approximately two-thirds the length of the long arm. Chromosome 3 is J-shaped, the short arm being slightly less than half the length of the long arm.

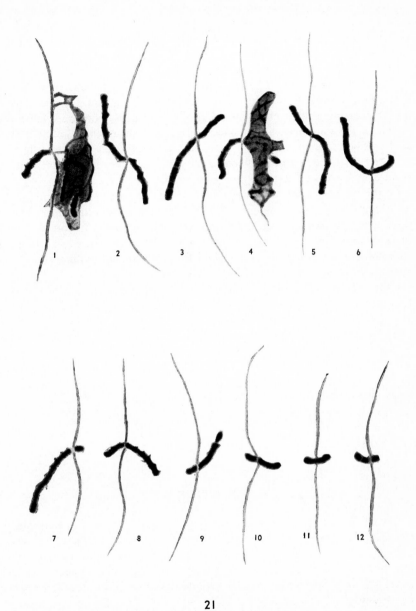

21

FIG. 21. Chromosomes of *Zelleriella elliptica*. A member of each chromosome pair is illustrated. For detailed explanation, see text. ×3680.

There are six chromosomes of intermediate lengths, one asymmetrical V and five J's. Chromosome 4 is V-shaped, with a submedian fiber attachment and two arms of nearly equal length. A large nucleolus is always associated with the mid-region of the longer arm. Chromosome 5 has a submedian fiber attachment, with a short arm a little more than half the length of the long arm. Chromosome 6 has a submedian fiber attachment, with a short arm a little more than one-third the length of the long arm. Chromosome 7 has a subterminal fiber attachment, with a short arm about one-eighth the length of the long arm. Chromosome 8 (much shorter than 7) has a sub-median fiber attachment, with a short arm approximately half the length of the long arm. Chromosome 9 has a subterminal fiber attachment, with a short arm less than one-third the length of the long arm.

Of the three short chromosomes, nos. 10 and 12 are J-shaped and no. 11 is symmetrical V. Chromosome 10 has a submedian fiber attachment, with a short arm approximately half the length of the long arm. Chromosome 11 has a median fiber attachment, with two arms of equal length. The shortest chromosome (no. 12) has a submedian fiber attachment, with a short arm approximately half the length of the long arm.

Other species of *Zelleriella* found in *Bufo valliceps* from New Orleans, Louisiana, appear to have the same chromosome configuration as *Z. elliptica*, except with regard to their association with nucleoli; they differ in the number of chromosomes associated with nucleoli, in the specific chromosomes associated and in the type of association. For example, *Z. louisianensis* apparently has the same configuration (Fig. 22) as *Z. elliptica* (Figs. 21) with the following minor differences: in *Z. elliptica* two pairs of chromosomes (nos. 1 and 4) are associated with nucleoli; in *Z. louisianensis* three pairs of chromosomes (nos. 1, 4, and 5) are involved. In *Z. elliptica* the nucleoli are formed outside the chromosomes; in *Z. louisianensis* the nucleoli are formed inside the chromosomes.

VIII. DIPLOIDY

The evidence for diploidy in these opalinids is conclusive. For example, in the several species of *Zelleriella* here under consideration there are twelve pairs of chromosomes with distinctive characters (size, location of fiber attachment, and structural peculiarities). The members of each pair are alike. The discovery of occasional haploid nuclei corroborates the evidence for diploidy.

The three species of *Zelleriella* that have been carefully studied (*Z. elliptica*, *Z. intermedia*, *Z. louisianensis*) are diploid. It is probable that the other two species studied (*Z. pfitzneri*, *Z. valliceps*) are also diploid. The criteria for chromosome homology within the species are size, form and structure.

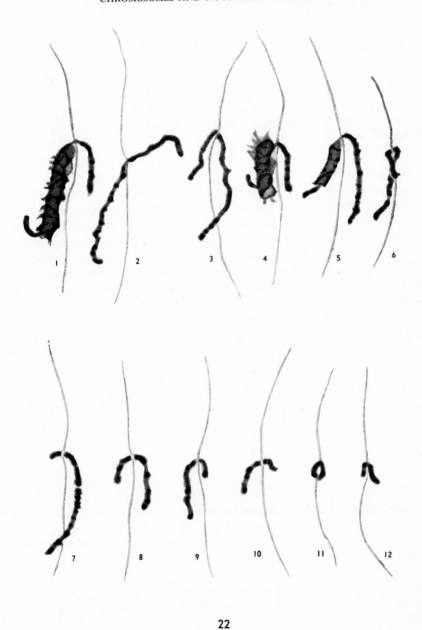

22

Fig. 22. Chromosomes of *Zelleriella louisianensis*. A member of each chromosome pair is illustrated. ×3680.

A. Size

In chromosome studies of Metazoa and Metaphyta, similarity in size is usually taken as evidence of chromosome homology within the species, heteromorphic chromosomes excepted. This seems to be also true in the opalinids. After prolonged study, it is clear that the twenty-four chromosomes in these three species of *Zelleriella* from *Bufo valliceps* can be arranged in duplicate series on the basis of their size alone, there being two chromosomes of each size.

B. Form

Chromosomes of the same size also show identical location of fiber attachment. Three chromosomes that are easily recognized (Fig. 21) will be described here as examples. Chromosome 2, one of the longest chromosomes, has a submedian fiber attachment; its short arm is two-thirds the length of the long arm. Chromosome 7 is a medium-sized chromosome with a subterminal fiber attachment and a short arm about one-eighth the length of its long arm. Chromosome 11, one of the shortest chromosomes, has a median attachment and two arms that are equal. Other individual chromosomes could also be cited.

C. Structure

Homologous chromosomes not only show similarity in size and shape, but also in structure. The nucleolus-chromosomes show this with particular distinctness (Figs. 75, 76). Each nucleolus-region exhibits intense chromaticity and structural modifications (Figs. 75, 76) and occupies a definite position on its chromosome. If one chromosome displays such structural peculiarities at a certain region, one other chromosome of corresponding size and form shows the same structural modification at an identical region.

IX. HAPLOIDY

Since the publication of my earlier reports, haploid animals having only twelve chromosomes have been found in *Zelleriella louisianensis*. The haploid animals are rare. They have been found in a single specimen of *Bufo valliceps* from New Orleans, Louisiana, and only fifteen individuals were observed among hundreds of animals which were examined from this host. Of the fifteen haploid animals, thirteen each have four resting nuclei, one animal has four prophase nuclei, and one other has two mid-telophase nuclei.

The haploid animals are of approximately the same size as the diploid animals, although each of the former has four nuclei, whereas the diploid animals have but two nuclei each. The haploid animal with two telophase nuclei is approximately the same size as a diploid animal with a single telophase nucleus.

The haploid nuclei are much smaller than the diploid nuclei (Figs. 23–26). In the haploid nucleus, there are only twelve chromosomes differing in size and location of fiber attachment (Fig. 24).

Each haploid nucleus contains three nucleolus-chromosomes (Fig. 24). Six is the characteristic number of nucleolus-chromosomes in the diploid nuclei. In the resting stage, each haploid nucleus shows three ribbon-shaped structures (the nucleolus-regions of certain chromosomes), whereas the diploid nuclei show six (cf. Figs. 25 and 26).

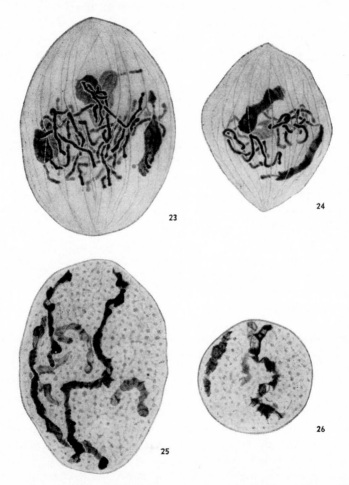

FIGS. 23–26. Diploid and haploid nuclei of *Zelleriella louisianensis*. All figures ×2640.

FIG. 23. A diploid late prophase nucleus.

FIG. 24. A haploid late prophase nucleus.

FIG. 25. A diploid resting nucleus.

FIG. 26. A haploid resting nucleus.

The relation of these haploid animals to the life cycle of the opalinids is not yet known. Some possibilities are: (1) The haploid animals may have their origin in the fusion of gametes without the fusion of their nuclei to form a synkaryon. (2) These exceptional animals may be haploid trophozoites. If this should be the case, it seems likely that the four nuclei may have been derived by two successive divisions of a single haploid nucleus, without a corresponding division of the cytoplasm. The four nuclei should then contain identical sets of chromosomes. (3) The four resting nuclei (or two telophase nuclei) in each haploid animal may be the immediate products of two meiotic divisions.

The discovery of these haploid nuclei each with twelve chromosomes confirms the previous statement (Chen,[6] that the usual type of nucleus found in these opalinids (containing twenty-four chromosomes) is diploid. It seems clear that in the trophozoite stage, which is the longest stage in their life cycle, the opalinids possess a diploid set of chromosomes, and that opalinids are diploid organisms, in contrast with some other protozoons, such as certain gregarines and coccidia among the Sporozoa, and certain flagellates, such as *Chlamydomonas* and *Protosiphon*, which are haploid organisms. If there is a sexual phase in the life history of the opalinids, as some previous investigators have reported, it is probable that the duplicate series of chromosomes present in the trophozoites is derived from two gametes and that the members of each chromosome pair are segregated at some stage of meiosis.

X. RELATIONSHIP BETWEEN CHROMOSOMES AND NUCLEOLI

As previously pointed out, within each nucleus of *Zelleriella*, there are, in addition to the chromosomes, conspicuous bodies—nucleoli. These stain intensely with hematoxylin, but not with Feulgen's reagent (Figs. 17–20), and are associated with specific portions of particular chromosomes.

A. Manner of Association

Depending on the species, there are four or six nucleoli found respectively on four or six (two or three pairs) of the twenty-four chromosomes. Only one nucleolus is found on each of the several chromosomes and it always occupies a non-terminal position. The location of nucleoli is identical on homologous chromosomes (Figs. 75–77). It is evident that the constant location of nucleoli at definite portions of particular chromosomes is an expression of the constancy in the internal organization of the chromosomes in these protozoans.

Each of the species of *Zelleriella* so far studied shows a characteristic configuration of these nucleoli and their associated chromosomes. Each

Figs. 27–43. Relation between chromosomes and nucleoli in *Zelleriella*.
All figures ×3680.
Fig. 27. A nucleolus-chromosome of *Z. valliceps*.
Figs. 28–39. Nucleolus-chromosomes of *Z. intermedia*.
Figs. 40–43. Nucleolus-chromosomes of *Z. elliptica*.

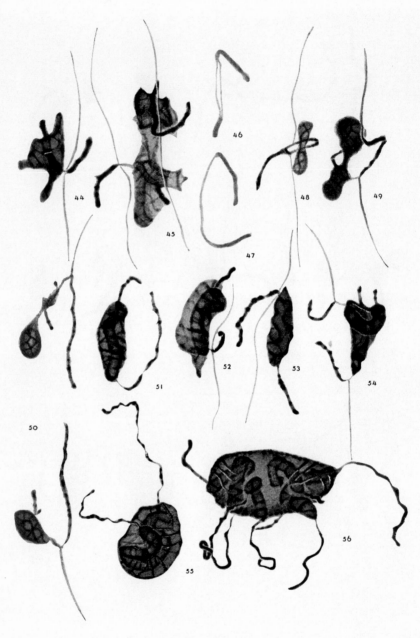

FIGS. 44–56

species differs from the others in the number of nucleoli, in the specific chromosomes associated with nucleoli, and in the type of association between chromosomes and nucleoli (Figs. 75–77).

As stated earlier, the association between chromosome and nucleolus is of three different types. In the first type, seen in *Zelleriella elliptica* and in *Z. intermedia*, the nucleolar substance is located outside the chromosome, forming an extra-chromosomal structure (Figs. 27–45). Such nucleoli do not have a definite shape and they vary in size from a small granule to one of considerable size. Some of the smaller nucleoli (Figs. 27–29) closely resemble the nucleoli ("chromomere vesicles") in Orthoptera, as has been pointed out by Chen[7, 10] and Corey.[12] The larger nucleoli seem to consist of a large reticular mass surrounding the chromosome segments (Figs. 33, 34, 35, 36, 38, 39–45). In some cases the chromosome segment on which the nucleolus is located is normal (Figs. 33, 36), having the same diameter as other portions of the same chromosome and as other chromosomes in the same nucleus; in other cases this segment may be distinctly constricted (Fig. 47) or precociously split (Fig. 46). At times it is difficult to trace the chromosome through the nucleolar substance. Such diverse appearances of the chromosome segment have been found in both the Feulgen-stained and hematoxylin-stained preparations.

In the second type of association, found in *Zelleriella pfitzneri* and *Z. valliceps*, the nucleolar substance is apparently located both inside and outside the chromosome (Figs. 51–56). The structure of the chromosome segment is much more disturbed and modified than in the first type of association; the diameter is greatly increased and its structure much modified. Instead of the presence of chromomeres, as found elsewhere on the same chromosome, this region seems to have a spiral structure which is surrounded by a reticular mass (Figs. 51, 52).

FIGS. 44–56. Relation between chromosomes and nucleoli in *Zelleriella*. All figures ×3680.

FIGS. 44–45. Nucleolus-chromosomes of *Z. elliptica*. (In Fig. 45 the two nucleoli are fused.)

FIGS. 46–47. Nucleolus-chromosomes of *Z. elliptica* stained by Feulgen technique. Note that the nucleolar substance is not stained. (In Fig. 46 the nucleolus-region of the chromosome is precociously split. In Fig. 47 the nucleolus-region of the chromosome is constricted.)

FIGS. 48–50. Nucleolus-chromosomes of *Z. intermedia* in various stages of division and separation of daughter chromosomes. (In Fig. 50 the daughter chromosomes have moved to opposite poles. Each daughter chromosome carries with it a daughter nucleolus.)

FIG. 51. A nucleolus-chromosome of *Z. pfitzneri*.

FIG. 52. A nucleolus-chromosome of *Z. valliceps*.

FIGS. 53–56. Nucleolus-chromosomes of *Z. pfitzneri*. (Fig. 53 shows an optical section. Fig. 54 shows separation of daughter chromosomes. Fig. 56 shows fusion of four nucleoli.)

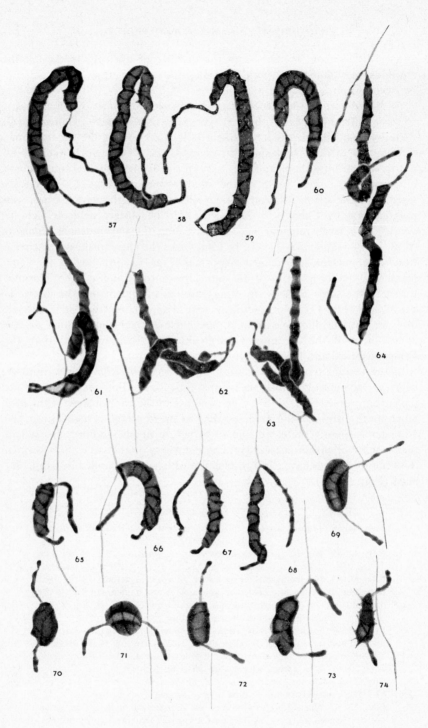

FIG. 57–74

In the third type of association, found in *Zelleriella louisianensis*, the nucleolar substance is apparently located inside the chromosome (Figs. 57–68). The chromosome segment becomes greatly modified and may be called the "puffed region." Its diameter greatly increases, and its structure changes much, particularly shown by the presence of chromatic spirals in place of the usual chromomeres. The nucleoli of this type are apparently similar to the so-called "puffed regions" of the salivary gland chromosomes, which are also believed to be nucleolar in nature (Poulson and Metz,[37].

In four of the five species of *Zelleriella* (*Z. elliptica*, *Z. intermedia*, *Z. louisianensis*, and *Z. pfitzneri*) only one type of association is found in each species. In *Z. valliceps*, on the other hand, two types of association (the first and second) may be found (Figs. 27, 52).

Studies on *Protoopalina caudata microhyla* show that nucleolus-chromosomes are also present in this species (Figs. 69–74). In each nucleus six nucleolus-chromosomes are found.† Only one nucleolus occurs on each of these six chromosomes. These nucleoli are similar to those present in certain species of *Zelleriella*. Judging from these observations, it seems clear that the phenomenon of association between chromosomes and nucleoli found in *Zelleriella* is not limited to a single genus. It has also been found in *Protoopalina* and possibly occurs throughout the family of Opalinidae.

B. Specific Chromosomes Associated with Nucleoli

1. *Zelleriella elliptica*. Nucleoli are located on chromosome pairs 1 and 4 (Fig. 75). The association between chromosomes and nucleoli belongs to type I. In chromosome 1 the nucleolus is located at the middle region of the long arm and occupies about 70 per cent of the total length of this arm. The nucleolar substance is located outside the chromosome segment, which may display precocious splitting, with the two chromatids close together, or may

---

† The protoopalinas found in most of the frogs (*Microhyla ornata*) examined have six nucleolus-chromosomes. (In the resting nucleus six elongated structures, the nucleoli, are found.) These animals with six nucleoli are probably diploid animals. In one frog, however, the majority of the animals have nine nucleoli instead of the usual six. These animals (as well as their nuclei) are larger than those with six nucleoli. It is possible that these animals with nine nucleoli are triploid animals.

FIGS. 57–74. "Puffed regions" of chromosomes in *Zelleriella* and *Protoopalina*. All figures × 3680.

FIGS. 57–64. Samples of chromosome 1 in *Zelleriella louisianensis*. Figs. 61–64 show separation of daughter chromosomes.

FIGS. 65–68. Samples of chromosome 4 in *Zelleriella louisianensis*.

FIGS. 69–74. Samples of chromosomes with "puffed regions" in *Protoopalina caudata microhyla*.

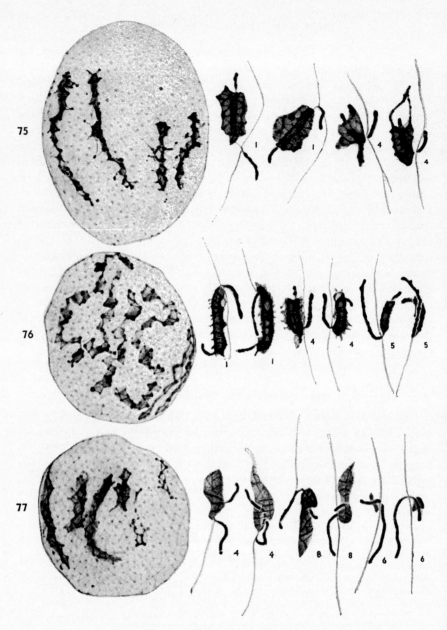

FIGS. 75–77. Nucleolus-chromosomes of three species of *Zelleriella*. All
figures ×2640.
FIG. 75. *Zelleriella elliptica*. Left, a resting nucleus; right, chromosome
pairs 1 and 4 with nucleoli associated with them.
FIG. 76. *Zelleriella louisianensis*. Left, a resting nucleus; right, chromosome
pairs 1, 4, and 5 with "puffed regions" (nucleoli).
FIG. 77. *Zelleriella intermedia*. Left, a resting nucleus; right, chromosome
pairs 4, 8, and 6 with nucleoli associated with them.

be constricted. In some nuclei it is difficult to trace the chromosome segment through the nucleolus.

In chromosome 4 the nucleolus is located at the middle portion of the slightly longer arm and covers about 80 per cent of the length of this arm. The nucleolar substance is located outside the chromosome segment which may be normal or modified. In case of modification, the segment may be constricted or precociously split. There are cases in which it is difficult to trace the chromosome segment through the nucleolus.

2. *Zelleriella intermedia.* Nucleoli are associated with chromosome pairs 4, 6, and 8 (Fig. 77). The nucleoli on pair 6 are regularly very small and may consist merely of one or two granules. The association in this species belongs to type I, the nucleoli being located outside the chromosome. The chromosome segment on which the nucleolus is located either appears as single or precociously split, with the chromatids close together. Unlike the nucleoli in other species of *Zelleriella* studied, the nucleoli in *Z. intermedia* do not appear as clearly defined and compact bodies. They are often irregularly shaped thin masses of material, with variable sizes, and are often fused together.

3. *Zelleriella louisianensis.* Nucleoli are located on chromosome pairs 1, 4, and 5 (Fig. 76). The association of chromosomes and nucleoli in this species is of type III—the nucleolar substance being apparently located inside the chromosome.

Chromosome 1, the longest chromosome in the nucleus, is J-shaped, having a submedian fiber attachment, with a short arm approximately half the length of the long arm. The nucleolus or the "puffed region" is located in the middle of the long arm. The "puffed region" is the most conspicuous part of the chromosome because of its great diameter and represents about 70 per cent of the total length of this arm. The "puffed region" seems to consist of lightly staining substance and chromatic spirals.

Chromosome 4 is a moderately long chromosome with the fiber attachment point near the middle. The middle portion of the long arm is the "puffed region", which is the most conspicuous part of the chromosome because of its large diameter, and equals approximately 70 per cent of the total length of the long arm. The structure of the "puffed region" of this chromosome is similar to that in chromosome 1.

Chromosome 5 is a moderately long chromosome, with two unequal arms. The "puffed region" is located at the middle portion of the short arm. The nucleolus-region of this chromosome, however, has regularly a smaller diameter and is less deeply stained when compared with those of chromosomes 1 and 4 in the same nucleus.

4. *Zelleriella pfitzneri* and *Z. valliceps.* The nucleolus-chromosomes in *Z. pfitzneri* and *Z. valliceps* are shown on Figs. 82–84. In the former species all six nucleoli are of considerable dimensions. In the latter species two of the six nucleoli are regularly very small.

C. Behavior of Nucleolus-chromosomes and of Nucleoli

Compared with that of other chromosomes in the nucleus, the behavior of nucleolus-bearing chromosomes is normal. Condensation and shortening in the prophase, arrangement at the equatorial region of the nucleus during the metaphase, appearance of longitudinal splitting, separation and movement of daughter chromosomes toward opposite poles take place regularly as in other chromosomes and at the same rate. When the daughter chromosomes move toward opposite poles, each carries half of the nucleolus.

The nucleolus-chromosomes show a tendency to occupy a peripheral position in the nucleus. In the resting stage they invariably lie in the periphery of the nucleus close to the nuclear membrane.

The behavior of nucleoli during mitosis has been studied in two of the three types of association of chromosomes and nucleoli—types I and III. As it will be recalled, in type I the nucleoli are located outside the chromosome; in type III the nucleoli are apparently located inside the chromosome.

The behavior of the nucleoli of type I has been studied in *Zelleriella elliptica* and to a lesser degree in *Z. intermedia*. (In the latter species, because of the frequent fusion of nucleoli into irregular masses, the conditions are not as clear as those in *Z. elliptica*.) In the prophase the nucleoli become gradually condensed and shortened. In the metaphase, when they are arranged at the equatorial region of the nucleus, their condensation and shortening reaches its climax. In the anaphase the nucleolus is divided and distributed to opposite poles. Division may be apparently equal or unequal. In the telophase the nucleoli become gradually lengthened, and this attenuated condition is retained throughout the resting stage. The parallel behavior of the nucleoli and of the chromosomes during mitosis may be accounted for by the very close association of these structures, as though the chromosomes take an active part in these changes, while the nucleoli are being passively carried along throughout the mitosis and mechanically torn apart during the anaphase.

The nucleolus-region of type III, as seen in *Zelleriella louisianensis*, behaves essentially the same as other regions of the same chromosome. In the prophase the nucleolus-region becomes condensed and shortened. In the metaphase the nucleolus-region reaches its climax in shortening and thickening; its diameter is greatest at this stage. The nucleolus, a more or less cylindrical structure, like other regions of the same chromosome, is intensely stained. In late metaphase, when other parts of the same chromosome are longitudinally split, the nucleolus-region, too, shows signs of duality and becomes longitudinally and equally divided. The longitudinal halves of the nucleolus often twist around each other (Figs. 62, 63) before they become entirely separated and move toward opposite poles. One interesting feature observed during the anaphase is the great amount of stretching at the nucleolus-region. Because of the unequal amount of stretching of different parts of the nucleolus-chromosome, the length ratio of the two arms becomes considerably altered in the anaphase. In chromosome 1 of *Z. louisianensis*, for example, during the metaphase the long arm is approximately twice as long as the short arm; whereas during the late anaphase the long arm is approximately three times as long as the short arm. In late anaphase and early telophase there is a gradual decrease in the chromaticity of the matrix substance of the nucleolus-region so that the chromatic spirals become clearly visible. The matrix substance continues to lose its chromaticity during the telophase. This is accompanied by the gradual lengthening of the nucleolus-region. As a result, the nucleolus-region becomes greatly attenuated in late telophase and in the resting stage. The six nucleoli are arranged at random and occupy the periphery of the resting nucleus.

D. Appearance and Arrangement of Nucleoli in the Resting Nucleus

As already described in previous pages, the nucleoli associated with different chromosome pairs in a given species are of different sizes, and such differences may be conspicuous. Evidently these relative sizes are retained during the resting stage.

In *Zelleriella elliptica*, the nucleoli associated with chromosome pair 1 are longer than those associated with chromosome pair 4 (Fig. 75, right). In each resting nucleus of this species two long nucleoli and two short ones are regularly found (Fig. 75, left). Judging from the study of the history of the chromosomes and nucleoli during mitosis, it is most probable that these two long nucleoli are those associated with chromosome pair 1, and the short ones are those associated with chromosome pair 4. It is, therefore, possible to recognize the individual nucleoli in the resting stage, and the positions of the nucleoli mark the location of the associated chromosomes in the resting nucleus.

In *Zelleriella louisianensis* the nucleoli associated with chromosome pair 1 are longer than those associated with chromosome pair 4 (these four nucleoli are of similar diameter). The nucleoli associated with chromosome pair 5 are short and distinctly narrower than those associated with the other chromosome pairs (Fig. 76, right). In the resting nucleus of this species two long nucleoli and two short ones (these four nucleoli are of similar diameter), as well as two short and particularly narrow nucleoli, are found (Fig. 76, left). It is evident that the two long nucleoli are those located in chromosome pair 1, the two short ones are those located in chromosome pair 4, the two short and particularly narrow ones are those located in chromosome pair 5.

In *Zelleriella intermedia*, six nucleoli are found associated with six chromosomes; two of these nucleoli are particularly small (Fig. 77, right). In some resting nuclei four large nucleoli and two very small ones are found (Fig. 77, left), but such clear-cut individual nucleoli are not often found either in the dividing or resting nuclei because the nucleoli in this species are usually irregular in shape and they are frequently fused together.

E. Basic and Apparent Numbers of Nucleoli

It may be recalled that the actual number of nucleoli associated with chromosomes is four in one species (*Zelleriella elliptica*) and six in the other four species. However, accidental fusion of two or more nucleoli gives rise to variations in their apparent number, not only in different individuals belonging to the same species, but also in the two nuclei of a single individual.

Another cause of variation in the apparent number of nucleoli comes from the detached nucleoli. In many nuclei of certain species of *Zelleriella* (e.g., *Z. intermedia*) there are, in addition to the nucleoli associated with chromosomes, detached pieces of nucleoli that may make the apparent number higher than the basic number.

It is interesting to note that the apparent number of these nucleoli has been used by some students of Opalinidae as a basis for species distinction and for establishing new species. It is now obvious that the apparent number is not a reliable basis for species distinction.

F. Fusion of Nucleoli

If two or more nucleoli are close to each other, they may clump together or even fuse into a single body (Fig. 45). Fusion does not take place in a definite pattern. There is no special attraction between the nucleoli of homologous chromosomes. The fusion of nucleoli appears to be accidental, and it may occur in only one of the two nuclei of an individual opalinid. The fusion may be only temporary, in which case the individual nucleoli separate in the anaphase.

Fusion of nucleoli is frequent in *Zelleriella intermedia* and *Z. elliptica*, which have type I of chromosome-nucleolus association (nucleolus located outside the chromosome); it is rare in *Z. louisianensis*, which has type III of chromosome-nucleolus association (nucleolus located inside the chromosome).

XI. CHROMOSOME DIFFERENCES BETWEEN DIFFERENT SPECIES OF ZELLERIELLA

Over a period of more than 10 years the writer has studied chromosomes in various species of *Zelleriella* (also some species of *Protoopalina* and *Opalina*) from different parts of this country, Bermuda, British West Indies, Panama, Brazil, Chile, and China. Four types of chromosome differences have been found among various species of *Zelleriella*: (1) difference in chromosome number, (2) difference in the number of chromosomes associated with nucleoli, (3) difference in the specific chromosomes associated with nucleoli, and (4) difference in the type of association between chromosomes and nucleoli.

As already stated, although twenty-four is the chromosome number commonly found in the various species of *Zelleriella*, a much higher number may be found. In an unidentified species of *Zelleriella* found in *Bufo marinus* obtained from Trinidad, British West Indies, for example, the chromosome number is close to thirty-eight.

The number of chromosomes associated with nucleoli varies among different species but is constant for each species. In the five species of *Zelleriella* found in *Bufo valliceps*, as already described, the number of chromosomes associated with nucleoli is four in one species and six in the other four species. The number of chromosomes associated with nucleoli in certain other species of this genus are six in one, eight in another and fourteen in still another.

The particular chromosomes associated with nucleoli may vary among different species but are constant within each species. For example, as already described, the chromosome pairs associated with nucleoli are nos. 1 and 4 in *Zelleriella elliptica*; nos. 4, 6, and 8 in *Z. intermedia*; nos. 1, 4, and 5 in *Z. louisianensis*.

Different species of *Zelleriella* may differ in the type of association between chromosomes and nucleoli.

XII. IDENTIFICATION OF SPECIES OF ZELLERIELLA IN BUFO VALLICEPS

In addition to *Zelleriella intermedia* which Metcalf[29] described, the writer has found four new species of *Zelleriella* in *Bufo valliceps* from New Orleans, Louisiana. The essential characteristics of these new species are given below.

A. *Zelleriella elliptica*. More or less elliptical in shape; a relatively large species; average length of well-grown trophozoites, 184 μ; average width, 90.8 μ; ectosarc alveoles, small and inconspicuous; average diameter of resting nuclei in well-grown trophozoites, 21.2 μ; twenty-four chromosomes (twelve pairs); nucleoli on chromosome pairs 1 and 4; nucleoli formed outside the chromosome. (See Figs. 2–7, 21, 75, 78).

B. *Zelleriella louisianensis*. More or less oblong in shape; average length of well-grown trophozoites, 154 μ; average width, 83 μ; ectosarc alveoles, large and conspicuous; average diameter of resting nuclei in well-grown trophozoites, 16.5 μ; twenty-four chromosomes (twelve pairs); nucleoli on chromosome pairs 1, 4, and 5; nucleoli formed inside the chromosome. (See Figs. 22, 76, 79, 80).

C. *Zelleriella pfitzneri*. More or less oval in shape; average length of well-grown trophozoites, 161 μ; average width, 96.2 μ; ectosarc alveoles, small and inconspicuous; average diameter of resting nuclei in well-grown trophozoites, 24.4 μ; twenty-four chromosomes; six nucleoli associated with six

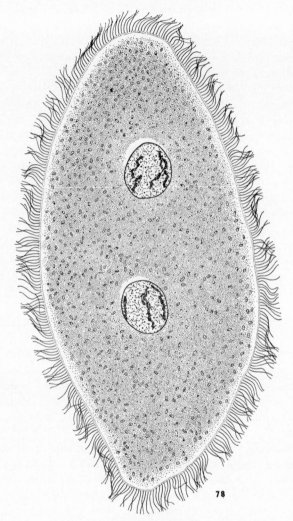

FIG. 78. *Zelleriella elliptica.* A vegetative animal. ×646.

chromosomes; all six nucleoli are of considerable dimensions; nucleoli formed both inside and outside the chromosomes. (See Figs. 81 and 82).

D. *Zelleriella valliceps.* Somewhat elliptical and generally slender in shape, posterior end distinctly pointed; a relatively small species; average length of well-grown trophozoites, 141 μ; average width, 70.5 μ; ectosarc alveoles, very fine and inconspicuous; average diameter of resting nuclei in well grown trophozoites, 17 μ; twenty-four chromosomes; six nucleoli associated with six chromosomes; two of the six nucleoli are regularly very small and may consist merely of one or two granules. (See Figs. 83 and 84).

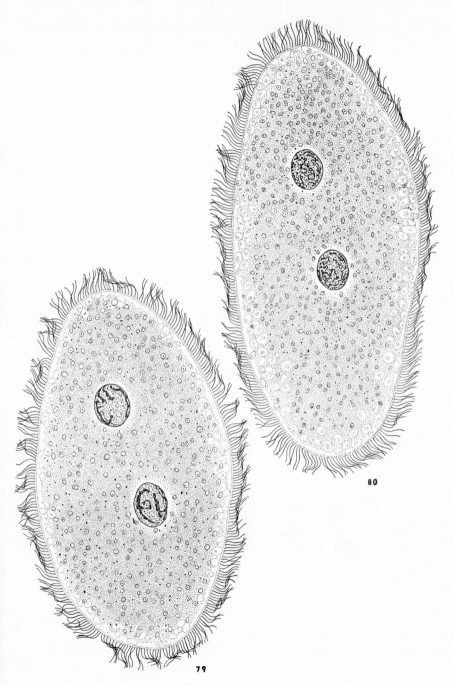

79

80

FIGS. 79–80. *Zelleriella louisianensis.* Two vegetative animals. Both figures
×646.

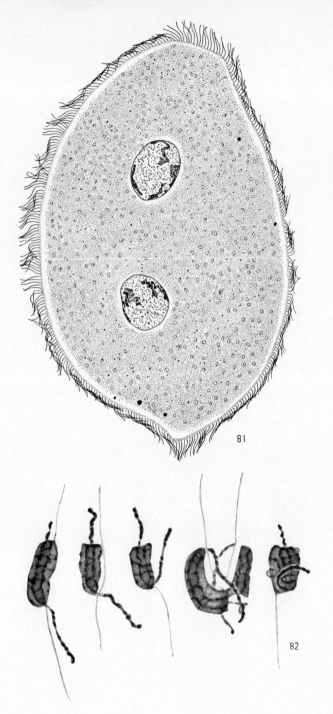

FIGS. 81–82. Vegetative animal and nucleolus-chromosomes of *Zelleriella*
pfitzneri.
FIG. 81. A vegetative animal. ×646.
FIG. 82. Nucleolus-chromosomes. ×2640.

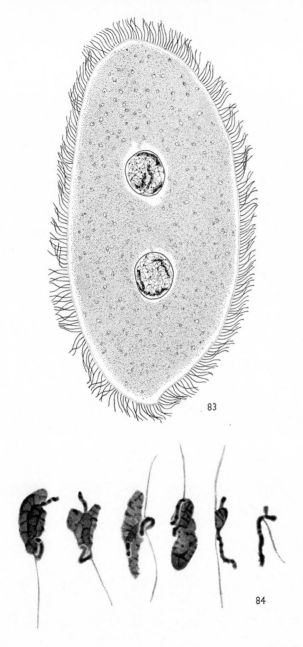

FIGS. 83–84. Vegetative animal and nucleolus-chromosomes of *Zelleriella valliceps*.
FIG. 83. A vegetative animal. ×646.
FIG. 84. Nucleolus-chromosomes. ×2640.

REFERENCES

1. AWERINZEW, S. (1913) Ergebnisse der Untersuchungen über parasitische Protozoen der tropischen Region Afrikas. II. *Zool. Anz.* **42**, 55–57.
2. BĚLǍR, K. (1926) Der Formwechsel der Protistenkerne. *Ergebn. u. Fortschr. der Zool.* **6**.
3. CALKINS, G. N. (1933) *Biology of the Protozoa.* Lea & Febiger, Philadelphia.
4. CHEN, T. T. (1932) Nuclear structure and mitosis in Zelleriella (Opalinidae). *The Collecting Net, Mar. Biol. Lab., Woods Hole* **7**, 270–1.
5. CHEN, T. T. (1932) Nuclear structure, mitosis, and chromosome individuality in an opalinid (Protozoa, Ciliata). *Anat. Rec.* **54**, no. 3 (Suppl.), p. 98.
6. CHEN, T. T. (1936) Observations on mitosis in opalinids (Protozoa, Ciliata). I. The behavior and individuality of chromosomes and their significance. *Proc. Nat. Acad. Sci.* **22**, 594–602.
7. CHEN, T. T. (1936) Observations on mitosis in opalinids (Protozoa, Ciliata). II. The association of chromosomes and nucleoli. *Proc. Nat. Acad. Sci.* **22**, 602–7.
8. CHEN, T. T. (1944) Staining nuclei and chromosomes in Protozoa. *Stain Techn.* **19**, 83–90.
9. CHEN, T. T. (1946) Chromosome studies in opalinid ciliate infusorians. *Year Book of the American Philosophical Society*, **1946**, 123–7.
10. CHEN, T. T. (1948) Chromosomes in Opalinidae (Protozoa, Ciliata) with special reference to their behavior, morphology, individuality, diploidy, haploidy, and association with nucleoli. *J. Morph.* **83**, 281–357.
11. CLEVELAND, L. R. (1938) Longitudinal and transverse division in 2 closely related flagellates. *Biol. Bull.* **74**, 1–40.
12. COREY, H. I. (1940) Chromomere vesicles in Orthopteran cells. *J. Morph.* **66**, 299–321.
13. CORLISS, J. O. (1955) The opalinid infusorians: flagellates or ciliates? *J. Protozool.* **2**, 107–14.
14. CORLISS, J. O. (1961) *The Ciliated Protozoa: Characterization, Classification, and Guide to the Literature.* Pergamon Press, Oxford.
15. DA CUNHA, A. M. and PENIDO, J. C. N. (1926) Nouveau Protozoaire parasite des Poissons: *Zelleriella piscicola*, n. sp. *C. R. Soc. Biol. Paris* **95**, 1003–5.
16. FANTHAM, H. B. (1929) Some parasitic Protozoa found in South Africa. XII. *S. Afr. Jour. Sci.* **26**, 386–95.
17. FANTHAM, H. B., and ROBERTSON, K. G. (1928) Some parasitic Protozoa found in South Africa. XI. *S. Afr. J. Sci.* **25**, 351–8.
18. GRASSÉ, P. P. (1952) *Traité de zoologie.* Tome I, fasc. I. *Phylogénie. Protozoaires: Généralités. Flagellés.* Masson et Cie, Paris.
19. HEGNER, R. W., and TALIAFERRO, W. H. (1924) *Human Protozoology.* MacMillan Co.
19a. HONIGBERG, B. M. et al (1964) A Revised Classification of the Phylum Protozoa. *J. Protozool.* **11**(1), 7–20.
20. IVANIĆ, M. (1933) Zur Aufklärung der Kernverhältnisse und der Kernteilung bei der in Enddarme der gemeinen Erdkröte (*Bufo vulgaris* Laur.) lebenden Opaline, *Cepedea dimidiata.* Stein. *Arch. Protist.* **80**, 1–35.
21. IVANIĆ, M. (1934) Ein Beitrag zur Kenntnis der im Enddarme des Laubfrosches (*Hyla arborea* L.) lebenden Opaline, *Opalina obtrigona.* Stein. *Zool. Anz.* **107**, 295–306.
22. KOFOID, C. A., and DODDS, M. L. (1928) Relationships of the Opalinidae. *Anat. Rec.* **41**, 51.
22a. KUDO, R. R. (1966) *Protozoology*, Fifth Edition. Charles C. Thomas, Springfield, Illinois.
23. LAVIER, G. (1927) *Protoopaline nyanza* n. sp., Opaline parasite d'un Reptile. *C. R. Soc. Biol. Paris* **97**, 1709–10.
24. LWOFF, A. (1923) Sur un Infusoire cilié homocaryote a vie libre. Son importance taxonomique. *C. R. Acad. Sci. Paris* **117**, 910–13.

25. Lwoff, A. (1936) Le cycle nucleaire de *Stephanopogon mesnili* Lw. (Cilié homocaryote). *Arch. Zool. Exp. et Gen.* **78**, 117–32.
26. Metcalf, M. M. (1909) Opalina. *Arch. Protist.* **13**, 195–375.
27. Metcalf, M. M. (1912) *Opalina mitotica. Zool. Jahrb.* Suppl. XV, **1**, 79–94.
28. Metcalf, M. M. (1914) Notes upon Opalina. *Zool. Anz.* **44**, 533–41.
29. Metcalf, M. M. (1923) The opalinid ciliate infusorians. *U.S. Natl. Mus. Bull.* 120.
30. Metcalf, M. M. (1940) Further studies on the opalinid ciliate infusorians and their hosts. *Proc. U.S. Natl. Mus.* **87**, 465–634.
31. Neresheimer, E. (1907) Die Fortpflanzung der Opalinen. *Arch. Protist.* (Suppl.), **1**, 1–42.
32. Nie, D. S. (1932) On some intestinal ciliates from *Rana limnocharis* Gravenhorst. Contrib. from the *Biol. Lab. Sci. Soc. of China, Zool. Ser.*, **8**, 183–99.
33. Nie, D. S. (1935) Intestinal ciliates of Amphibia of Nanking. Contrib. from the *Biol. Lab. Sci. Soc. of China, Zool. Ser.*, **11**, 67–95.
34. Pätau, K. (1937) SAT-chromosomen und spiralstruktur der Chromosomen der extrakapsulären Körper (*Merodinium* spec.) von *Collozoum inerme* Müller. *Cytologia, Fujll. Jub. Vol.* **2**, 667–80.
35. Pfitzner, W. (1886) Zur Kenntnis der Kerntheilung bei den Protozoen. *Morph. Jahrb.* **11**, 454–67.
36. Pitelka, D. R. (1963) *Electron-Microscopic Structure of Protozoa.* Pergamon Press, Oxford.
37. Poulson, D. F., and Metz, C. W. (1938) Studies on the structure of nucleolus-forming regions and related structures in the giant salivary gland chromosomes of Diptera. *J. Morph.* **63**, 363–95.
38. Reichenow, E. (1929) *Lehrbuch der Protozoenkunde.* 5. Aufl. Jena.
39. Sugiyama, T. (1920) Studies of the structure and the nuclear division in a Japanese species of Opalina, *O. japonica*, nov. sp. *J. Coll. Agric., Tokyo* **6**, 361–90.
39a. Swarczewsky, B. (1928) Beobachtungen über *Spirochona elegans* n. sp. *Arch. Protist.* **61**, 185–222.
40. Tönniges, C. (1919) Weitere Mitteilungen über die feineren Bauverhältnisse und über die Fortpflanzung von *Opalina ranarum. Sitz.-Ber. d. Ges. Z. Beförd. d. ges. Naturw., Marburg*, no. 6, 32–47.
41. Tönniges, C. (1927) Die Karyokinese von *opalina ranarum*. Ein Beitrag zur Zahlenkonstanz und Individualität der Chromosomen der Protisten. *Sitz.-Ber. d. Ges. z. Beförd. d. ges. Naturw., Marburg* **62**, 345–80.
42. Valkanov, A. (1934) Die Kernverhältnisse der Opaliniden. *Arch. Protist.* **83**, 356–66.
43. Wenyon, C. M. (1926) *Protozoology.* William Wood & Co., New York.
44. Wessenberg, H. (1961) Studies on the Life Cycle and Morphogenesis of Opalina. *Univ. of Calif. Publications in Zoology* **61**, 315–70.

INDEX OF SCIENTIFIC NAMES

CONTENTS OF VOLUME 1

CONTENTS OF VOLUME 2

CONTENTS OF VOLUME 3